Science and the Soviet Social Order

Science and the
Soviet Social Order

Edited by
Loren R. Graham

Harvard University Press
Cambridge, Massachusetts
London, England
1990

Library of Congress Cataloging-in-Publication Data

Science and the Soviet social order / edited by Loren R. Graham.
p. cm.
Includes bibliographical references.
ISBN 0-674-79420-6
1. Science—Social aspects—Soviet Union. 2. Technology—Social
aspects—Soviet Union. I. Graham, Loren R.
Q175.52.S65S35 1990   89-78235
303.48'3'0947—dc20   CIP

Contents

Preface

Most Western authors interested in Soviet science and technology have concentrated on international security issues and military competition or, less frequently, the history of science. The authors in this volume have a different goal. They wish to understand the Soviet Union better by analyzing the role that science and technology have played in shaping Soviet culture and politics. They share the assumption that Western analysts of the Soviet Union have usually ignored or underestimated science and technology in their efforts to understand the evolution of Soviet society.

Within two generations the USSR has made the transition from a peasant society to an industrialized superpower; today it has the world's largest scientific and technical establishment, surpassing that of the United States by almost one third. Nonetheless, anyone visiting the Soviet Union will probably conclude that this transition to modernity is strikingly incomplete. In rural areas and even in some aspects of urban life, the Soviet Union displays many characteristics of an underdeveloped nation.

The uneven modernization of the Soviet Union that is so evident in a physical sense when one walks down village roads or attempts to buy goods even in large stores in the capital cities has invisible intellectual, cultural, and political correlates. Science and technology have heavily impinged on Soviet society, but the results in politics and culture often are distinct from what one observes in modernized societies in the West. The modernization process has been accompanied by severe cultural and social strains. And even where modernization has not caused social disruption, it has often produced effects that can be explained only in terms of the context of Soviet society, including its political, ideological, and religious traditions. This volume is a first attempt at identifying and analyzing some of those effects. Its title, *Science and the Soviet Social Order,* indicates that its subject is not the technical details but rather the social consequences of Soviet science and technology.

This book is the product of a four-year collaboration of scholars from universities in the United States and Canada who met regularly in Cambridge, Massachusetts, to discuss the social and political implications of science and technology in the Soviet Union. The members of this core group had actually begun to meet much earlier, in the late seventies, to study the strengths and weaknesses of Soviet science and technology for a project supported by the Ford Foundation. During the Ford project, chaired jointly by Mark Kuchment and myself, the group concentrated on the nuts and bolts of Soviet science and technology, relying heavily on the reports of scientists and engineers who had recently emigrated from the USSR to the United States. At the conclusion of that project it was apparent to us that the most interesting issues involving Soviet science and technology are not technical ones but social and political ones. We decided to deepen our knowledge of the role of science and technology in Soviet culture and, in particular, the conflicts over values that have accompanied scientific and technological change in the Soviet Union.

With this goal in mind we applied to the National Endowment for the Humanities for a three-year grant aimed at creating a Coordinated Research Center on the Humanistic Dimensions of Science and Technology in the USSR, which I would chair at MIT, with the strong support of the Russian Research Center at Harvard University. NEH decided to support this effort, but asked that we raise half the necessary money through matching grants from other sources. We were fortunate to receive such grants from the MacArthur Foundation, the Rockefeller Foundation, the Carnegie Corporation, and IBM. We are grateful to NEH and these four matching donors for enabling us to pursue this research project and to produce this book.

From an early point in the project Howard Boyer, Science Editor at Harvard University Press, became deeply involved. He attended most of the seminars, made suggestions for revisions of papers, and became a full member of our group. This unusually fruitful relationship between authors and editor is an important explanation of the fact that the book is not merely a collection of individual chapters, but is instead the product of a group who from the beginning worked toward a common goal, applying both intellectual and editorial criteria.

The Program on Science, Technology, and Society at MIT was the principal home of the project, but Harvard's Russian Research Center provided cooperation and support. Gregory Crowe, a doctoral student in the history of science at Harvard, and Jennifer Haywood of the STS program at MIT coordinated the project, with the help of Rhea Epstein, also of the STS program. Kenneth Keniston, director of the STS program, and Adam Ulam, director of the Russian Research Center, helped us to obtain financial support and gave much encouragement.

The book is divided into several parts. It starts out with an overview of the impact of science and technology on Soviet society and politics. The first specialized part deals with the effects on Soviet society of communications technologies, and includes one chapter written by a historian, S. Frederick Starr of Oberlin College, and another by a computer scientist, Seymour Goodman of the University of Arizona. The second part discusses the social effects of developments in the biological sciences, both chapters of which were written by historians, Douglas R. Weiner of the University of Arizona and Mark B. Adams of the University of Pennsylvania. The third part, concerning the place of engineers and "big technology" in Soviet society, was written by a historian and a historian–political scientist, Harley Balzer of Georgetown University and Paul R. Josephson of Sarah Lawrence College, respectively. Part Four, on philosophical questions of science and technology, combines the efforts of a philosopher, Richard T. De George of the University of Kansas, and a political scientist, Bruce J. Allyn of Harvard. The final part, on literature and art, was written by researchers in four different specialties. They are Katerina Clark, a scholar in Russian and Soviet literature from Yale University; Richard Stites, a social historian from Georgetown University; Mark Kuchment, a historian of science from Harvard University; and Peter Nisbet, a curator and historian of art at Harvard University.

L.G.

Science and the Soviet Social Order

Introduction: The Impact of Science and Technology on Soviet Politics and Society

Loren R. Graham

The revolutionaries who established the Soviet regime believed that they were not only opening a new epoch in human history but also creating a new intellectual world. According to Soviet Marxist theory, scholarship and learning were a part of the ideological superstructure of society deriving its characteristics from the economic base. Once the transformation to a socialist economy was accomplished, the whole world of scholarship would also be reformed, with Marxist approaches to social and natural reality supplanting capitalist ones. Early Soviet Marxist theorists derided the achievements of bourgeois scholars in Western countries and called for uniquely Soviet developments.

From the first years of the regime, however, science presented a special problem. Revolution is based on discontinuity, but science relies on continuity. Marxist literature and art might be different from bourgeois forms, but what about Soviet physics or biology? It might be possible to dismiss from the universities all the unreliable prerevolutionary professors of philosophy and history and replace them with young Marxists favoring different interpretations; but was it likewise possible to replace politically suspect mathematicians, physicists, and other technical specialists? Answering these questions proved to be far more difficult than a Western observer might have suspected. The most militant Soviet ideologists argued that the principle of Soviet uniqueness, or exceptionalism, extended to science itself. Even though Lenin defended the older "bourgeois technical specialists," especially engineers, who could help with practical tasks, the idea of a revolution in science comparable to that in other areas of culture lived on for many years.

The concept of a distinctly Soviet science was the subject of a lengthy

1

debate among Soviet philosophers in the 1920s.[1] The rejection of Western scientific theories as "bourgeois," however, did not become a political possibility until Stalin stoked the fires of ideological passion during the Cultural Revolution of the late twenties and early thirties. In the thirties and forties Stalin supported the rise of Trofim Lysenko, a poorly educated agronomist who had some ideas about growing plants. Lysenko learned that when he described his biological theories as an ideologically superior rival to the Mendelian genetics of the West, he received praise from many journalists, politicians, and philosophers. Soon the theory of two biologies—one Marxist and Soviet, the other bourgeois and Western—became known throughout Soviet society. Lysenko's forcible imposition in 1948 of his Marxist version of biology on all Soviet scientific institutions marked the extreme point in the Soviet commitment to exceptionalism in science. This catastrophe for Soviet scholarship was not overcome until Lysenko's downfall in 1965, and some of its effects linger even today.

Most Soviet scientists, philosophers, and political leaders have learned the lesson of Lysenkoism, however, and they now recall the theory of two sciences with considerable distaste. Even Soviet Marxist theorists have surrendered their claim to a unique Marxist natural science, a change officially recognized when the Party program declared that science was becoming "a direct productive force"—in other words, a part of the economic base.[2]

The implications of this long and tortured transition are enormous. Science proved to be an element of continuity between Soviet and non-Soviet intellectual life. It reduced the exceptionalism of the Soviet Union and brought its intellectuals closer to the rest of the world and the rest of history. Temporally, a link was created between the prerevolutionary and the postrevolutionary periods, since scientists educated before the Revolution could be celebrated for their contributions to international science even if their political views were unacceptable. Geographically, links were created between Western and Soviet natural scientists, who had grounds for believing that their subjects of study were less fraught with political difficulties than those of humanists or social scientists.

The Soviet recognition in the post-Lysenko period of the international character of natural science has allowed Soviet and Western scientists to cooperate much more fully in many joint projects. Not surprisingly, scientists on both sides have often used the allegedly apolitical character of science as a bridge to discussing decidedly political questions such as human rights and arms control. This psychological bridge was especially useful in the years of early contacts.[3]

An irony can be found in the fact that Soviet Marxists have accepted the internationalism of science at the same time that quite a few modern Western historians and sociologists of science have come increasingly to see science as a "social construction," a body of thought heavily influenced by society.[4] Taken to its extreme, this approach might seem to be a form of the Marxist superstructure theory, and one would be tempted to say that Soviet and Western analysts of science are exchanging places. The situation is, of course, not nearly so polarized. Most scholars studying science in all countries accept the fact that it is an international endeavor and that there exists in the world one physics, one biology, not several. The degree to which social and cultural forces mediate science is, however, still an open question on which some of the most interesting Western work in the history and sociology of science is currently being done.

Yet another area where science and technology have had a genuine impact on Soviet intellectual attitudes is in literature and art. In the early years of the Soviet regime the leftist members of the cultural intelligentsia were often enthusiastic about the official promotion of a positive connection among the fields of science, technology, art, and literature. In paintings, films, plays, poetry, short stories, novels, and even musical compositions the metallic world of the machine and the clangorous environment of the factory were major motifs. One group of proletarian poets after the Revolution was known as "The Smithy"; its members wrote odes to the machine as a "friend and a deliverer."[5] Aleksei Gastev, a promoter of industrial acceleration on the basis of time-and-motion studies of work habits, was also a poet who wrote verses celebrating machines and technology.[6] Two of the most popular novels of the twenties were the technological communist utopias *Red Star* and *Engineer Menni*, written by Lenin's erstwhile Marxist colleague Aleksandr Bogdanov.[7] Many other writers, artists, sculptors, and architects followed the same path, including Vladimir Tatlin, who promoted "machine art," Vladimir Maiakovskii, who called his study a "word workshop," and the constructivist artist V. P. Krinskii, who produced drawings showing approvingly the conversion of churches into factories.[8] In paintings the image of the spume of black smoke coming from a smokestack did not evoke fears of pollution—this would come decades later—but instead it honored the progressive march of technology into rural areas, the elimination of the difference between city and countryside that Marxist ideologists promoted. The village was seen by writers and artists sympathetic to the regime as a backward area to be transformed into a rural factory. Several musical composers and conductors in their search for proletarian inspiration went straight to industry itself, creating "proletarian symphonies" in which the dominant sounds were steam whistles, drop-hammer clangs, turbine roars, and the grinding, cutting, whining, and banging of various types of machinery.[9] In films of the twenties, the gears, lathes, wheels, and smokestacks of industry were often featured, just as in Charlie Chaplin's *Modern Times,* but embedded in the opposite message of the beneficence of technology. Dziga Vertov, a leading Soviet documentary director, wrote:

> Revealing the souls of machines,
> enthusing the worker with the lathe,
> the peasant with the tractor,
> the driver with his engine,
> we bring creative joy to every mechanical labor,
> we join men with machines,
> we educate the new men.[10]

It would be an exaggeration to maintain that all these trends have been reversed in recent years, as there are still many admirers of science and technology in the Soviet Union. Nonetheless, a striking shift has occurred in the predominant and most popular images of science and technology in cultural works. A conflict between two styles of life is taking place in the Soviet Union—a clash between a pastoral tradition and a campaign for a technological future.[11] Although similar conflicts have occurred in other countries experiencing rapid industrialization, including the United States, the division is particularly deep in the Soviet Union.[12] The split is exacerbated by nostalgia and the romantic recall of ancient Russia's thousand-year history, by the denigration of much of this history by the Soviet government, and finally by the unrelenting demand of the Soviet regime to catch up with and surpass the economic power of its Western competitors. This combination of factors has never before existed in history in such a potent brew, and the reaction against technocratic culture by a substantial portion of the creative intelligentsia is strong.

In recent years one of the most powerful trends has been the emergence of the *derevenshchiki* (rural writers), authors who celebrate not the machine but the virtues of pretechnological life. In many of their writings an antitechnological theme is evident. One of the best known of the *derevenshchiki* is Valentin Rasputin, whose novel *Farewell to Matyora* tells the story of an ancient Russian village which is submerged when the river is dammed in order to construct a hydroelectric power station.[13] The flooding of the village with its old Orthodox church is used as a metaphor for the general destruction of rural values in the Soviet Union by onrushing technology. Rasputin graphically demonstrated that this theme was intended as a political statement when he became one of the most vocal opponents of the Northern Rivers Project, a river diversion plan promoted by the Soviet government until widespread protest among the intelligentsia caused it to be shelved, at least temporarily.[14]

In art, a corollary to the rural writers can be found in the paintings of Glazunov, who portrays old villages in a nostalgic style somewhat reminiscent of that of Andrew Wyeth in the United States.[15] A revealing contrast in attitudes toward technology of the cultural intelligentsia can be seen by recalling that, whereas the early Soviet poet Maiakovskii called his study "a word workshop," the later artist Glazunov has made his studio into a religious and nationalistic retreat, complete with icons, candles, and portraits of Nicholas II.[16]

In films, an example of the emergence of criticism of science and technology in Soviet culture is Tengiz Abuladze's sensational work *Repentance,* seen by many millions of Soviet citizens in 1986 and 1987. Throughout much of the film the scientific and religious approaches to human issues are contrasted. Whereas in the twenties Krinskii approvingly portrayed the use of old churches as industrial shops, Abuladze depicts a protest against the destructive effects of machines being used in an ancient church. In discussing the film with a Western reporter Abuladze remarked, "Be sure to notice the woman with a book on her head and a rat on top of it. She is the medieval symbol of overreaching science which is destroying us."[17]

In architecture, one of the most popular movements in the Soviet Union in recent years has been the restoration of prerevolutionary churches, village huts, tsarist castles, and the homes of nineteenth-century cultural figures in such areas as the old Arbat near downtown Moscow. In several instances the restorations have taken place over the objection of Soviet officials still clinging to the modernization ethos of earlier times, a message that has lost much of its popular attraction.

The interest in the past in contemporary Soviet cultural expression goes far beyond the restoration of churches and old buildings or the praise of rural Russia in literature and painting. It is a powerful yearning for an older, more variegated and sensually rich culture than the Soviet-style machine age permits. In the name of industrial strength and the scientific-technological revolution, Soviet citizens in the last two generations have been so harangued and coerced with calls for industrialization and acceleration that many of them have become immune. A constant sermon urging modernization can be as deadening as a constant sermon on repentance.

The subject of ethics concludes my consideration of the impact of science and technology on changing Soviet intellectual attitudes. Many commentators on classical Marxism have noted that it fails to provide an adequate treatment of ethical concerns.[18] According to Marx and Engels, the ethical systems of capitalist society are based on idealistic, ultimately religious principles rather than on an analysis of objective reality. Marx believed that in a communist society the "is" and the "ought" would come together as the sources of exploitation and greed disappeared. He did not, therefore, see any reason for a separate and autonomous system of ethics. Many of his followers interpreted this stance to mean that anything that brought communism closer was moral.[19]

The inadequacy of this approach becomes particularly clear in the fields of biomedicine and biotechnology. Soviet scientists, along with those in other countries, have developed very powerful technical means of altering, prolonging, and reproducing life. What are the ethical rules that should hold here? If a biologist wishes to insert human DNA into the embryo of an ape and continue its development in the ape's womb with the goal of producing a hybrid between ape and man, should the experiment be permitted? Or if, like Pierre Soupart in the United States, he or she proposes to fertilize human embryos in the laboratory and then freeze and preserve them indefinitely, should the government fund the effort?[20] Is it permissible to "pull the plug" on life-support equipment for a terminally ill patient, and if so, when? What sorts of rules, if any, should govern surrogate motherhood? Is it permissible to apply genetic engineering to human beings, either to cure individual genetic deficiencies or to "improve" the human race in future generations? Can genetic principles and technologies be applied to produce better mathematicians, musicians, or athletes to excel in international competitions? These questions have in recent years moved from the speculative to the possible to, in some cases, existing reality.

In Western societies, answering these questions has been very difficult, just as it is in the Soviet Union, and many of them remain controversial. Western society, however, is far ahead of the Soviet Union in developing systems for discussing and coping with these issues. The predominant approach in the West is to form ethical advisory committees or institutional review boards with broad public representation and membership. These committees typically include scientists, ethicists, community representatives, and religious leaders.[21] Often the last three types of members have served as "brakes" on the application of biotechnology to human beings, slowing the process until the ethical implications have become clearer. To scientists like Soupart, who lost his federal research grant because of such resistance, this retardation has sometimes been frustrating, but many recognize the necessity to proceed cautiously and to consult the public.[22]

Finding both institutional mechanisms and intellectual rationalizations for such a braking process has been difficult in the Soviet Union. The official opposition to religion as a source of meritorious values has so far kept the Russian Orthodox church and other religious groups off the few scientific councils that have been formed to deal with these questions, despite the request of some religious leaders to be involved.[23] At the same time, the lack of a developed ethical theory within Soviet Marxism has prevented the Marxist philosophers—some of whom are being consulted, although thus far at a distance—from playing effective roles in these discussions.

Perhaps recognizing the practical inadequacy of their ethical views, reformist Marxist philosophers such as I. T. Frolov have joined with American bioethicists at the Hastings Institute in New York and the Kennedy Institute in Washington in international exchanges and discussions of these issues.[24] In some instances the Soviets seem simply to be adopting the guidelines of Western countries on biotechnology, worked out with considerable controversy and with the involvement of groups unacceptable by Soviet standards, such as religious leaders. The Soviet Union adopted, for example, the guidelines of the National Institutes of Health on recombinant DNA research.[25] In this sense the Soviet Union is engaged in both technology transfer and ethics transfer with the West, a concept that would have shocked early Soviet Marxist philosophers, who believed that bourgeois ethics had nothing to offer the Soviet Union.

Such borrowing of Western experience in the area of ethics will probably continue until the Soviet Union finds a way to create truly representative bodies (including religious leaders) to answer these crucial questions involving life, death, and human values. Biotechnology is, then, a powerful example of the way in which science and technology are influencing the Soviet Union, reducing its exceptionalism among other nations, and creating forces calling for greater public representation.

State Policies

The historian who takes the long view of the evolution of Soviet attitudes toward the rest of the world since 1917 will see, in my opinion, a fundamental shift in its underlying assumption, a transformation in which technology has played a leading role. The original assumption of Soviet foreign policy was that the antagonism between socialism and capitalism was so intense that cooperation between the two systems could only be episodic, used for temporary strategic purposes. Eventually one of the two world systems would triumph, and Marxism left no doubt about which one it would be. Beginning with Nikita Khrushchev, however, a different principle based on some degree of long-term cooperation began to emerge implicitly in the speeches of Soviet leaders. Mikhail Gorbachev has pronounced that principle explicitly and prominently. The new view is that the two systems must cooperate on common long-term problems. It is striking how many of these problems have a technological component: avoiding nuclear war, reducing pollution, controlling atomic energy from fission reactors and attempting to obtain it from fusion reactors, dealing with the greenhouse effect, combatting international health threats such as AIDS, and joining forces for expensive projects such as space exploration and high-energy physics research.[26]

The area where technology has had the greatest impact on Soviet attitudes toward the rest of the world has been on the question of the likelihood and consequences of war. Lenin's position was that wars are inevitable among capitalist powers and that when they occur, socialists should try to convert them into class wars.[27] Early Soviet leaders believed that international conflict would bring about the emergence of successful socialist revolutions. The Soviet Union was born, after all, amid the destruction and violence of the First World War, and the entire East European bloc of socialist countries arose in the aftermath of the Second

World War. In its purest ideological form, the belief that war leads to revolution and positive social change did not mean that the Soviet Union should aim for war; on the contrary, the orthodox Leninist belief was that capitalist states would wage wars against one another which the proletariat would convert into revolutionary struggles. Regardless of the phrasing of the ideological reasoning, however, early Soviet foreign policy saw conflict, both among nations and among classes, as the seedbed of positive social change.

At the Twentieth Congress in 1956 Khrushchev noted that "there is a Marxist-Leninist principle which says that while imperialism exists, wars are inevitable," but he went on to declare that this principle was no longer valid.[28] Khrushchev believed that the advent of deliverable atomic weapons required the revision of this Leninist principle, for now even wars among capitalist states would likely involve socialist ones as well and were therefore to be avoided. In 1959 he emphasized the need for peaceful coexistence. But, he added,

> it would be too late to discuss what peaceful coexistence means when the talking will be done by such frightful means of destruction as atomic and hydrogen bombs, as ballistic rockets which are practically impossible to locate and which are capable of delivering nuclear warheads to any part of the globe. To disregard this is to shut one's eyes, stop one's ears, and bury one's head as the ostrich does when in danger.[29]

Not believing in the inevitability of wars left plenty of room for their occurrence, however, and Khrushchev and his immediate successors continued to maintain that, horrible though modern war was, it was still possible for the Soviet Union not only to survive a war but to win. To do so it was necessary for the Soviet Union to be adequately armed; during Leonid Brezhnev's tenure as leader of the Soviet Union, that country achieved rough parity with the United States in nuclear armaments.

Gorbachev has since renounced the view that the Soviet Union could emerge victorious from a world war, and atomic weapons are the main reasons he cites for the change. In his book *Perestroika* he expresses his fear that nuclear war might break out even accidentally, continuing, "Everyone seems to agree that there would be neither winners nor losers in such a war. There would be no survivors. It is a mortal threat to all."[30] And in his speech to the United Nations General Assembly on 8 December 1988, he said that nuclear weapons had revealed the "absolute limits" to military power.[31]

The skeptic might say at this point that the Soviet shifts over the years on the subject of war are more rhetoric than substance, and point to the Soviet support of conflicts in Africa, Latin America, and Afghanistan. But the most important fact to note is that although the Soviet Union has continued on occasion to support local wars (as has the United States) by both word and deed, it has demonstrated its fear of major ones and its abandonment of the hope that armed conflicts among capitalist nations will result in revolution.

Avoiding nuclear war requires cooperation between what the Soviet Union calls the two major world orders, as the series of agreements involving communications, major military maneuvers, arms testing, and arms control and verification attests. These agreements began and over the years have extended the "new thinking" that is so much discussed in the Soviet Union.

That new thinking is progressing into many other areas, such as co-operation in controlling threats to the environment and in scientific research. As in arms control, however, the two nations have a very long way to go in these areas. Many of the scientific exchange agreements between the USSR and the United States in recent decades were so hobbled by bureaucratic regulations and so restricted by Soviet fears of free contact, as well as by American fears of technology drain, that their results were meager. With the advent of Gorbachev and glasnost in the Soviet Union, this situation is changing. The prospects for fruitful work on common problems are brighter now than ever before.

In summary, the creation of atomic weapons and the emergence of a broad spectrum of scientific and technological issues requiring joint work have been essential factors in moderating Soviet hostility toward the rest of the world. While it would be a mistake to assume that science and technology have played exclusive roles in this growing moderation, their significance is unmistakable.

The impact of science and technology can also be seen on Soviet modernization policies. Throughout most of Soviet history those policies have displayed a narrow focus on technical criteria and a neglect of social and economic ones. One reason for the overlooking of social considerations in Soviet modernization policies may well have been the predominance of people with narrow engineering educations in the top leadership. Between 1956 and 1986 the percentage of members of the Politburo, the most powerful political body in the USSR, with educations in technical specialties rose from 59 percent to the remarkable level of 89 percent.[32] Although it is not possible to prove that their educational background has influenced their management style and policy preferences, one can point to strong circumstantial evidence that this is the case.

For decades Soviet leaders promoted enormous construction projects that were seriously flawed from the standpoint of investment choices, environmental considerations, and social costs. Many of the top administrators were former engineers who admired such projects but knew little about economics and cost-benefit analysis, not to mention sociological factors. Even today the profession of social work does not exist in the Soviet Union. One of the results is inadequate awareness of the effects of government priorities and policies on underprivileged citizens.[33]

The large-scale Soviet construction projects of the past included the most ambitious programs in hydroelectric power and canal building in the twentieth century, as well as the largest nuclear power plants ever constructed. Even more breathtaking projects have been discussed, but not yet approved, such as the Northern Rivers Project, mentioned earlier, under which the direction of flow of several of Siberia's largest rivers would be partially reversed in order to provide irrigation for Central Asian agricultural regions. It has been called the largest civil engineering project in history. Favored by Central Asian political leaders, it has been vehemently opposed by environmentalists, Russian nationalists, and several leading economists. Soon after Gorbachev came to power the Northern Rivers Project was shelved, and many Western observers considered it dead, although some articles report that Gorbachev subsequently authorized feasibility studies of this gigantic project.[34]

Soviet agricultural policies are also attempts to find a technological fix for what is essentially an economic and social problem. The original preference for collectivized agriculture was based not only on the principle of socialist ownership of the land but also on a conviction that the full potential of modern agricultural machinery, such as tractors and combines, could not be realized as long as the land was divided into small private plots. There was a certain amount of justification for this belief, as the increase in the average size of farms all over the world in the last fifty years indicates, but it was too narrowly grounded in a reliance on technology and not sufficiently attuned to the economic and psychological factors that make the difference between a hardworking private farmer and a listless state employee. When it became clear to Khrushchev in the late fifties and early sixties that Soviet agriculture was in deep trouble, he reached once again for a technocratic solution: the extension of massive mechanized state farms (which he called agricultural cities) to virgin lands not previously cultivated. This program compounded the flaws of the original technocratic vision of collectivized agriculture with the difficulties of raising crops on arid land. Only in the last few years have Soviet leaders begun to recognize that all the agricultural machinery in the world will not solve the problem of motivation that lies at the heart of their low productivity in agriculture.[35]

Internal Political Development

Soviet authorities have traditionally feared the political effects of communications technology. Evgenii Velikhov, a vice president of the Soviet Academy of Sciences in charge of computer science, admitted that personal computers have been subject to unusual restrictions in the Soviet Union, but he believed that situation was changing. In the past, he maintained, Soviet citizens were so politically immature, so vulnerable to Western propaganda, that they needed to be protected by censorship. During those years foreign books, newspapers, and radio broadcasts were blocked by the Soviet censors. This task was feasible because of the relatively primitive communications technologies involved. At first, similar restrictions were applied to personal computers. But now, crucial changes have occurred in both communications technologies and Soviet society itself.[36] "Two curves have crossed," he observed, "the curve of communications technology capability, and the curve of increasing political maturity of Soviet citizens." Communications technologies have developed so extensively that to block the information flow would not be possible now even if the Soviet government wished to do so. Fortunately, the maturity of the Soviet population has developed to such a degree that the earlier controls are "no longer needed."[37]

As the observations of a leading Gorbachevian reformer, Velikhov's remarks are more an indication of his desired program than of established policy.[38] Certainly not every Soviet politician or ideologist agrees with Velikhov that the time has come to drop all censorship controls.[39] Andrei Sakharov warned during his visit to the United States in late 1988 that several proposed laws in the Soviet Union would once again tighten censorship, and one of them might be aimed at personal computers.[40]

Nonetheless, Velikhov's observation nicely poses the dilemma that new communications technology presents to the Soviet government. Because of the ubiquitousness of technologies like personal computers, communications networks, fax machines, and satellite broadcasts, the leaders probably cannot maintain the traditional censorship barriers. They may be able to hold on for a while longer, but one thing is clear: if they impose heavy restrictions on such technologies as personal computers, they will retard the growth of computer literacy and technical innovation, thereby causing their nation to fall further behind other advanced industrial states.

Computers are by no means the only example of how science and technology are changing the internal political development of the Soviet Union. They are also deeply involved in the recent dramatic rise of interest groups in Soviet politics. These groups are made up of independent clubs and associations with their own political agendas.

These movements are quite different from one another, often centering on ethnic, religious, economic, and political problems.[41] The oldest, largest, and most visible of them, however, was a response to technology, a protest against the damage to the environment caused by industry. It was the campaign in the 1960s and 1970s to save Lake Baikal—the largest body of fresh water in the world and one possessing many unique features—a movement that galvanized environmentalists throughout the post-Stalin Soviet Union.[42]

In its gathering of public support against official policies, this early campaign was in many ways an antecedent of the glasnost of the late eighties. Defense of the environment was politically possible in the Soviet Union of Nikita Khrushchev and Leonid Brezhnev in a way in which calls for political and religious freedom were not.[43] The environmentalist movement served as a model for groups with other causes that emerged after Gorbachev declared his policy of glasnost.

While many other associations have appeared, the Soviet environmentalist movement continues to be the largest and most successful of the new interest groups. Within recent years environmental protests in the Soviet Union have resulted in the cancellation of plans to construct several nuclear power plants, the improvement of pollution controls on lakes and rivers, the closing of a plant manufacturing synthetic additives for livestock feed, and the delay or shelving of plans for river diversion. Many other protests have been unsuccessful but are continuing nonetheless. They include demonstrations against a chloroprene rubber plant in Armenia and against a flood-protection dike around the harbor of Leningrad.[44] And these are only the best known of the protests; by now almost every large city in the Soviet Union and many small ones as well have witnessed environmentalist demonstrations or petition-gathering campaigns. To Americans or West Europeans these developments may not seem remarkable, but within the context of Soviet history they are momentous. It is still not accurate to describe the Soviet Union as a pluralistic society, but these developments show how inadequate the old totalitarian model is for describing the Soviet Union today.

As we look back over the various influences of science and technology on Soviet society, what general conclusions can we draw? Mainly that science and technology have helped to make Soviet society more like the rest of the world, eroding the revolutionary and exceptional ethos in which the USSR was born. The early Soviet leaders who, like Lenin, admitted that science and technology presented the Bolsheviks with a special challenge were right. How could they embrace these refined products of bourgeois civilization without being contaminated by the social and political values of that culture?

In subsequent decades we have seen alternating moments of acceptance and rejection of some of those values, but the trend toward the former seems clear. Gorbachev has asserted that he does not have "any intention to be hemmed in by our values." That would result in "intellectual impoverishment, for it would mean rejecting a powerful source of development—the exchange of everything original that each nation has independently created."[45] At least partially as a result of science and technology, Soviet leaders have permitted a relaxation of censorship controls; they tolerate more heterogeneity in politics; they no longer insist so strongly on their unique intellectual characteristics; they admit that they can learn something about ethics from the West; and they have dramatically changed their views on the consequences of war. With experience they have also learned about the limitations of the technocratic vision and are paying more attention to the social and political consequences of science and technology. These are momentous shifts, and the role of science and technology in helping to bring them about has been underappreciated.

Nonetheless, in none of the areas would it be accurate to say that the Soviet Union is now just like other modern societies. The impact of science and technology on all societies is heavily mediated by the culture, politics, and economy of each particular society. Thus, although controls on communications technologies in the Soviet Union are currently much looser than they were even a few years ago, there are no signs yet of the dropping of all controls. In literature and art, the earlier apotheosis of technology as a means of achieving socialism has been replaced not just by the ambiguity toward technology that is common in the West, but even by expressions of hostility toward it that can be understood only within the context of prerevolutionary Russian literary and religious traditions, which are distinct from those of Western societies. In scholarship, the desire to flee from the political interference of Stalinism has led many Soviet intellectuals to shun the discussions of science as a social construction that currently engage Western historians of science, placing themselves outside an important debate. We see, then, that although the most obvious effect of science and technology on Soviet society has been to reduce its exceptionalism and bring it closer in its characteristics to Western societies, the results that we witness are still quite distinct from what we see in other societies.

Whether in terms of the emergence of interest groups in domestic politics or the evolution of attitudes toward international cooperation in foreign policy, the impact of science and technology on the Soviet Union has not been unidirectional, resulting in behavior changes that can be understood outside of the social context; instead, the relationship between science and Soviet society has been a complex interaction in which social and technical influences intermingle. It is entirely possible that with the advent of Gorbachev, the USSR has already passed through the period of maximum development in this direction. The Stalinist regime of the forties and fifties, and even the Brezhnev regime of the sixties and seventies, were striking deviations from the predominant economic and political patterns of other industrialized nations. In a basic sense Stalinism was an irrational political and economic order; science and technology are based on rationality, and therefore they were corrosive elements within the Stalinist system, eating away at its most aberrant features. As the Soviet Union reduces its most irrational characteristics under Gorbachev, the potency of science and technology as transforming factors is reduced. Other factors in Soviet society, such as cultural, religious, and national traditions, are now emerging and playing larger roles. We have seen this development already, with nationalistic movements in Armenia and the Baltic republics and religious movements within Russia itself gaining prominence in place of demonstrations for scientific freedom and basic human rights.

It is therefore not at all certain that science and technology will continue to drive the Soviet Union toward a pattern common to all other industrialized nations. The influence of the unique history and culture of that country is probably too pervasive for such a homogenization to occur. In order to understand what a powerful influence science and technology has been on Soviet society and politics, we should not rely on a model of universal technological determinism. Instead, we should see simply that at a certain moment in Soviet history—Stalinism and its aftermath—science and technology emerged as major factors in reducing the exceptionalism of the Soviet Union. That period will not last forever, and may already be passing.

· PART ONE ·

Communications and Computers

· 1 ·

New Communications Technologies and Civil Society

S. Frederick Starr

Few aspects of Soviet life today are untouched by change. Social organization, administrative structures, basic principles governing the economy, cultural values, and media of expression are all undergoing apparently fundamental transformations. The rapid pace at which all this is occurring, combined with the participatory character of the process, suggests that the very nature of change in Soviet life is changing.[1]

Communications stand prominently among those areas of transformation. Taking both complex and simple technologies into account, it is evident that in communications in general the USSR lags far behind other advanced industrial societies, especially in computerization but also, to a lesser extent, in telecommunications.[2] It is undeniable that the effects of this deficit will be felt widely in the future. Yet to concentrate on it to the neglect of other developments in communications, let alone of the inevitability of eventual computerization in the USSR, is to undervalue severely the changes that have occurred. Telephone, radio, television, photocopiers, print journalism, audio and video cassette recordings, automobile transport, and international travel and transmissions of various sorts are among the many areas of Soviet communications that have seen rapid development. My purpose in this chapter is to identify those changes and determine their likely impact on the political system.

A rich body of theoretical literature can be brought to bear on this topic. As early as 1957 Karl W. Deutsch studied the process by which communications stimulate the integration of societies.[3] Lucien W. Pye subsequently presented his own theoretical writings on *Communications and Political Development;*[4] Marshall McLuhan stimulated thought on the mass media through "The Medium Is the Message";[5] and literally hun-

19

dreds of writers have pondered the question, posed by Oswald H. and Gladys Ganley, of whether the tendency of the new media is *To Inform or To Control?*[6]

Nearly all of these writers tend toward deterministic views on the impact of communications on politics. Few, however, bother to analyze closely the question of just how deterministic communications technologies might actually be.[7] Daniel Lerner offers a warning in his essay "Toward a Communications Theory": "The mass media, as a distinctive index of the participant society, flourish only where the mass has sufficient skill in literacy, sufficient motivation to share 'borrowed experience,' sufficient cash to consume the mediated product."[8]

Many forces besides communications are fostering political change in the USSR. Indeed, the capacity of that country to assimilate and exploit new conduits of information is arguably as much the effect as the cause of change in other areas of the society. Undeniably, communications and overall social change are closely bound up with each other. At the least, developments in communications are a good index of social transformations.

We shall therefore consider a range of questions, by no means all of which can be answered conclusively. Is the Soviet communications system made up of multiple simple systems, or is it moving toward fewer more complex and integrated systems? How interactive are Soviet communications? Are the new technologies as readily controlled by the state as the old? Do they protect or erode Soviet notions of national sovereignty? Above all, does the evolution of communications foster vertical or horizontal human networks in the USSR?

This last question, posed by Deutsch a generation ago, provides the backbone of the present analysis.[9] It presupposes that autocratic and authoritarian regimes one-sidedly develop vertical communication links ("transmission belts," to use Lenin's phrase), while democratic societies require elaborate horizontal networks as well as vertical ones. These requirements are not absolute, since all societies need multiple links in both directions, and since both types of linkage are more fully developed in complex societies than in simple ones. Our objective, then, must be to determine whether vertical or horizontal integration is proceeding more rapidly in the USSR.

The evolution of communications in Western Europe and the United States provides an inevitable context for such a study. Yet the level of development in such countries is so far in advance of that in the USSR that comparisons minimize the significance of incremental change on the Soviet side. To avoid this problem, developments in the Soviet Union should be seen in the context of the earlier history of communications in Russia.

The Vertical Tradition of Tsarist Communications

Beginning in the eleventh century, written chronicles recorded and standardized the deeds of Russia's church and state leaders. Because they were maintained for centuries, these chronicles systematized history over time and, since copies were made and preserved in various towns, over geographical area as well. At the most local level village church bells provided a simple signal system, while in the ancient Russian city of Novgorod birchbark papyri were employed to document commercial transactions. The latter are particularly useful as an early example of horizontal communication. The fact that channels for such nonofficial communication did not significantly expand until the advent of modern technologies attests to the extent to which vertical communication dominated in both Kievan Rus and Muscovy.

Movable-type printing and hand-carved wood-block broadsides (*lubki*), both of which appeared in Russia in the sixteenth century, present an interesting contrast between vertical and horizontal linkages. In western Europe, as Marshall McLuhan reminds us, movable-type printing fostered pluralism, individuation, and autonomy.[10] The Muscovite state's exclusive patronage of Ivan Fedorov, Russia's first printer, and its subsequent suppression of all publishing outside of the central Printing Court (Pechatnyi dvor) indicates the very different function the same technology fulfilled in Russia. It is revealing that one of the first uses to which movable-type printing was put in Russia was not to publish locally edited bibles for a literate public, as occurred in Germany, but to issue authorized service books in great number so that priests in the isolated parishes across the newly conquered Tatar areas of the Upper Volga basin would not fall into heresy.[11] Notwithstanding this effort, freshly edited scriptural texts issued in the seventeenth century by a handful of independent presses in the Ukraine gave rise to a major schism in the Orthodox church. By the end of Peter I's reign, however, these presses too had been muzzled, and print technology was limited to the dissemination of acts of state, official documents, scientific treatises, and Orthodox Christian liturgical books in forms approved by the state church.

In contrast to the state's domination of the "high" technology of movable-type printing, independent firms in Moscow and elsewhere dominated the "low" technology of wood-block printing. Technologically primitive, *lubki* by the late seventeenth century were nonetheless established as a major conduit of horizontal communication in Russian society, used to disseminate the first printed satires, alphabet books, folk stories, popular religious tales, and pornography.[12] Thanks to its technological simplicity and portability, *lubok* technology was virtually uncontrollable and came eventually to flourish in the very shadow of the Kremlin, at the Lubianka.

Postal service was established in the late seventeenth century with the help of Swedish and German experts. While postal messengers were able to transmit letters between Moscow and Kiev or Arkhangelsk in just over a week, their services were used exclusively by the court and bureaucracy.[13] By contrast, the development of roads and canals facilitated autonomous economic and social intercourse. Following the French pattern, the Russian government established a state engineering school to prepare specialists in bridge, road, and canal construction.[14] The canal system begun by Peter I linked the major European Russian waterways and was designed according to the needs of commerce at the time. Roads, by contrast, were designed first to meet the state's military needs, and only secondarily for the purpose of private communication.[15] Typically, the first macadamized road in Russia was built in 1816 by Count Arakcheev as a purely military venture.

Military considerations also figured prominently in Nicholas I's decision to engage American engineers to build the first railroad link between St. Petersburg and Moscow.[16] The objective in this case was to move troops quickly between the two capitals should further crises like the 1825 Decembrist revolt occur. To be sure, the first Russian railroad between St. Petersburg and the Summer Palace had been privately constructed, and the St. Petersburg–Moscow line itself was built by foreign concessionaires. Nonetheless, the state's deep suspicion of this new channel of communication—both Baron Toll, supervisor of the Directorate of Communication, and Count Kankrin, the minister of finances, opposed railroads as "democratic"[17]—ensured that railroads would remain firmly under state control, if not ownership. Military considerations also figured in the design of the rail grid, even if the decision to use the broader American gauge was made to facilitate speed rather than security, as is often claimed.[18] The slow development of steamboat transport in Russia (there were only ninety-seven steam-propelled craft in 1850)[19] can probably be traced to the military's lack of interest in this technology and to the slow development of internal commerce.

No substantial and autonomous medium of communication developed in Russia before the mid-nineteenth century. Pressed by a depleted treasury, Catherine II had opened the door to private publishing in the 1760s.[20] But even the nominally independent entrepreneur who responded to her call used mainly state-owned presses and was subjected to heavy censorship. Further progress was slow. When private printing began to expand in the early nineteenth century, censorship laws were extended in order to regulate it. Moreover, publishing devoted to lateral communication—for example, private printing—remained technologically backward. Whereas in Great Britain the first steam press had been introduced by the *Times* of London, it fell to the tsar's Ministry of Internal Affairs to introduce that technology in Russia.[21]

Thus, down to the mid-nineteenth century the Russian state provided the main locus for technological innovation in communications. Naturally its primary aim was to provide systems to meet its own military and administrative needs, and only secondarily to develop society locally or to link its components horizontally. Suffice it to say that the *Provincial News* (Provintsialnye vedomosti) published by the government in each administrative district was a conduit mainly for official information, much to the chagrin of the local populace.[22] Only when urban society began to develop in the late nineteenth century did pluralism and horizontality in communications begin to flourish.

New Technologies and Horizontal Communication

The communications technologies that dominated the late nineteenth and early twentieth centuries continued to foster the increase of centralized control, but they also stimulated more decentralized initiatives. Postal services, for example, had initially been designed to meet official needs. The Russian postal system began selling stamps to the public for domestic mail in 1857 and for foreign mail by 1864. The number of all letters mailed in 1854 had been only 34 million, but soared to just under 200 million per annum by 1878.[23] Since this increase far outstrips the growth of governmental services, one assumes that private communications made up the bulk of the growth.

Aside from a few such successes, however, Russian communications lagged. The Russian road network by midcentury was less than a fifth as long as that in France or Germany.[24] Shipbuilding expanded fitfully, with only 15 percent of the tonnage passing through Russian ports being carried in Russian bottoms.[25]

Such instances of retardation can be traced to the slow development of private commerce and, more important, to the severe shortage of capital. Since the government could not fund projects on a large scale, it had no choice but to turn to private and foreign investors. This occurred even in the militarily vital area of railroads. The General Company of Russian Railroads was established in 1857 in order to tap Dutch, British, and French banks at a time when the tsar's finances were in disarray. The resulting concessions brought about a twelvefold increase in railroad mileage by 1880. In that year the state, having recovered from earlier fiscal crises, began repurchasing domestic and foreign concessions from their owners. By 1912 only a third of Russian railroad mileage was privately owned.[26]

Even if ownership of this important channel of communication remained largely in state hands, the actual movement of people and goods in Russia was increasingly determined not by the state but by myriad private and individual—that is, horizontal—decisions. For example, the peopling of western Siberia, while not discouraged by the government, occurred largely because hungry peasants used the new rail network to escape famine and communal control.[27]

Publishing also felt the effects of public initiative, with the state-owned sector decreasing steadily as a proportion of the whole after the 1860s. Autonomous publishing houses strove to meet the interests of the public. Revised censorship laws instituted in the 1860s defined the limits of glasnost (the term entered the Russian political vocabulary in the course of these debates), yet they did not attempt to reinstitute the degree of vertical control that existed formerly.[28] Independent forces rushed in to exploit the situation. The great Moscow publisher Aleksandr Suvorin first introduced the rotary press to Russia for his newspaper *Novoe vremia,* while the entrepreneur Ivan Sytin pioneered the exploitation of linotype and rotogravure presses after 1900, enabling his newspaper, *Russkoe slovo,* to achieve the largest circulation in Russia between 1900 and

1917.[29] The kinds of mass entertainment literature that had earlier been produced only in broadsides now spewed forth from presses in the form of penny newspapers and fugitive journals, with little or no effective state control.[30] Only when local self-governing councils (*zemstva*) tried to link their separate printing activities did the government intervene harshly by imposing strict censorship.[31] In much the same way, "societal organizations" today frequently enjoy extensive freedom to publish but have only recently gained limited rights to disseminate their magazines and journals beyond the immediate district in which they are licensed.

The telegraph and telephone are among the nineteenth century's most sophisticated communications technologies, and Russians played a prominent role in the development of both.[32] P. L. Schilling, a German from Russia's Baltic provinces, invented electric telegraphy before Morse; B. S. Jacobi in 1839 invented the "writing telegraph"; in 1858 E. Ia. Slonimskii was the first to send two telegraphic messages over the same line; S. M. Berdichevskii-Apostolovyi invented the first automatic telephone switch in 1895; and Aleksandr Popov constructed a working radio telegraph in 1895. The Russians also laid the longest optical telegraph line in the world in the 1840s and the longest telegraph line in the world in 1871.[33]

Notwithstanding these achievements in research, the practical development of both telegraphy and telephones was retarded in Russia. Government offices in Moscow and St. Petersburg could not communicate with one another by telegraph at the time of the Crimean War, and in 1863 there were fewer than three hundred telegraph stations in the entire empire.[34] As late as 1900 the Russian telegraph system was only half as long as Germany's and a third as long as England's.[35] Again, the cause was a shortage of capital, which also accounts for the decision to grant private telegraph concessions to the public. Seeking to maintain control over what it did not actually own, the government passed a telegraphic charter which imposed strict punishments against those transmitting anything deemed threatening to life and health, mandating the death sentence for telegraph agents who willfully violated the code. The Directorate of Communications also hosted an international convention in 1875 which endorsed punishments against those transmitting across national borders telegraphic messages "hostile to the interests of states, against the laws, the social order, and morality."[36] By such means the state tried its best to regulate strictly the individuating aspects of telegraphy, even when it did not own the systems.

A similar process occurred with the telephone, but in the decades after 1880 in which that technology developed the state was willing to allow concessionary firms to dominate the field.[37] It was widely held that privatization sped the development and lowered the cost of telephone service. Such arguments no doubt helped to justify the fact that entire local systems in Odessa and other cities were privately owned.[38]

Railroads, telegraphy, and telephones developed in chronological sequence. Comparing them, one notes the nationalization of railroads before 1913, the steady but not increasing role of the state in telegraphy, and the prominence of private and concessionary ownership in telephones. In addition to the growing privatization of their ownership, all three technologies increasingly served horizontal communication. Usage soared when semiconstitutional rule was instituted after the Revolution of 1905. Between 1903 and 1913 the number of telegraph stations grew by almost as much as it had in the entire forty years previously, while the number of telegrams transmitted increased even more.[39] Between 1900 and 1910 the number of intercity telephone lines quintupled, with still more growth in the following half-decade.[40] The new technologies assumed a role in politics. The reactionary politician Konstantin Pobedonotsev listed his telephone number in the St. Petersburg directory in 1900, as did the newly formed political parties a few years later; during the revolutions of both 1905 and 1917 the public at large used telegrams to communicate its demands to the government.[41] Private publishing also grew phenomenally in these years, the number of titles nearly quadrupling between 1907 and 1913 alone.[42]

A Yiddish proverb reminds us that "an example is not a proof." Nonetheless, such instances, multiplied by hundreds, suggest the way in which Russia's developing society seized on new technologies to enhance both horizontal communication as well as vertical communication upwards from society to the state. The evidence does not permit us to ascribe the rise of constitutional rule in Russia to a prior growth in horizontal communications, nor does it prove the reverse. What is clear is that they arose together before 1917 and that each fostered the other.

The Vertical Structure of Bolshevik Communications

Lenin, asked why bourgeois ideology prevailed whenever there existed an open competition of ideas, responded that it "has at its disposal *im-*

measurably more means of dissemination."[43] Faced with this, Lenin, like other authoritarian rulers in the twentieth century, seized control of the vertical conduits of communication and used them to transmit Bolshevik ideas to the public.[44] In addition to this positive step, he also systematically suppressed horizontal communication, thus isolating individuals and groups from one another and atomizing the society as a whole. All this left individuals more readily subject to control from above.[45]

This process was all the easier owing to the virtual collapse of electronic communications, printing, and railroad transport after the Bolshevik Revolution. The number of telephones in use in Russia shrank from 232,000 in 1917 to 127,000 by 1921.[46] The mass evacuation of cities reduced drastically the number of people with access to telegraph stations, and the combination of Menshevik domination of the printers' unions and plummeting paper production after 1914 led to drastic declines in the publication of books and newspapers.[47] By the end of 1920, in the words of a student of the subject, "even the smallest private printing shops had disappeared,"[48] while by 1923 three-quarters of all Russian bookstores and daily newspapers existing in 1917 had closed.[49] Rail transport too was severely disrupted, although surviving photographs showing hordes of people clambering onto those few trains still running indicate that public demand had, if anything, increased.

In its effort to reestablish the priority of vertical channels of communication, the Soviet state pursued policies reminiscent of those of the tsarist state in the seventeenth century—namely, seizing the channels of communication, focusing the production of information in the capital, and closely regulating its dissemination. In printing this meant, in addition to the abolition of private printing and the establishment of the state press, the concentration of printing facilities in a few readily controllable locales, the elimination of autonomous distributors, and the nationalization of existing inventories of books.[50] In telegraphy and telephones it led to the creation of the state telegraph agency (ROSTA) as an instrument of top-down communication and propaganda. Internal passports were eventually introduced as a means of controlling access to railroad transport.

In addition to taking hold of existing channels of communication, the new Bolshevik state suppressed the development of potentially individuating new technologies. Private automobiles, which had been produced in small numbers in the last years of tsarist rule, virtually ceased to exist in Russia at the moment they were becoming ubiquitous in the West.

International telephone communication, first considered by the Soviet state in 1923, grew very slowly and was limited to a few official calling points. Direct telephone lines linking Moscow with Warsaw and Berlin were opened only in 1927, and the line to Paris, opened in 1930, was not direct.[51] All international telephone lines from the USSR were subjected to close surveillance. In a burst of utopian enthusiasm, free postage was established in 1919 but was quickly discontinued.[52] As controls over mail were tightened, the volume of mail began to fall.

The result of these various policies was to restrict severely all lines of horizontal communication. It is worth noting that this process was well advanced even before Stalin's Cultural Revolution completed the task. The growth of urbanization required an absolute expansion of communication facilities in the 1930s, but the USSR ended that decade even further behind in relation to the West than it had been ten years before. In the postwar era the decline became absolute as well as relative. The number of both letters and packages sent by Soviet citizens in 1950 was smaller than in 1940, while the slight increase in intercity telephone calls can be attributed to official rather than private use. By contrast, since the content of books and newspapers could readily be controlled, their production was allowed to increase.

Along with controlling existing technologies of communication, the Soviet regime tried to exploit new technologies to enhance top-down communication. Loudspeakers, introduced in the late 1920s, were well suited to this purpose and were produced in quantity. Lenin had a keen appreciation for the potential of film, but insisted that this technology too be closely overseen from above. Private filmmaking collapsed during the Civil War,[53] to be replaced by the State Film Agency (Goskino, later Sovkino).

The Bolshevik government also seized on radio technology. Introduced first by the Imperial Navy to improve communications during the Russo-Japanese War, radio remained a military monopoly up to the Revolution, by which time there were twenty stations in Russia, all under the navy's control. By the end of the 1920s nearly sixty stations were broadcasting in the USSR, and plans were being made to produce millions of receivers.[54]

Authoritarian regimes in the twentieth century have been said to attach special importance to controlling and developing communications technology. This certainly occurred in Hitler's Germany and in Mussolini's Italy.[55] Russia's centralizing leaders were also determined to place the various new technologies of communication at the service of their cause. Lenin and a host of practitioners in various media developed an impressively detailed body of theoretical writings to undergird their plans.

State control of existing communications developed rapidly and steadily throughout the 1920s, exploiting new technologies that strengthened the regime's ability to transmit messages to the populace and suppressing potentially individuating technologies. The result was a thoroughly authoritarian, even totalitarian, system of communications in which the state controlled both the conduits of information and the messages carried by those conduits.

Acknowledging this, one cannot help but be struck by the relatively primitive fashion in which the Soviet state developed and exploited communications technologies. For all their monopoly in film, the regime's filmmakers achieved far lower levels of public saturation than were achieved by Hollywood or the leading studios of the major Western nations. Not surprisingly, Goskino was chronically underfunded and had to rely on receipts from popular foreign films for its revenue. Moreover, there existed only nine hundred film projectors in the entire country in 1925, and half of these were broken and hence idle.[56] Not until the 1930s did the production and distribution of Soviet films begin to meet public demand, and even then only imperfectly.

Having gained a monopolistic position in radio, the regime likewise failed to exploit its new position. Notwithstanding a 1932 plan to build 14 million receivers, only 3.5 million were in operation in 1937, or a mere twenty-five receivers per thousand population.[57] A key obstacle to radio communications was the USSR's inability to produce vacuum tubes in the quantities needed. As a result, production of popularly priced models like the EChS-4 (1934) and SUD-9 (1939) fell far short of targets.[58] This, along with the desire to restrict access to the open airways, led to the extraordinary development of cable ("wired") radios, with tuning fixed to the two official stations. As late as 1952 two out of three radio receivers in the USSR were of this type, with fewer than 6 million radio-wave receivers available for the entire Soviet population.[59]

Only in the technologically less innovative areas of book and newspaper publishing did the regime achieve high levels of production. Hence, Peter Kenez does not exaggerate when he concludes that "Soviet leaders had much to learn from Westerners in the field of mass communications and almost nothing to teach them."[60]

Although state-dominated top-down communication was vastly strengthened under Soviet rule, this was done at the expense of developing horizontal communications. It is striking that in the years between the Bolshevik Revolution and the death of Stalin in 1953, Soviet citizens achieved no breakthroughs in communications technology comparable to the earlier achievements of Jacobi, Schilling, and Popov. Lacking them, the regime became a consumer of other nations' technologies rather than an innovator itself. This stands as evidence of the relatively conservative record of the Soviet government in the field of communications, its claims to the contrary notwithstanding.

It goes without saying that the messages transmitted over the vertical media strongly supported the regime. Two qualifications must be introduced, however. First, a cursory review of the Soviet press and of Soviet films of the 1930s and 1940s suggests that while virtually no anti-Soviet messages were transmitted, only a part of the production focused directly on the regime's goals. Far from being the relentless bombardment of propaganda anticipated in *Brave New World,* much of the content was composed of ideologically bland and even unassimilable data. Second, at least as much attention was devoted to what was *not* communicated as to what was. In other words, Soviet communications policy under Stalin was aimed at suppressing data judged harmful rather than effectively disseminating positive messages. As in the communications system as a whole, far more attention seems to have been devoted to eliminating autonomous horizontal channels than to fully exploiting vertical channels. Closer comparisons with fascist Germany would be instructive on this point.

For all the force Stalin devoted to suppressing horizontal communications, he never managed to destroy the ideal of a more pluralistic culture like that which had begun to appear on the eve of the Revolution. As soon as the harshest controls began to be relaxed in the 1950s, horizontal channels, both official and unofficial, were once again put to use.

Toward a Horizontal Information Culture

The post-Stalin era has been the victim of hyperbole. Dubbed "The Thaw" after the title of a novel written before any thaw had occurred, the early years of dramatic change are said to have given way to torpor and "stagnation," to use Mikhail Gorbachev's self-serving phrase. In terms of social change, however, the evolution was both more steady and more basic than either supporters or critics admit. Collective farmers made up almost half the population on the eve of the Second World War but had shrunk to a fifth by 1971, a smaller percentage than that constituted by members of the white-collar intelligentsia.[61] The number of postsecondary students soared, from 6.2 million in 1957–58 to 25 million in 1964–65.[62] Corresponding changes occurred in the rates of literacy and urbanization as the population grew younger and geographically more concentrated. Such shifts, accompanied by the USSR's steadily improving technological capacity, prepared the way for a fundamental change in social communications. The fact that the law governing communications was extensively revised as early as 1954 suggests that the leaders themselves understood change to be impending.[63]

As we shall see, changes in communications occurred through both the addition of new technologies and the expansion and alteration of older technologies to make them capable of fulfilling new functions. Together these shifts brought about a transformation far more extensive than is evident from an examination of only the separate parts. On the one hand, they expanded and strengthened vertical channels of communication in Soviet society. But on the other hand, they also rendered those channels more interactive than formerly and gave them a stronger role in horizontal communications. More important, they vastly increased the ability of individuals and groups to communicate directly with one another, unmediated by the state. All of these changes presupposed a reduction, albeit partial, of the Stalinist controls on horizontal communications. As soon as these controls were cut back somewhat in the 1950s, Soviet society showed itself eager to exploit existing and new technologies of communication, as indeed it has been ever since.

We will consider the implications of these changes for the Soviet polity in the concluding section of this chapter. For now, let us review the elements contributing to the new horizontality of Soviet communications.

The Expansion and Alteration of Old Technologies

The Soviet postal system provides a good example of the impact of social change on communications. Between 1940 and 1974 the number of letters handled grew from 3 to 9 million per annum.[64] The number of packages quadrupled in the same period. Most of this expansion was concentrated in the late 1950s and early 1960s, coinciding with a phase of rapid urbanization, increased literacy, and greater openness.[65] Increased efficiency also stimulated public use of the mails. Today, when 60 percent of Soviet mail is shipped by air, the volume of letters has grown so rapidly as to cause a shortage of postal carriers and an increase in theft.[66]

Communication by telephone also soared. Twice as many new phones were installed between 1965 and 1974 as between 1940 and 1965, with the number of urban telephones trebling in the same period.[67] Nearly all the new urban phones were automatic and thus increased privacy. By the mid-1980s there were 24 million telephones in the USSR, half of them in urban apartments. By contrast, only 2 million private rural residences had phones.[68] The nearly 2 billion intercity calls being made annually by then and the sevenfold increase in international calls in the decades before 1974 attest to rapidly changing public access to this medium.[69]

As the USSR became less of an "information-poor" society, the content of communications grew less readily controllable. The growth in the sheer number of phone calls makes it difficult, if not impossible, for the state to monitor their content, just as the quantity of private mail has rendered it impossible for the KGB to maintain former levels of surveillance over that medium. It is no surprise that as early as the 1970s persons in many fields came regularly to use both domestic and international telephone lines for unofficial and purely personal purposes. Among such users were those with agendas different from the state's. As one student of the subject put it: "The international telephone, despite continued control that amounts to persecution, has given Russia's dissidents the means for immediate direct contact with the outside world, something quite unthinkable not much more than twenty years ago."[70]

The increase in mail and telephone usage has facilitated horizontal communication, while the rapid growth of publishing and the press, by contrast, has benefited both vertical and horizontal linkages. The number of periodicals nearly doubled between 1958 and 1965,[71] with *Pravda* going from a four-page format to six pages in 1970. The central press grew with particular speed, with nearly all major Moscow newspapers being printed simultaneously in thirty-five cities by 1966.[72]

If such changes served uniformity and top-down communication, other changes in traditional print media enhanced interaction. The much-heralded rise of letters-to-the-editor columns indicates that Soviet newspapers are becoming vehicles for interactive communication from bottom to top, providing feedback to the government in the process. Moreover, the appearance of job ads, lonely hearts announcements, and other forms of personal notices in various local newspapers reflects the public's growing interest in exploiting traditional print technology to improve horizontal linkages among individuals.

Radio, too, gradually became more interactive. Rare is the student of Soviet affairs who cannot regale friends with a few "Radio Armenia" jokes. Few realize that these have their origin in programs begun in the 1960s in which listeners were invited to call in their questions. Such programs, aired on most Soviet domestic stations, constituted the first bottom-up use of the vertical medium of radio, and provide the same kind of feedback to the regime that letters columns do in newspapers.

So much has been written about the USSR's failures in the mass dissemination of personal computers that it is easy to forget the dramatic increases registered in many other electronic media of communication, particularly in the 1960s. Nowhere is this more striking than in radio.[73] For all the emphasis on top-down communications in the Stalin era, there were only 17.5 million radios in the entire USSR in the year before Stalin's death.[74] By 1968 this number had risen to 89.5 million.[75] While the ratio of cable to wave radio in 1952 had been approximately two to one, by 1968 the ratio slightly favored wave sets.

The proliferation of wave radios during the 1960s made it all but inevitable that the public should become interested in receiving international as well as domestic broadcasts. Shortwave transmissions had greatly multiplied since the early 1950s, with stations in the United States, Great Britain, West Germany, Sweden, Luxembourg, and Iran beaming broadcasts to the USSR. Receivers capable of tuning in such broadcasts were constructed in large numbers by amateurs, while others were imported unofficially through diplomatic channels. Transistors enabled such equipment to be miniaturized during the 1970s and made it readily importable through informal channels. By the end of the 1960s Radio Liberty could claim that 27 million radios in the USSR were capable of receiving its broadcast.[76] Even if this figure is exaggerated, as seems likely, the number was large enough for the Soviet government to decide that it should manufacture such equipment itself so as to at least be able to co-opt what it could not control. Selective jamming limited access to certain foreign transmissions, but the manufacture of shortwave radios indicates the government's acceptance of broadcasts across its borders as an unavoidable feature of modern communications.

New Technologies of Communication

No less significant than the expansion and transformation of existing channels of communication are the major new media introduced in the past twenty years. Among these television is the most prominent. Developed by a government confident in its ability to control the social impact of the medium, Soviet television burgeoned quickly, expanding from 2.5 million sets in 1958 to 30 million a decade later.[77] By the end of the 1970s television was all but universal in Soviet households. During the decade ending in 1974 the number of transmitters trebled.[78] Cable television, by contrast, has made very slow progress in the USSR, partly because it requires such a large initial investment but doubtless also because it introduces a greater element of choice than the government is yet prepared to allow.[79] That the latter consideration is significant is suggested by the fact that the USSR did not shrink from the expenditures required to transmit its few channels by satellite, which it has done since 1967.[80]

In contrast to the Soviets' wholehearted acceptance of television, their attitude toward the private automobile has been more ambiguous. On the one hand, production grew from 64,000 cars in 1965 to over 1.3 million in 1982, and would have grown still more had the Kama River Truck Works not gobbled up more than half the rubles designated for the motor vehicle industry in the late 1970s.[81] On the other hand, retail prices were set extraordinarily high, and were not reduced until 1985.[82] Frequent articles in the press have warned of the negative social impact of private automobiles, leading to charges that more than simple inefficiency lies behind the refusal of ministries to provide the necessary infrastructure for private automobile ownership. Only in 1984 did the government announce plans to increase the number of gas stations for private cars from 1,200 to 3,000.[83] This more positive attitude has spread rapidly, however, extending to the expansion of partial credit programs for car buyers,[84] reductions in the prices of certain models,[85] and even to discussions of the possible production of sports cars for Soviet citizens.[86]

Since the retarded growth of computerization in the USSR has been widely discussed in the Western press, it is not necessary to repeat the story here. Suffice it to say that while microchip technologies have made substantial progress in the military sphere and in certain areas of industrial planning, they have made little headway at the crucial level of desktop personal computers. With no modems, few printers, and inferior floppy disks, the situation in the USSR will not change rapidly.[87] Networking of all sorts is proceeding slowly at best,[88] even though a system linking institutes of the Academy of Sciences in three cities is now in place.

The introduction of a single "gateway" for all computerized data entering the USSR reflects the government's concern with controlling information transmitted for use by this new medium; 80 percent of all data bases, after all, originate in the United States.[89] Resistance to demand-based systems stems from a similar concern to maintain at least some central control. But the adherence of the USSR to international architectural standards in computing and the rapidly declining costs of transmitting data within the USSR,[90] along with the tremendous positive publicity given to the interactive nature of the Academy of Sciences' new network, suggest that an environment more hospitable to the computer revolution is beginning to emerge, albeit slowly. Even the ideal of a single gateway in Moscow for computerized data from abroad may prove so clumsy or so difficult to enforce that it will eventually have to be abandoned.

The record of the USSR in adopting the major communications technologies of private automobile, television, and computing is mixed. Television, the most vertical and hence controllable technology, has progressed most rapidly, while the private automobile and personal computer have made only slow advances, development of the former having been held back by more than half a century and the latter by at least a decade. Yet this is not to say that the advance even of these technologies will be permanently thwarted. The Soviet government has officially committed itself to rapid progress in both automobile and computer production, which will have the effect of stimulating public demand. In the conclusion of this chapter I will show that such demand is becoming increasingly difficult to resist.

The Inexorable Advance of "Small" Technologies

No journalistic account of Soviet life today seems complete without tales of VCRs, home movies, and black market audio and videotapes. Rarely, though, do such accounts go beyond the level of anecdote. Yet the "small" technologies of the past generation are uniquely suited to foster horizontal communications, just as film, radio, and loudspeakers represented new means of facilitating vertical communications in the 1920s and 1930s. The history of such technologies dramatically highlights the fundamental changes occurring in Soviet communications over the past decades.

The rise of such minor technologies as home photography, cassette recording, ham radio, and video cassettes shows certain common features. All benefited greatly from public demand, which in turn was stimulated by the public's knowledge of how the given medium was being exploited abroad. All gave rise to simple networks of aficionados, and all became the object of official efforts at co-optation. Eventually all gained a legitimate place in Soviet society as a whole. To see these patterns in action, let us review more closely the copying and transmission of static visual images, reproduction of sound, and the replication of movie images.

Various stencil, xerography, and ditto systems existed in the USSR prior to the 1960s. All were considered printing presses in law, however, and hence could not be owned privately. In practice, access to stencils was widespread, and materials as diverse as music and architectural drawings were being unofficially reproduced for select private audiences as early as the mid-1950s. As is well known, the USSR maintains strict controls over all photocopying machines, including the cumbersome domestically produced models. In the 1960s and 1970s, however, a number of samizdat publishers in various fields gained access to such machines and used them extensively. The example of the Voronezh engineer Iurii Vermenich is typical, in that he succeeded in reproducing translations of several dozen books on jazz on primitive machines owned by his institute.[91]

Many voluntary (*obshchestevnnye*) organizations beginning in the late 1960s gained official permission to issue informal newsletters and magazines for local distribution; most of these publications, such as the Leningrad quarterly *Kvadrat,* were reproduced on photocopying machines. The independent Ukrainian journal *Ukrainsky visnik* and the religious journal *Vybor* are both reproduced in the same semilegal fashion.

Attempts to control access to these machines have failed to repress the demand for horizontal print communication. Private photography was always available to fill the gap. An article in the autonomous journal *Svobodnaia mysl'* in 1971 presented detailed instructions on how inexpensive and widely available photographic equipment could be used as a surrogate printing press.[92] Such techniques were made readily accessible by the excellent and inexpensive single-lens reflex cameras manufactured in the USSR with equipment taken from the Zeiss factories in Jena, the Zenit-E being the model of choice for unofficial printing on account of its excellent close-up resolution. Negatives were easily transmitted by mail and could be read with the help of a lens for viewing filmstrips, available in children's stores for thirty-five kopeks. Countless manuscripts, reports, poems, lyrics, and other documents were independently transmitted throughout the USSR by this means.

The spread of radio stimulated interest in recording. Wire recorders were manufactured in the USSR in the 1940s but were rarely available to private citizens. Instead, amateurs constructed simple machines capable of recording sound, using the emulsion of discarded X-ray plates. Such recordings were of poor quality and had a short life expectancy but offered the double advantage of being inexpensive and readily transmitted through the mails. By the early 1950s this *Roentgenizdat* was widely exploited for recording both music and voice, leading eventually to a 1958 law making it illegal "to produce home-made records of the criminally hooligan trend."[93] Meanwhile, Soviet-made open-reel tape recorders appeared in the 1950s with the large El Fa-6 model, which was followed before 1960 by the lumbering Dnepr-3 and Spalis models. More compact foreign-made cassette machines entering the country in great numbers in the early 1960s forced the authorities to choose between losing all control over the technology or attempting to co-opt it by producing a home-grown portable product. They chose the latter course. Sales of Yauza series tape recorders reached half a million by 1965 and over 1 million by 1970.[94] The social impact was enormous. The late Anatolii Kuznetsov described the situation:

> Soviet ideological organs, busy in the field of radio production . . . completely failed to pay attention to such a seemingly innocent technical branch as the production of tape recorders. A demand existed and it was satisfied, and when at last ideological firemen discovered the catastrophic breakthrough, it was too late. Now it is a rare home without a tape recorder, and an evening party or get-together without one is unthinkable.[95]

Cassette tape recordings, shipped through the domestic and international mails, provided a channel of horizontal communication that was at once inexpensive, legal, and virtually beyond official control. Ham radio operators seized on another means of sound transmission that was equally efficient, equally inexpensive, and nearly as difficult to control.[96] It is estimated that up to twenty thousand licensed radio amateurs were operating in the USSR in the late 1960s. According to Gayle Hollander, the number of illegal operators increased dramatically in the 1960s, when a do-it-yourself handbook for amateur radio was published. While details of this medium are lacking, it is known that ham radio operators in the Ukraine warned of the Soviet troop buildup on the eve of the invasion of Czechoslovakia in 1968, that hams in the Ukraine spread lurid reports at the time of the Chernobyl disaster and helped force the government to release authoritative information, that a ham operator in Vilnius was given three years' incarceration in the early 1970s, and that more than a thousand hams in the Donetsk region were detained in 1974.[97]

Photography, tape recording, and ham radio were all exploited by Soviet citizens to create more adequate horizontal conduits for information than official media could provide. Much the same process is going forward today with video cassette recorders. Large numbers of these inexpensive and compact instruments were being unofficially imported into the Soviet Union by the late 1970s. Crew members of a Soviet cruise ship that made frequent stops in New Orleans were known to purchase several hundred VCRs at a time from dealers in that city, to be resold on the Odessa and Leningrad black markets. Dubbing machines, essential if the medium is to respond to market demand, were bringing 1,000 rubles at Riga commission stores in September 1986.

What *Izvestiia* termed the "currently fashionable passion for videotapes" led police in Riga to confiscate 415 imported and domestically produced videos depicting "cruelty, violence, mysticism, and superstition," which were being shown by independent operators to paying audiences of local students. The operators of this library were charged under an article of the Latvian civil code that banned the distribution of videotapes "harmful to the state or to public order, health, or morals."[98] The analogous law in the Russian Republic was invoked to punish a Moscow piano teacher caught trading in videotapes and equipment.[99]

VCRs by 1986 had spread so far that it would have been impossible to rein them in completely. Instead the government limited its intervention to co-opting the medium and policing its most objectionable excesses.[100] The worst danger lay in the seemingly uncontrolled nature of communications across the country's borders. Dish receivers have until recently been all but nonexistent, and any that might find their way into private hands could easily be controlled. Videotapes, by contrast, are as disrespectful of national borders as audio cassette tapes. Because they are so readily imported, reproduced, and disseminated, they effectively destroy the state's autarkic control over both television and film production.[101] Whether or not Soviet citizens produce their own original videos, the exercise of independent choice over what is imported and disseminated creates a kind of video samizdat. It is for this reason that the Soviet government began producing its own Elektronika VM-12 VCR. Reportedly costing from 1,200 to 1,400 rubles, the Soviet machines may be less expensive than imports but have the overriding disadvantage of being unable to play standard Western tapes without modification. It is doubtful that more than ten thousand Elektronika VM-12 units had been manufactured before the end of 1986.[102]

A second attempt to preempt the video import boom was the decision in 1985 to produce large numbers of video cassettes in the USSR. Manufactured at the same Elektronika plant in Voronezh that produced the VM-12, the Soviet video cassette library consists mainly of mainstream popular music (Pugacheva, Vysotskii, and so on) and old films, mainly Soviet. By the end of 1985 the library included 450 titles, which were distributed mainly at electronics stores in such ports of entry as Riga, Moscow, Odessa, and Tallinn, where the black market in foreign tapes was most active. Production remained low, however, because the only source of tape was the Soviet film industry (Soiuzkinofond), which jealously hoarded all videotape to meet its own needs.[103] Moreover, the Soviet press candidly admitted that many customers were buying the local product solely to rerecord imported films and programs for their own use.[104] No wonder private video traders have concluded, as the official press acknowledges, "that, for the time being, there is no threat of competition."[105]

With the exception of audio tape-recording, all of the small technologies of communication that have appeared in the USSR remain, by Western standards, fairly limited in their reach. Yet together the VCRs, ham radio stations, audio cassettes, photographic labs, and xerographic machines touch the lives of tens of millions of Soviet citizens. Responding to market demand, these media have expanded rapidly in recent years and will doubtless continue to do so. Inevitably this activity has produced a strong reaction in the form of efforts to co-opt and control. None of these attempts has met with success, however, for the small technologies are too decentralized for their use to be more than marginally shaped from above.

Toward an Information Revolution

The USSR's stagnant economy, coupled with its stumbling approach to personal computers, has caused observers there and in the West to conclude that in the 1970s and 1980s it missed out on the information revolution. The foregoing overview of the expansion and transformation of old technologies, the emergence of major new large-scale conduits of information, and the rise of small technologies suggests this generalization is overstated. However stagnant the Soviet economy may be as a whole, the realm of communications has been steadily, radically, and irreversibly changed.

To be sure, different groups and regions of the USSR have sharply different levels of access to the transformed or new media. As I have noted, urban families are three times more likely than rural families to have telephones,[106] while major cities and international points of entry have far greater access to new communications than secondary and interior cities. Overall, access to public media correlates closely with the differing levels of economic development among the republics.

Whatever their unevenness, the changes are profound and show every sign of continuing. Repeated statements by Gorbachev from his arrival in office heralded his hope of increasing investment in telecommunications and computing. Moreover, there is ample evidence of a suppressed demand for communications so large that it can scarcely be ignored. Twelve million citizens were waiting for telephones to be installed in their homes in 1985, with a quarter-million more waiting to receive long-distance service.[107] The total of 25 million civilian phones in the Soviet Union compares with 170 million for the less populous United States, suggesting that even the addition of 12 million more phones may eventually not be enough.[108] With only 32 automobiles for every thousand Soviets, as compared with 471 for Americans and nearly the same ratio for West Germans, demand in that area too seems likely to continue to rise.[109] Only in computing has the Soviet state escaped market demand, and this is bound to change as a core of civilian computer buffs is formed.

Together these many changes are beginning to create a horizontal information culture in the Soviet Union, supplementing but not replacing the vertical structure inherited from the Stalin era. At the same time, that vertical structure itself is being revived and altered as more messages flow both downward and upward through it and as the number of interactive or feedback elements increases. Indeed, one of the most significant innovations that can be traced directly to Mikhail Gorbachev is the infusion of new vitality into the previously moribund sphere of vertical communications, both downward and upward.

Needless to say, the strengthening of horizontal communications has evoked concern in some quarters. V. M. Chebrikov of the KGB denounced the exploitation of Soviet citizens by foreign media conspirators,[110] while he and other Soviet commentators have singled out as evidence of such manipulation the nationalist demonstrations held in the Baltic republics as well as the protests in Armenia.[111] To check such untoward occurrences, Stalinist traditionalists mounted efforts to influence the drafting of new laws so as to limit the right of assembly and suppress independent publications as well.[112]

In light of the extraordinary tenacity and initiative shown by Soviet citizens seeking greater access to modern communications, such accusations and measures seem quite tame—mild rearguard actions rather than a serious campaign of suppression. The failure of efforts to maintain the old controls raises the question of whether horizontal communications could actually have been suppressed in the late 1980s. Of course they could, but, as we will see, only at a very high price. For now it is worth noting that the Gorbachev government through mid-1989 took no drastic measures against any medium deemed subversive, even though it moved against particular publications in several instances. Until the government makes such a counterthreat, and until it succeeds, it is reasonable to conclude, first, that a kind of communications revolution is under way in the USSR; second, that that revolution is modifying the received communication culture by stressing horizontality and interaction across levels where top-down verticality once reigned unchallenged; and third, that the new communications order in the USSR benefits from the government's acquiescence, if not approval.

Technotronic Glasnost and Civil Society

The Soviet newspaper *Literary Gazette* in 1987 carried a long article entitled "The American and the Computer," in which the author charged that Americans want nothing better than for the USSR to wallow in the same "technotronic openness" (*glasnost'*) that exists in the United States.[113] Information, he admitted, is power. For the USSR to suspend all controls on information would be to weaken the country for no better purpose than to satisfy the demands of Americans. In spite of such grumbling, a kind of technotronic glasnost already exists in the Soviet Union and will have profound implications for the political culture of that country.

I have characterized this new information culture in terms of the rise of horizontal links and systems. While acknowledging that vertical conduits not only continue to exist but have been strengthened by new technologies and the new leadership, I have stressed the relatively greater impact in the USSR of the new horizontal communications in recent years. In many ways this recalls the situation in late-nineteenth-century Russia, when fresh technologies also stimulated horizontal communication within society. Today's developments in horizontal communication outstrip those of the past both in the diversity of new channels and in the number of people affected. It is therefore necessary to evaluate the impact of these developments on political life. This impact can be detected in at least six areas.

Privatized Information and the Stimulation of Public Opinion

Far more information is available to the Soviet public than ever before. The capacity to acquire, preserve, and transmit information has grown sufficiently to enable one to speak of at least partial privatization in this area. Stated differently, improved horizontal communications and advances in education have almost certainly increased the percentage of all Soviet information that is now generated outside the Party and state and circulating freely in society.

This means that the regime must reckon with more numerous and more diverse sources of input than formerly. At the least, this more pluralistic situation places greater burdens on "the attention-giving, information processing, and decision-making capabilities of administrators, political elites, [and] legislatures."[114] No wonder that in 1988 the Gorbachev government moved to establish two new institutes for the systematic study of Soviet public opinion.

The Increasing Internationalization of Information

Technologies both high and small foster communication across the borders of the USSR. This is true of both unofficial and official channels. At the level of popular culture, contraband songs by the émigrés V. Tokarev and A. Rozenbaum gained great popularity even during the late Brezhnev era through tapes widely distributed at sanatoriums and vacation spas.[115] Similarly, nearly 40 percent of all films shown in the provinces are foreign made, while the percentage of VCR films that came from abroad is even higher.[116] Telephone calls, letters, and international radio transmissions all attest to this internationalization of information.

A century and a half ago that notorious French traveler the Marquis de Custine wrote, "The political system of Russia could not survive twenty years' free communication with the west of Europe."[117] It is clear that his observation overstates the case today. But if the regime has survived greater international communication, it has increasingly had to respond to information from abroad, the importation of which it can no longer control. No longer willing to pay the price necessary to control international conduits completely, the state attempts merely to minimize the negative impact of the information they convey. Implicitly it acknowledges that the internationalization of information is inevitable.

Communications Technology and Social Individuation

Much has been written about the way cassette recorders, VCRs, photography, and other small technologies not only privatize communications but individuate the communicators. Such individuation is one of the strongest currents in Soviet society today, helping explain phenomena as diverse as the rising prestige of careers in writing and the burgeoning fashion industry.

Existing small technologies in the USSR foster this individuation because they enable people to exercise choice among the aural and visual sources from which they draw information. Desk-top personal computers have the same impact, since they allow people to choose and, if necessary, generate data pertinent to their interests.

Individuation extends even to such top-down media as television. Viewing a movie in a theater places limits on one's response. Seeing the same movie at home frees the individual to react actively and independently. While it is true that all three Soviet television stations still air the news program "Vremia" (Time) at the same time, this practice has been attacked publicly in the Soviet press on the grounds that it suppresses choice.[118] Similarly, the state controls nearly all newspapers and periodicals, but their sheer proliferation enables readers to seek out what interests them, again expanding the realm of choice.

The exercise of choice over information emancipates the individual from his surroundings. A cassette tape of a foreign pop tune that finds its way into the hands of some provincial teenager may conjure up the existence of an alternative life, of some other world where freedom and eros are untrammeled. Suddenly his immediate environment becomes nothing more than the drab setting from which the taped tune emancipates him.

Choosing among the welter of information carried by new technologies, a subject is transformed into a citizen, eager to exercise broader choice over all the decisions that affect his life. Eventually the political system must accommodate that citizen and the individuated personality that is his essence.

The Growth of Networks and Groups

Amateur builders of outlandish homemade aircraft held a convention at an airport outside Moscow in September 1987. Brought together at the urging of scientists in the capital, these inventors and their craft attested to the existence of a nationwide network of Soviet Rube Goldbergs, most of them known to one another and communicating through the mail, telephone, and personal travel.

Such networks exist in hundreds of fields in the USSR. People interested in unusual sports, various forms of collecting, virtually every marginal field of culture have organized themselves into informal lateral networks with little or no support from the state and often wholly independent of it. Hundreds of groups are chartered as societal (obshchestvennye) organizations. Others thrive without official recognition. While less institutionalized than the major formal organizations, they have the advantage of being sustained by the members' genuine enthusiasm. The proliferation of such organizations owes much to social and educational change, but it could not have occurred without vastly improved conduits of horizontal communication.[119]

This mode of self-organization is ideally suited to those promoting special interests. When Moscow's city planner M. V. Posokhin proposed to cut the new Kutuzovskii Prospekt through the historic core of the city, opponents organized the now notorious Memory (Pamiat') group. Over the fifteen years of its existence, Memory has gained branches in Leningrad and Novosibirsk and maintained informal communication on issues pertaining to historic preservation through intercity telephones and open mails.[120]

Similar groupings in the ecological field have existed for years, only the best known of which deal with the problems of Lake Baikal. In a typical effort at co-optation, the Leningrad Komsomol organized the association BER, which quickly aligned itself with a coalition of unofficial youth groups publishing a samizdat journal and advocating, among other projects, a monument to the victims of Stalin. The Moscow Perestroika Club made similar demands, and in August 1987 had the opportunity to express them at a convention of similar organizations held in the capital under the patronage of the Moscow branch of the Community Party.[121]

Unlike the nineteenth-century *zemstva,* whose efforts to federate nationally were easily thwarted, the new groupings can proliferate and federate easily, albeit informally, simply by using the networking potential of the new communications media. In their informality, their horizontality, their openness to all supporters of a given cause, and in their participatory character, made possible by the telephone, such groups contrast sharply with both the Communist Party and the organs of state. They thus pose a fundamental problem to the Soviet leadership. In the autumn of 1987 V. M. Chebrikov, chairman of the KGB, delivered an astonishing and measured assessment of these organizations: "A characteristic feature of our time is the marked increase in the Soviet people's social activeness, clearly manifested, in particular, in the creation of independent associations whose participants seek to contribute to the development of this or that aspect of public life. The CPSU regards the activity of such associations as a concrete manifestation of socialist democratism."[122] The KGB chief then went on to decry the fact that "extremist elements" have penetrated the leadership of certain of these associations, "taken to the streets to make unwarranted protests in public, advanced provocative demands, and fulminated against those who disagree." Yet while he charged that these extremists were under the sway of "foreign subversive centers," Chebrikov, like the Leningrad Komsomol, seems to have accepted the inevitability of autonomous organizations. Indeed, by mid-1988 Communist officials advocating Gorbachev's reforms were themselves proposing the establishment of mass organizations independent of the Party as a means of strengthening their cause. Such entities were actually created in Latvia and elsewhere and represent the Party's acknowledgment of the existence of change in the nation's political culture.

Proliferating Communications Technologies and Surveillance

Governmental surveillance of private communications was simple in a society in which potentially significant communications were limited to a few educated people using a limited number of public technologies. Now the number of communicators has soared, and numerous private technologies serve their individual and group needs.

Even before Chernobyl there was ample evidence that an autonomous and internationally linked communications culture had grown up among the Soviet people. To be sure, this culture has not broken through a number of barriers which in Poland were penetrated early by the Solidarity movement. It has not, for example, created an autonomous radio beyond the level of ham operators; it has not launched publishing efforts on the scale of Poland's NOWA enterprise; it has not exploited videotape and film to the extent done by Video NOWA; and it has not managed to establish an independent newspaper on the scale of Poland's *Robotnik*, with a national circulation of twenty thousand.[123] Nonetheless, the autonomous communications culture of the USSR has shown sufficient strength for officials to deem it unwise to attempt to destroy it.

For such an effort to succeed, it would have to cut back much of the officially sanctioned communications system as well. Since jamming cannot blot out all international broadcasts, legally acquired shortwave receivers would have to be banned. The use of intercity telephones and mails would have to be severely restricted and intercity travel sharply reduced so as to thwart the transmission of independently reproduced sound, video, and print data. All this could be done, but it would require vast expenditures in money and manpower to reach anything like the former level of surveillance. The economic cost would be staggering, while the price the regime would pay in terms of public support would be greater still, particularly if it resorted to force, as would probably be necessary.

Undermining the Party's Role as Culture Maker

Such considerations suggest that the new communications culture is largely irreversible, even if the Soviet regime did wish to abolish it. And who would staff a party or government that would undertake such an effort? The same process of individuation and pluralization that has affected society at large has been felt among those running the official media. When the volunteer civil defense organization DOSAAF recently decried the erosion of communist values, it attacked not the independent small technology media but all the television, film, and radio industries of the Soviet state.[124] In effect, it acknowledged that the masters of these official conduits had come to share the same individuated and pluralistic values that permeated the broader culture. This being the case, there would appear to be too few Bolshevik traditionalists—Stalinists, in the reformers' terminology—to staff the input end of Lenin's conveyor belts today.

No careful reader of the Soviet press in recent years would be surprised by this assertion. As early as 1982 Soviet cultural leaders were publicly debating "mass culture." It is clear, declared the staunchly Leninist head of the Moscow Union of Writers, that mass culture is unrelated and even hostile to socialist culture and the socialist way of life: "It is its polar opposite."[125] Yet in the course of the 1982 debate it became apparent that mass culture was already a reality in the Soviet Union, and that this more independent and market-related phenomenon represented a loss of the party's cultural leadership.[126] By no means did everyone consider this a bad thing. One writer saw the freer operation of market mechanisms in publishing as likely to benefit good literature as much as bad, since it provided an alternative to the moribund bureaucracy in publishing.[127]

Through such debates Soviet commentators struggled toward accepting the new reality of public opinion. Their conclusion can be easily summarized: that mass culture is not controllable from above; that many, if not most Soviet citizens are drawn to it; and that such attraction is the obverse of the public's alienation from those cultural values promoted by the Communist Party.[128]

This 1982 debate came increasingly to focus on the new technological media as such. In the process, the position of the old intelligentsia came very close to that of conservative Party leaders, for both feared the way their status as shapers of public values was being eroded by television and film. Both understood that the vanguard role that the Russian revolutionary movement had assigned variously to the intelligentsia and the Party was being diminished by the new technologies. Writer Andrei Bitov's fulminations against mass culture thus paralleled those of Party apologists, although they began from radically different premises. Both look to the age of democratization with deep skepticism.[129] This is not to deny that intellectuals, especially those of the generation that reached maturity in the late 1950s, have played a central role in Gorbachev's reform movement. But the very nature of the changes they advocate will eventually broaden the degree of public participation in political life and hence weaken their own role, which has in fact occurred among the younger generations. This helps explain the frequent attacks on the young by reformist intellectuals who realize that popular culture is incompatible with their own role as an independent source of values.

That the realm of culture and values has gradually gained independence from Communist Party edicts in the Soviet Union is evident from recent developments in virtually every field of expression. What remains to be seen is the extent to which the Party will accept this reality by reducing its expectations.

What if it fails to do so? It can attempt to reimpose Stalinist controls on horizontal communication, which I have acknowledged to be possible, but only at an exceedingly high price. Alternatively, it can simply adapt received institutions. This too seems unlikely, for such a policy, carried to its logical limit, would deeply undermine the position of the Communist Party in Soviet society. Admittedly this is the effect of various proposals put forward by Gorbachev at the June 1988 Party conference, but he balanced them by calling for the strengthening of the central executive power. Finally, it can choose to move neither backward nor forward, in which case state and society will remain at loggerheads, as was the case prior to Gorbachev.

Given both the need for change and the strong opposition to it in some quarters, some combination of the first two variants seems most likely, with a strong movement toward accepting the new realities, limited by the Party's commitment to maintaining as much initiative and power as the changed circumstances allow.

This overview of communications in Russian history suggests several conclusions. At the least, it demonstrates the close relationship in Russia between political development and the state of communications technology. In most eras the two have been closely connected, with progress in one inseparable from progress in the other. Many anomalies in Russian social development—the slow appearance of an urban elite in the seventeenth and eighteenth centuries, the isolation of the peasantry from politics in the nineteenth century, and the diminished role of the bourgeoisie from 1917 through the 1950s—are reflected in and amplified by the communications system. This history suggests an answer to the question of whether the Soviet Union's large new technical and managerial class will develop communications technologies capable of serving its own needs, as distinct from those of the Communist Party. Against the background of earlier history, the burden of proof would lie on the side of anyone claiming it would not.

This overview of Russian and Soviet history from the standpoint of communications technologies suggests the need to revise accepted notions about several key moments in this century.

First, the march of communications technologies in the late imperial era contrasts sharply with historians' arguments about the internal decay of the social structure that supported semiconstitutional government in Russia. There is no evidence that the emerging communications system of the period 1900–1917 was collapsing from within and ample evidence that it was burgeoning. The present era appears as the lineal descendant of the late imperial phase, after skipping two generations.

Second, it is hard to view the era of the New Economic Policy as representing something wholly separate from the Stalin era in the sphere of communications, as is often claimed in the area of political philosophy. The abolition of private printing, film, and record production, the cessation of private automobile production, and the thwarting of telecommunications at both the intercity and international levels all went forward as rapidly as the Party could promote them, the process beginning under Lenin himself. The pace at which the Party severed horizontal communications was defined less by philosophical or legal limits than by raw power. While a careful review of Lenin's writings may reveal differences between his and Stalin's approach to communications technologies, their actions differ more in degree than in kind.

Third, the reconstruction of horizontal communications and development of feedback systems and interactive media after 1953 proceeded steadily throughout the Brezhnev era. Whatever stagnation might have occurred in the broader economy, modern horizontal communications continued to develop rapidly up to Brezhnev's death. Indeed, the pronounced breakdown of vertical communications in the late Brezhnev era actually stimulated the development of horizontal links within society and hastened the creation of the situation that exists today. What is taking place at present can thus be seen as the fulfillment of changes begun in the 1960s and 1970s rather than their refutation.

What, finally, is the essence of this fulfillment? The expansion of communications technologies in the USSR has fostered both horizontal and vertical links. In terms of social impact, however, the horizontal ties have predominated, and are reinforced by the increasingly interactive nature of the old vertical ones. This has created a kind of pluralism of information in the Soviet Union quite unlike anything else since 1917. Old monistic models of Soviet politics seem less and less appropriate as new patterns of communication deepen.[130] Each network and group arising from the new pluralism boasts its own body of information, and each is therefore capable of providing an independent input into the political process. Together these changes are creating what is recognizably a "civil society" in the USSR.

As I have noted, this technotronic glasnost still lags far behind what exists in Poland, which in turn remains far removed from the style of communications prevailing in the parliamentary democracies of Western Europe, North America, and Japan. Nonetheless, it is far closer to these prototypes than to anything seen in Russia since 1917 and may eventually lead to a very different type of political order than has heretofore existed. Gorbachev acknowledges as much when he speaks of democratization not as a goal but as a fact. He also affirmed it when in the spring of 1988 he appealed to the public to support his reforms in the face of opposition from many in the Party and state. Under such a new order society may remain partially controlled, but it in turn exercises a control of its own, thanks to the existence of autonomous channels of communication. Such circumstances impose absolute boundaries on absolute power. They limit the government's ability to shape society and introduce the possibility of society shaping government.

This situation exists today only in embryo. Even in its present form, however, it exhibits many characteristics commonly associated with the notion of civil society—for example, the free flow of information within society, the ability of individual groups to articulate their demands, a government subject to control by the governed, and the existence of rights against the state as well as duties to it. Technotronic glasnost does not itself create these conditions, but it provides fertile soil in which they can grow, and therefore represents a profound source of change in Soviet politics in the waning years of the twentieth century.

· 2 ·

Information Technologies and the Citizen: Toward a "Soviet-Style Information Society"?

Seymour Goodman

The Soviet Union will be compelled to develop and apply the information technologies (IT) on a large scale if it aspires to be a twenty-first century superpower.[1] These technologies are fundamental to solving problems concerned with the location and efficiency of political and economic controls, with industrial and military modernization, and with the need to present the image and substance of a progressive society to the world and to its own people.

How is the Soviet citizenry likely to see and be affected by these technologies? At one extreme is the possibility that the state could exploit these tools to institute Orwellian surveillance and control over the people. At the other extreme is the possibility that information technologies could greatly expand the freedoms of the Soviet citizen, producing an open, widespread information society not unlike our own. Neither scenario is realistic.

If the appropriate technologies had existed in Stalin's day, the Orwellian nightmare might well have come to pass. The internal security organs were so dominant under Stalin that the elimination of massive and unpredictable terror has been one of the greatest tests of Soviet stability since his death. Although the system of coercion has never been dismantled, it is now more refined, more discreet, and more predictable; many Soviet citizens may even regard it as an integral part of a reasonable social order.

More than likely, the KGB and the MVD do use information technologies for sociopolitical control. Applications such as large centralized data bases on mainframes, microelectronics-based surveillance, and personal computers for case officers are of obvious value to these agencies, and are almost certainly within their means. There also appear to be few constraints to prevent such applications. In time, the security agencies may even undertake selective surveillance through large collective networks such as the telephone system or—if they come into common use—computer networks for communication, vehicle registration, or travel lodging.

On the whole, however, Soviet citizens probably have less to fear from the use of IT for surveillance and other intrusions into their lives than do Americans. There are far fewer entry points for such intrusion. Soviet citizens are "protected" by the relative backwardness and simplicity of their society, economy, and private lives.

The state could deliberately create such entry points by setting up elements of *1984*-style control. But these are not at all likely to materialize in this century. Any nationally pervasive system for controlling people and information would strain the limits of current technology and would be outrageously expensive. Moreover, no such system could be imposed on today's Soviet population by anything short of a return to Stalinist oppression, and any effort to do so on a large scale would likely be resisted to a degree that would threaten the stability of the USSR.

Perhaps more important, the security agencies do not *need* to go to Orwellian lengths to control the population. In the transition from the mass arrests and prison camp sentences under Stalin to the punitive use of psychiatric wards under Brezhnev and his successors, the organs of internal security have shown themselves to be flexible and capable of dealing with political deviance. In response, most Soviet citizens have developed to a high degree the kind of self-discipline necessary for avoiding conflicts with the authorities. They have also been conditioned to believe that Soviet surveillance technology is already both potent and pervasive, although this conditioning may have eroded a bit over the last few years.

But just as information technology in the Soviet Union is not likely to become an instrument of enslavement, neither does it seem destined to spread with the same abandon as in the United States. Various attempts have been and will continue to be made to control the spread of IT in the Soviet Union and to relate the development and application of these technologies to nationally determined goals. Moreover, in their pursuit of these goals, the state and the Party will be limited by the need to maintain the stability of Soviet society.

This stability derives from a variety of sources.[2] Among others, they include upward mobility (through education, achievement on the job, and Party membership), full employment, a relatively low level of economic stratification, shared cultural values and interests between the political elite and the working class, an ideology that has produced required and pervasive forms of official culture, and a history of subservience to authority. With some exceptions (for example, the use of the broadcast media to establish and disseminate the official culture and the fact that certain high-tech consumer goods have become the object of privileged acquisition), the information technologies by and large have had almost no impact on these institutions. But what about the future?

The same technologies that must proliferate in order to strengthen the Soviet economy have the potential to weaken the pillars of stability. For example, factory automation is good for productivity but bad for job security; personal computers and networks are good for engineering education and communications but bad for keeping dissidents in check; VCRs are good for mollifying consumers but bad for preserving traditional Soviet values. In each case the Soviet authorities will attempt to weigh the benefits against the risks. But is it likely that the Soviet authorities will have less control and influence over the ultimate disposition of benefits and risks than has been the case in the past? Will the spread of IT lead to an "informatization" of Soviet society, bringing with it a form of democratization that will be essentially irreversible?

This chapter considers some of the conflicting pressures to expand the use of IT in the Soviet Union and to preserve the status quo. It concludes with a necessarily tentative and speculative discussion of some of the possible humanistic features and prospective changes in the Soviet social order that may accompany the more widespread introduction of the information technologies.

Pressures for Expanding and Developing the Information Technologies

Over the past several years deliberations among the Soviet Union's political leadership and technocratic elite seem to have produced the following four principal IT-dependent goals:

1. To attain real gains in productivity and to modernize the industrial base.

2. To improve and modernize the economic planning and control mechanism.
3. To support both military and internal security needs.
4. To present the image and some of the substance of a progressive society both to the people of the USSR and to the outside world.[3]

It is not apparent to what extent these goals can be attained by the end of the century, but any effort to meet them will require a massive infusion of IT. As a result, the Soviets are under tremendous pressure to step up the development, effective distribution, and application of IT systems.

Other forces will push them in the same direction. One such force is the rising expectations of Soviet citizens. There is a growing gap between East and West in the production of consumer technology. In the industrially advanced Western countries most people now live far above the subsistence level for housing, food, and education. Standard of living is increasingly defined by the range and availability of services and products, by career and leisure opportunities, by personal communications, and by the quantity and quality of opportunities for entertainment. The developed and developing countries of the West and Far East (hereafter loosely referred to together as "the West") are meeting these needs with an unprecedented variety and volume of electronics-based products and services. The USSR, meanwhile, is producing little of this technology and exporting even less. Imports are reserved for the elite and near-elite and for high-priority or high-profile applications. And even with the rise of glasnost, the country remains largely outside a world that is increasingly able—and increasingly inclined—to communicate among its members.

In the past this East-West gap has had little effect on the expectations of most Soviets. The average citizen still has rather modest criteria for standard of living and quality of life. As Seweryn Bialer writes:

> What is salient for the Soviet citizen—and hence crucial for the stability of the system—is not how this level of material well-being compares with Western society, which he has never seen, or even how it compares to that of the political elite, since their conspicuous consumption is not reported in the media. The benchmark by which the citizen assesses his standard of living is that of his parents and of his own past.[4]

Since the end of the Second World War, this benchmark has consisted mainly of the basics: food, housing, health care, public transportation, and so on. But as citizens become more aware of Western-style information societies—often through IT channels of one kind or another—their perceptions of what contributes to the quality of life are likely to change. Increasingly, happiness may be equated with owning telephones, VCRs, and personal computers. And the state will not be able to get its citizens to contribute adequately toward its four principal IT-dependent goals unless it provides or permits more of what they want.

But improving the performance of both blue-collar and white-collar workers will be more than a matter of raising their standard of living. The average Soviet worker is simply not up to the task of helping to transform the USSR into a robust economic power. Life under the Soviet system has produced a being known as New Soviet Man. He is depicted in posters as a heroic soldier of the Revolution, but a more accurate portrait is painted by Mikhail Heller and Aleksandr Nekrich:

> The founding fathers of the October revolution took upon themselves the task of creating a "new man" . . . At the Twenty-fifth Party Congress, in 1976, Brezhnev called Soviet man the "most important result of the last sixty years." The general secretary of the CPSU was exactly right. During those sixty years the party's main efforts were concentrated on the Sovietization of its people . . .
>
> "The people of our country submit uncomplainingly to all the shortages of meat, butter, and much else. They put up with the gross social inequality between the elite and the ordinary citizens. They endure the arbitrary behavior and cruelty of the authorities . . . They do not speak out—sometimes they even gloat—about the unjust retribution against dissidents. They are silent about any and all foreign policy actions" [attributed to Andrei Sakharov, 1980].
>
> This is the portrait of Soviet man, the product of seventy years. A population that has lost hope for a better future and lives in fear of tomorrow is an essential factor in the stability of the Soviet system.
>
> A new human community has come into existence in which no one has rights, but each possesses a tiny share of power: he can work poorly, mock the customer if he is a sales clerk, denounce his neighbor, and be arrogant towards little people if he is a civil servant. He can steal, and give and take bribes. This bit of power is always gained by an abuse or infraction of official legislation, to which the state closes its eye . . .
>
> The regime's stability is explained by a new kind of "social contract"; the citizens surrender their freedom to the state, and in exchange the state gives them the right (under its supervision) to abuse their positions and violate the law. At the same time the state guarantees minimal conditions for survival.[5]

In the past this New Soviet Man might have been necessary or desirable to "preserve the Revolution," to maintain internal political control, and to enable the country to achieve military superpower status. He may suffice for a long time to come if the USSR continues to be socially and economically isolated, or at least to deal with the rest of the world on its own terms. But this form of New Soviet Man is not up to achieving the four principal IT-dependent goals of perestroika. A New New Soviet Man may be necessary.

The need to cultivate this better disciplined, better motivated kind of worker or manager will place additional pressure on the Soviet authorities to expand the use of information technologies. The days of Stalinist coercion are over. To impose discipline in the workplace, the Soviets will probably need to adopt methods similar to those that prevail elsewhere in the world: positive incentives such as better working conditions and equipment, more professional challenges and opportunities, and improved information channels—as well as negative incentives such as demotion, job redefinition, and perhaps even some automated monitoring of work. All of these measures will entail far greater exposure to information technology than the state has so far been able to provide or been prepared to allow.

To add to this pressure, the heightened technological awareness that workers gain on the job is likely to spill over into their private lives. They will soon start to notice a vast and unnecessary technology gap between the workplace and the home. And unlike the more elemental gaps between Eastern and Western standards of living, this will not be a gap that can be filled with simple material provisions like meat and milk.

Finally, the very size of the Soviet Union exerts its own pressure on the authorities to loosen their control over the flow of information. So much information is necessary to manage a large, advanced industrial (not to mention postindustrial) society that an information monopoly among a small leadership elite is no longer possible if such a society is to be run effectively. The regime can retain or refine many forms of power, but strong social control does not necessarily require a rigid and pervasive hierarchy. Soviet society has enough sources of stability for the political order to survive a certain amount of decentralization and dissemination of information controls. Of course this implies the need for trust in, and participation by, a much larger number of people.

The history of the USSR since 1953 shows cautious but steady movement in this direction. Under Gorbachev, more than at any other time since the end of the Second World War, international developments and pressures are such that this movement will have to be continued and intensified. And increasingly the pace will be dictated by developments in information technology.

The Power of the Status Quo

Responses to these pressures have been, and will continue to be, constrained by the inertia and self-interests of the people and institutions that have characterized the USSR for decades.

Among the strongest of these counterpressures is the authorities' desire to maintain control over Soviet society. Technologies that give people more control over their own lives—enabling them to see, hear, read, and disseminate whatever they want—present an obvious challenge to the regime. And there is no guarantee that the citizens' hard-learned ability to police themselves will prevent abuses, especially on the part of dissidents and other factions.

Despite over seventy years of efforts to impose cultural and political unity, the Soviet authorities still have to contend with a number of troublesome groups. These include not only the political, religious, and nationalist dissidents but also those groups that practice cultural experimentation and innovation, such as writers, musicians, and artists. At present these elements are small and are regarded mostly with apathy or hostility by much of the rest of Soviet society. But suppose they had easy access to PCs, printers, modems, electronic bulletin boards, and extensive telecommunications networks. Would this longer technology-aided reach allow them to acquire much greater influence? Would they evolve into a sizable intelligentsia of the sort that essentially disappeared along with the tsars?

And what of the rest of the population? Conceivably information technologies could bring large numbers of people from many classes out into the open, with intellectual, spiritual, and entertainment demands that would be better met by groups outside Party control. The state tries hard to "protect" various ethnic minorities, such as Jews and Chinese, from better communication with, and cultural influence from, their brethren abroad. Even the peasantry could become a source of worry. How would rural life change if the younger people in the boondocks (imagine what they do at night now) had access to a greater volume and variety of entertainment?

For the time being, at least, the Gorbachev administration is in no rush to test any of these conjectures on a very large scale. Certainly the current form of glasnost falls far short of undertaking such a test.

A related concern is that a proliferation of information technologies could threaten the already attenuated hold of ideology. Marxism-Leninism is an important official form of Soviet culture. To some extent it helps provide stability by encouraging morality, respect for the law, and obedience to and recognition of the existing political hierarchy. But over the years this has become a stability of rote and suffocation. Although the ideology has been somewhat updated since the 1960s—modifications under the so-called Scientific Technological Revolution have included additional flexibility for technological modernization and a more lenient view of Western progress[6]—it is doubtful if these changes have done much to revitalize widespread interest in ideology.

Whether most students and Communist Party (CPSU) members actually devote much time and effort to ideological studies is debatable. It is safe to say, however, that if access to the entertainment and other consumer applications of IT is greatly expanded, attention to ideology will suffer. Of course the authorities may try to take advantage of these new media to present the ideology in more palatable forms. Nevertheless, both the time and the effort most people put into absorbing ideology will probably decrease. Information technology provides means for developing and disseminating alternative cultural perceptions and tastes; thus, as it has done with religion in the West, it could weaken the position and influence of Soviet ideology. This, in turn, could lead to changes in morality and to Westernization.

Yet the demands of maintaining social control are only part of the reason why the Soviets are reluctant to embrace IT without restraint. They are also worried about job security at all levels. In spite of outpourings of "progressive" rhetoric and recognition that modernization is necessary across a broad spectrum of domains, the political and managerial elites and ordinary workers are greeting the information technologies with caution and concern. The jobs of many senior officials in the Gorbachev administration will depend on how well they implement programs for advancing the state's IT-dependent goals. The pressure will be particularly heavy on people at the ministerial and enterprise management levels. Unlike those "liberal" academics who are among the most vocal advocates of perestroika, these officials will be visibly accountable and blamable, and will have to work with new technologies and procedures that are poorly suited to their backgrounds.

Lower down, among the middle-level professional classes, IT seems to be regarded as both an opportunity and a threat. On the one hand, it presents these people with substantial opportunities for career advancement, higher profile, greater professional sophistication, and a higher standard of living. On the other hand, it has the potential to change broadly and deeply the way they do things. Thus it poses a serious risk to people whose careers have developed in very conservative industrial and research and development environments, where risk-free, steady state work has been the way to survive and prosper. The information technologies also make possible more timely, direct, and effective communication from above. This may mean more control and interference on the part of high-level management, as well as the elimination of many middle-level management and engineering jobs. Not everyone in the middle-professional classes may be happy with those prospects.

It is not in the interests of most of either the political-technocratic elite or the middle-level professional classes to push restructuring too far and too quickly. To be sure, they appreciate the need for somewhat greater decentralization and for entrusting information technology to larger segments of society. But at the same time, all of them have prospered within the current system, and all would have something to lose—and certainly a lot of adjustments to make—if the system were to change very much very fast. In the short term, at least, perestroika appears unlikely to do away completely with such basic features of the system as centralized planning, price and currency controls (which can be seen as strong forms of protectionism for inefficient industries and people), organizational rigidity, or many of the other factors that have inhibited the effective and widespread use of computing at the enterprise level in the Soviet economy over the last twenty years.[7] Nor do many technocrats and workers seem prepared to press for spiritual or cultural innovations and freedoms that go far beyond the current limits of glasnost.

The Soviets are also plainly worried about the "psychological barriers" to the absorption of the information technologies in the workplace. This term has been used as a catchall to cover every problem from technoshyness to the workers' and managers' desire to continue various forms of questionable economic and social practices. For example, a Bulgarian study of the reasons for worker resistance to the introduction of new technologies revealed the following seven serious concerns:

Fears of a cutback in personnel and the loss of bonus remuneration guaranteed by the old working conditions.

A reluctance to retrain as required by scientific and technical innovation.

A fear of the need to take on more of the work that had previously been allotted sparingly; a fear of risk and a lack of confidence.

Fear of increased demands in the profession.

A conviction that only the showcase value of the introduced innovation was being sought.

The view that the money used on scientific and technical progress could have been more effectively spent.

Skepticism over the qualities of the introduced innovation.

Another general conclusion of the Bulgarian survey is also worth repeating:

> All of this has not only a psychological sense but also a real reflection in the life and activities of the collective, in the conduct and actions of individuals and in their work and social activity. Albeit very rarely, there still are individual members of a collective [who] more actively resist innovations. Usually the resistance is not apparent but is expressed in so-called "cosmetic measures," that is, a fictitious, external change while maintaining the old essence or, in other words, only "the reporting is new." But there are also instances of overt resistance, of purposeful and well-founded actions. Certainly it is another question of whether these actions are valid or not.[8]

In one form or another, the same fears have surfaced in the USSR. Some of those who "more actively resist innovations" go so far as to physically sabotage robots.[9]

Beyond all the specific concerns of the leadership and the general population, the information technologies face another, more general impediment: a widespread aversion to open communication. The society of New Soviet Man is highly cellular, and often highly suspicious of outsiders. The predominant social unit is the small group of people who all know one another well, and there is little trust across units. In the workplace, where a bureaucratic, insider-outsider, finger-pointing mentality prevails, the Soviets have never been forthcoming with information they were not forced to share. This applies from top officials down to the lowest-level clerks and civil servants. When this general problem is compounded by severe technical deficiencies, it is not surprising that it is so difficult to find large, successful, advanced, nonbroadcast IT systems in widespread use in the Soviet Union.

Toward "Enlightened Collectivism"?

In the coming years the Soviets will have to strike a balance between the conflicting quests for modernization, as defined by the four principal IT-dependent goals, and for the preservation of past forms of stability and control. What emerges will no more resemble a Western-style information society than Gorbachev's emerging Soviet-style democracy will resemble Western-style democracy. The CPSU has no use for the turmoil, pace, and freedom (or, in the view of some Soviets, near-anarchy) of either. Not surprisingly, past and current Soviet initiatives and experiments with the not unrelated problems of proliferating information and democratization exhibit strong, controlled top-down features.[10] It is doubtful that they can achieve a broad, stable balance for any long period.

The most likely changes over the next several years range from modest increases in technology applications narrowly related to achieving the state's four main goals to the attainment of some more ambitious vision of "enlightened collectivism" that has yet to be defined by Soviet academics or technocrats. In neither case are most Soviet citizens going to reap many of the benefits that the information technologies have been bringing to their counterparts elsewhere in the world. In fact, even if the USSR undergoes social and economic reforms as radical as those in Hungary—a scenario that is not inconceivable—information technology will still be more tightly controlled and less pervasive than in the West.[11]

Under any circumstances short of a massive neo-Stalinist regression, we can expect to see a further opening up of information flows in the Soviet Union. For many Soviet citizens much of this will consist of the low-tech forms described by S. Frederick Starr (see Chapter 1 of this volume). These low-tech flows, however, are likely to generate increased demand for high-tech means. Conversely, the already rapidly growing demand for IT communications and consumer and commercial products has enormous potential for further opening up the flow of information, thus forming what might be seen as a demand feedback loop.

The "technology controllers" among the Soviet authorities have various tools for dampening the growth of such demand feedback, and of consumer demand more generally. These tools include technical weaknesses (for example, products with very poor reliability); shortages (whether contrived or owing to the honest inability to make and distribute suitable products); high prices (or, effectively in some cases, non-ruble prices); infrastructure problems (for example, the lack of widely available service); co-option (for example, the production of ideologically sound Soviet-made VCR tapes in an attempt to decrease the demand for Western tapes); incompatible standards (for example, Western tapes cannot be played on Soviet-made VCRs without some technical modifications); and administrative controls (see the example of photocopying machines in note 10). While one might argue that these controllers are fighting a rearguard action against the wave of globalization of IT and the need to make changes to achieve the IT-dependent goals, it is also important to note that they have been fairly successful so far, giving ground only slowly.

For decades Soviet citizens have invested enormous amounts of time and energy acquiring assorted goods and services through black or gray markets, as well as through various lawful forms of hustling. They know that they can get some of the things they want if they try hard enough. For their part, Soviet authorities have generally been much more tolerant of such deviant entrepreneurial economic practices than they have been of deviant political behavior. Many observers have argued that widespread permissiveness in this area has been a "substitute for political aspirations and to some extent performs the function of a safety valve for the pent-up dissatisfactions in Soviet society."[12] It has also served as a useful surrogate for other, more common (in terms of Western consumer practices) means of improving one's standard of living.

Consumer electronics products have been in enormous demand in the USSR for several years. Demand is so high and intense that these products sometimes leverage ten to twenty rubles to the dollar. In this domain unofficial private entrepreneurial economic practices (such as charging fees for home showings of Western videos) have been fairly widespread, imaginative, and difficult to control completely or to co-opt. Some of this activity is now in the process of being expanded via the new, officially approved private cooperatives. In fact at times it seems that almost every young programmer in the USSR has some vested interest in one or more of the emerging cooperatives. Nevertheless, the number of consumer products like VCRs and personal computers in the USSR remains a tiny fraction of the number in the United States.[13]

Not only will the information technologies be at least partially controlled, but it is also likely—given Soviet capabilities, goals, and concerns—that their absorption into the economy will be very uneven. There are prospects for substantial Soviet achievements in some application areas in industry and the military. Among others, these include process control, computer-aided design, and avionics and space systems. But large parts of the Soviet economy may fall even farther behind than they are today in comparison with industry's leading and "show" elements, with the West, and with parts of the Far East and even Eastern Europe. In effect, it is very possible that islands of moderately advanced facilities and applications will develop in what will continue to be a sea of backward work environments.

This uneven distribution of IT may have to hold true not just for industry but for society as a whole. For the Soviets to meet their principal goals, a larger segment of the population must be enlisted to contribute directly to economic revitalization; the state will need to foster a New New Soviet Man—someone with sufficient discipline and motivation, and who can be relied on not to abuse the freedoms conferred by new technologies. Yet it will not be necessary to offer unbridled trust to all elements of Soviet society. Significant progress can be made by placing additional trust in those relatively small but expanding groups that already have a strong stake in, and contribute to the strength of, the system, and that seek greater participation in the system for their own benefit. The Old New Soviet Man, necessary to preserve some of the basic features of stability and control in Soviet society, may continue to make up most of the population.

The actual composition of New New Soviet Man remains to be seen, but it will probably include upper-level managers and intellectuals such as academics and journalists. Compared with Old New Soviet Man, these people will be better educated, more responsible, more trustworthy, and no less nationalistic. It will be possible to buy their loyalty with a share of power and privilege. Some of this power and privilege may come from progress toward a more participatory type of democratic centralism— one in which the upper political and technocratic elites continue to dominate but in which these rising groups have a voice. If they help bring economic success, there will be a larger stock of power and privilege to share.

As this new class of citizens emerges, access to the information technologies will become the substance and symbol of their participation.

They will acquire assorted high-tech consumer products and will be exposed to better and more interesting entertainment and cultural possibilities, including significantly loosened censorship. We can expect to see personal computers and printers showing up in their homes and offices. For the short term, however, and perhaps even over the longer haul, these gains in standard of living will be quite modest by Western standards. And most of them will come along gradually, in a way that is not likely to result in the disintegration of the political order.

For the active participants in economic revitalization, the spread of information technology promises to raise the quality of life and work appreciably. But what of everyone else—the majority of citizens who will not be in the ranks of New New Soviet Men (and what will probably be a much smaller number of New New Soviet Women)? Their prospects appear to be mixed.

As the high-tech sector of the economy grows, so will the need for skilled, computer-literate workers. This may provide the average citizen with more opportunities for upward mobility. In general, exposure to the information technologies is likely to increase. People will come in contact with them through educational efforts such as the secondary-school computer-literacy program and through the new cooperatives. They will also acquire somewhat more in the way of consumer products.

But even with these improvements, the standard of living of the average Soviet citizen will still be far lower than that of the average citizen in the advanced Western countries. The emerging Soviet-style information society has no place for many truly pervasive applications of IT (and especially for the complex, decentralized, and massive supporting infrastructures that are characteristic of such applications in the West) or for most of the Soviet population to become consumers on the scale of their Western counterparts. Moreover, Old New Soviet Man is bound to notice that New New Soviet Man is getting the lion's share of the benefits from information technology, as it will be more difficult to keep this inconspicuous than differences in living standards in the past (automobiles being a notable exception). As a result, the average citizen's expectations may rise in ways that could be hard for the state to fulfill.

The workers may also have to adjust to the job displacements that could accompany industrial automation. For example, in March 1986 Radio Moscow announced that there would be some 16 to 18 million automation-related layoffs by the end of the century.[14]

Workers can scarcely be expected to welcome this by-product of modernization. But from the standpoint of Soviet economic health, a redistribution of labor could be one of the few major short-term benefits to emerge from the technological revolution. Various forms of automation may permit the Soviets to redefine or eliminate the positions of many troublesome middle managers and workers whose lack of skills and discipline is hindering the pursuit of the principal goals. Automation might provide a convenient excuse for remedying some of the slack and underemployment in the Soviet workplace and redirecting workers elsewhere:

> The Soviet workforce has developed into a generally underutilized asset while management has continued to hide its excess labor capacity, as it does with excess material resources and production capacity. In a society where management (Party and Government) mostly work their way up from the shop floors, this deception is universally understood. The resulting situation has become the most pervasive conspiracy in Soviet economic life—and the reason for many of the economic "crimes" (absenteeism, bribing of supervisors, stealing of work materials, etc.) that Andropov and now Gorbachev have been fighting.
>
> But how to turn such a universal situation around? With negative or at best slightly positive population growth in the Slavic population of the Soviet Union and ambitious plans for modernization and industrial expansion, where will the expanded workforce come from? Stalin ripped these assets free; his successors must be subtler.
>
> Technology has presented the first perfect opportunity to extract this hidden labor resource and redistribute it. By introducing "labor saving technologies," Soviet organizations can be ordered to surrender the capacity which has been (theoretically) freed. If the technologies are never used, there will be no real effect on [these enterprises], save the requirement . . . to better utilize [their] remaining workforce.[15]

On what basis did the Soviet authorities predict the displacement of 16 to 18 million jobs? If this conjecture is correct, the figure may correspond to the number of additional workers the Soviets expect to need for industrial modernization projects over the next dozen years. Of course some of the dislocated workers may find productive jobs that are unrelated to these needs—making consumer products, building roads, or working in expanded service or private sectors. But others may find themselves pushed to join the Slavic workforce wanted for rebuilding the Soviet industrial plant or for expanding industrial facilities in the more remote areas of the USSR.

Either way, the workers dislocated by automation will probably be found new jobs. In this regard the Soviet-style information society may turn out to be more humane than the U.S. version, in which workers enjoy no such guarantee. It is easy to fault the Soviet system for the low quality of many jobs and for de facto underemployment. Yet in no small way the policy of full employment contributes to the stability of Soviet society and to whatever positive image the USSR has abroad. Full employment is fundamental to Soviet ideology and is regarded as a basic civil right.[16] Thus, managers of Soviet enterprises where automation is being introduced will probably try harder to maintain employment levels than will most of their Western counterparts—both out of compassion for the proletariat and out of self-interest. Even if these policies change somewhat under perestroika, unemployment owing to automation will not become the problem it is in the West.

There is also some genuine concern about techno-stress in the Soviet Union, and lower-level employees are likely to be subjected to less job pressure and redefinition than may often be the case in the United States. Despite gradual modernization, Soviet society will still be simpler and less technologically advanced than Western society. Just as this way of life offers some protection against electronic intrusions, it may also be better for the human nervous system. Soviet citizens will probably continue to enjoy a less hectic pace of living, less intense competition, and less economic risk in their personal and work lives.

There is so much room for revitalization, and so many opportunities for visible and real improvements, that the information technologies will have at least some initial effect over the next few years. A modest pace of fairly easy improvements (for example, the current form of glasnost and the precollege educational program) would also give the regime time to think about what it is doing and to consider those changes that may be coming about in ways that it has not consciously promoted. It is possible that revitalization may be successful enough—however that success might eventually be defined or rationalized—that it can be sustained to the end of the century. Even if it is not so successful, it may be adequate if Western pressure is decreased—if, for example, the Western economies go into a serious tailspin. One might conjecture that, ideally, the Soviets would like to slow the pace of the rest of the world to something that they can control and be more comfortable with.

It is at least likely that, after three or five or ten years, the Soviets will experience diminishing economic and social returns from perestroika. Meanwhile, rising internal expectations and Western progress with the information technologies will have increased the pressure for additional change.

In either case Soviet society will change. A process of change fueled in part by developments in the information technologies will be under way, and the Soviets will have to deal with this on a nearly continuous basis for a long time to come. New balances of discipline, control, and trust will have to emerge between the state and Party and different groups within society. This will be partly the result of controlled responses and experiments forced by foreign pressures, Soviet goals, and the need for modernization, and partly a result of various forms of the porously controlled spread of products, thoughts, and attitudes. Soviet society in the year 2000 may be transformed to an extent that would have been considered remarkable by a citizen in 1970, perhaps almost as much as Soviet society in 1970 was transformed from the standpoint of a citizen living in 1940.

· PART TWO ·

Biology

· 3 ·

Prometheus Rechained:
Ecology and Conservation

Douglas R. Weiner

It is still less than a century since ecology has become a recognized branch of biology. Among the international community of ecologists there has been little consensus during this brief period regarding the proper scope of their science, its methodology, or its governing principles and paradigms. Because so much of ecological doctrine has remained speculative, ecologists have enjoyed greater freedom than researchers in other areas of science to choose theories and approaches according to taste. As has been the case in Western Europe and in the United States, extrascientific considerations, including political ones, have played a major role in influencing the particular ecological doctrines embraced by Soviet scientists and elites. Accordingly, this chapter seeks to explain the persistence through the 1970s of the belief held by many Soviet ecologists in the existence of closed ecological communities. I shall argue that this ecological belief was embraced in good part because it provided a scientific rationale for a conservation program hostile to rapid development and supportive of wilderness preservation. In particular, such a program was aimed at providing tangible and symbolic islands of autonomy from the state—under the control of the scientific community— while at the same time propounding the view that scientists, owing to their ecological expertise, should play a central role in economic planning and resource management. I shall also explore some of the reasons for the emergence of a new style of ecological thinking, and, correspondingly, some new arguments for conservation, in the Soviet Union during the last decade. "Social ecology," with its new agenda emphasizing the need for attaining a balance between human desires and needs on the one hand and the environment's capacity to provide for them on the

71

other, has arisen to challenge the old belief in the idealization of pristine ecological communities as well as the equally magical conviction that science can attain a complete mastery of ecological processes and environmental problems, solving them through technological "fixes." I shall suggest that the profound changes introduced by Mikhail Gorbachev, which include unprecedented official recognition of the importance of scientific advice and of such values as diversity, aesthetic preferences, and human psychological needs, have undercut the need for Soviet nature reserves to serve as vehicles for the intelligentsia's expression of those values; they can now be defended outright. While scientists and others will continue to support the preservation of biotic and landscape diversity, it will not be under the scientifically shaky banner of the closed ecological community as model of nature theory. Finally, I shall address the question of the comparative history of ecology in the West and the Soviet Union, pointing out how and why the Soviet experience might be considered distinctive.

The Origins of Soviet Ecology

Perhaps as much as anywhere, ecology developed in Russia in intimate connection with conservation values.[1] Leading ecologists from the 1890s held the belief that nature was made up of discrete, self-regulating subsystems called biocenoses,[2] which were characterized by two chief properties: a total, holistic interdependence of their components and a tendency to reach a state of equilibrium.

These notions formed the basis for the zoologist Grigorii Aleksandrovich Kozhevnikov's development in 1908 of the *etalon* concept.[3] This entailed the creation of *zapovedniki,* or inviolate preserves of virgin nature, dedicated to the study of the biocenoses, or ecological communities, which they were intended to incorporate. In addition to serving as centers for ecological research, *zapovedniki* were to serve as *etalony,* or models of healthy, primordial nature. Because economic activities required the alteration of virgin nature, causing pathological changes, Kozhevnikov argued that society needed to know how to reverse that process and recreate a natural equilibrium as an antidote to a potentially fatal dose of progress. The pristine *zapovednik-etalon* could serve as a reference point, preserving parcels of healthy nature as models for the rehabilitation of degraded lands. Moreover, on the basis of their knowledge about the ecological carrying capacities of different types of biocenoses, ecologists could offer expert advice on appropriate land use and other questions of economic development.

Although the network of ecological *zapovedniki* prospered during the 1920s, by the mid-1930s their unique status as inviolate ecological research centers was critically impaired. Economic ministries, notably the People's Commissariat of Agriculture, derided the ecological reserves for pursuing "science for science's sake," and sought to incorporate them into their own networks of *zapovedniki*, which served the more narrowly utilitarian goals of maximizing the propagation of selected, economically valuable species of wild animals. Cultural revolutionaries denounced the reserves as havens for that despised species the bourgeois academic. Finally, Isai Izrailevich Prezent, a close collaborator of the notorious charlatan agronomist T. D. Lysenko, accused the *zapovedniki* of permitting a counterrevolutionary resistance to such key economic programs as collectivization and acclimatization (introduction of exotic animals and plants) under the cover of scientific argument. While we may deplore Prezent's political thuggery, we must acknowledge that there was more than a grain of truth to his charges.

Owing, perhaps, to Prezent's potent political connections, as well as to Stalin's desire to favor only those scientific findings that supported his economic and social policies, Prezent's attacks proved the most telling. Through his direct involvement, the highly innovative work in trophic dynamics of Vladimir Vladimirovich Stanchinskii at the Askania-Nova *zapovednik* was brought to an abrupt end (he was arrested in 1934), and Prezent served notice that holistic ecological doctrines that asserted limits to humans' ability safely to transform nature were now to be regarded not only as flawed but as devised by the class enemies of Soviet socialism. For an unrelated reason (Prezent's discomfort with mathematics), he also proclaimed the unsuitability of attempts at the formal, mathematical description of biological phenomena. The result of these measures, in the words of two of Stanchinskii's students, was that "theoretical research in biology, including ecology and biocenology, was excluded from the work plans not only of Askania-Nova but also of all scientific institutions for two decades at the very least."[4] While this assessment seems to be somewhat exaggerated, especially for the period after the Second World War, there is no doubt that severe damage was done.

The *zapovedniki*, although they continued to increase numerically, could not avoid becoming bases for the selfsame radical transformation of nature that they had originally been intended to prevent. Acclimatization of all manner of exotic fauna and flora proceeded apace, together with such other aggressive management techniques (biotechnics) as predator and pest elimination, winter feeding of certain species, and the introduction of measures designed to change the mix of tree and shrub species to a more economically advantageous one. The organized conservation movement, in addition, was correspondingly vitiated, with membership in VOOP (The All-Russian Conservation Society) falling from 15,000 in 1932 to 2,553 by 1939.[5]

While some initially justified these measures with a theory holding that nature was rife with unexploited opportunities for new niche creation— a refutation of the older view among many ecologists that biocenoses were *closed* systems that represented the most efficient possible utilization of the productive potential on their territory—by the late 1930s and 1940s the great transformation of nature was simply accepted as a given, with no attempt at theoretical justification.

Ironically, just before the development of theoretical ecology was cut short, a tendency arose among the most mature ecological thinkers to emancipate themselves from a priori judgments about the nature of the ecological community. Indeed, it can be argued that such seminal thinkers as Stanchinskii and the geobotanist Leontii Grigor'evich Ramenskii had arrived at the sophisticated view that the biocenosis was just another useful category we impose on nature to make it comprehensible and manipulable. At first Ramenskii waged total war on what he believed to be the idealistic conception of the natural community. In his earliest view vegetation was seen as an unbroken continuum, whose patterns of species distribution could be explained on the basis of environmental gradients rather than on the basis of the structure of some mystical community.[6] Ramenskii admitted the conditional utility of the community concept, but only so long as its arbitrary nature was clearly recognized:

> In connection with the multifaceted inexhaustibility of phenomena there does not exist nor can there be a single all-embracing classification of them, fitting for all times and situations. In fact, such a classification system is not desirable. We need taxonomies firmly linked to specific objectives, helping to solve definite scientific and economic tasks.[7]

Stanchinskii, for his part, in the years just prior to his arrest, had been moving steadily away from the view of the biocenosis as closed. Migratory animals and birds participate in multiple systems, he pointed out, precluding absolute closure. Instead, his trophic pyramids (food webs) were only "loosely ordered systems," to use R. H. Whittaker's felicitous term.

Neither ecologist's colleagues were receptive to these deidealized views of nature.[8] Furthermore, they were too sophisticated to be pressed into serving the regime's crude policy of nature transformism. Consequently, for over a decade a theoretical vacuum reigned over a sullen and silenced camp of holists and a tiny, ignored band of antiholists. The policing of biology by Prezent and Lysenko, combined with the pervasive fear among scientists and editors, all conspired to impose an unnatural silence over ecology, a field so recently brimming with discussion.

Liquidation

Vasilii Nikitich Makarov, nominally only deputy director of the Main Administration for Zapovedniki but in fact its real leader since late 1930, had managed not merely to preserve but to expand impressively the network of nature reserves in the USSR; by 1950 there were 128 reserves with an aggregate area of 12.5 million hectares, .6 percent of the total territory of the country.[9] He had paid a price for that achievement, however: the sacrifice both of the principle of *zapovednik* inviolability and, linked with that, of the reserves' unique role as institutions dedicated to the study of undisturbed ecological communities. Now, in the charged atmosphere following 1948, Makarov suddenly discovered that this compromise was about to be nullified. His new supervisor, Aleksandr Vasil'evich Malinovskii, a Civil War veteran, CPSU member, and former chief of the State Forest Inspectorate,[10] had submitted an astonishing proposal to eliminate 85 percent of the protected territory of the USSR. Taking the existing utilitarian ethos of the *zapovedniki* to its logical conclusion, Malinovskii offered to transfer over 11 million hectares of land from his reserve network to the Ministry of the Lumber Industry and to state farms, reasoning that those institutions would be able to exploit the areas in question more intensively than his own small agency. In return, he requested only the upgrading of the truncated remainder of the Main Administration to the status of an All-Union State Committee; in all, only twenty-eight reserves were to be spared.[11]

Despite a desperate attempt to rally elite support behind the preservation of the reserves,[12] the liquidation went through on 29 August 1951.[13] Mass dismissals of the old ecologist-activists accompanied this reorganization; and with Prezent occupying the dean's chair in the biology faculties of both Moscow and Leningrad universities, there was a real danger that the ecologist-conservation tradition of activism would irrevocably pass from the scene.

Recovery

Just as it seemed that all was lost, the Academy of Sciences stepped decisively into the breach. Losing practically no time, the new president, A. N. Nesmeianov, on 28 March 1952 signed an order in the name of the Presidium establishing a new commission for *zapovedniki*.[14] With an impressive membership headed by V. N. Sukachev, director of the Academy's Institute of Forests, and including Makarov, the committee at once sought to undo the damage of the preceding year.

Malinovskii's days of glory were disappointingly short. On 15 March 1953 his state committee was again downgraded to the Main Administration for Zapovedniki and Hunting (Glavpriroda) within the Ministry of Agriculture and Supplies, and by that same decree fifteen *zapovedniki* were transferred to the All-Union Academy of Sciences or to its republican affiliates.[15] In April, with Stalin's body barely cold, the Academy geared up for a major showdown.[16]

May 1954, the very beginning of the thaw and roughly at the time of Sukachev's first attacks on Lysenko, was the moment the ecologist-activists chose to inaugurate their struggle to liberate the *zapovedniki*. A conference on *zapovednik* matters convened in Moscow, jointly organized by the Moscow Society of Naturalists, the Geographical Society, the All-Russian Conservation Society, the Academy of Sciences, and the Ministry of Agriculture's Main Administration for Zapovedniki and Hunting, now isolated by an impressive display of scientific solidarity. The opening speech, by Professor Varvara A. Varsanof'eva, a geologist and naturalist at Moscow University, set the tone by demanding a return to fundamental ecological research in the reserves. One Soviet historian characterized the meeting as "an arena of uncompromising struggle by scientific public opinion for the restoration of the *zapovedniki*."[17] As Mark B. Adams has shown for genetics during this period (see Chapter 4 of this volume), the old academic ethos, emphasizing the right of scientists to be able to determine their fundamental research agendas independently of state meddling, had lost little of its former vigor despite the prior two decades of intermittent pressure and terror.

Within four months Nesmeianov, meeting with Sukachev and other major conservation leaders, decided to reorganize the Academy's commission into a conservation commission.[18] In conjunction with this, the newly reorganized commission made speedy plans to restore and expand the network of *zapovedniki* as models of "healthy" nature (*etalony*), with Academician E. M. Lavrenko, the botanical ecologist, heading a high-powered policy study group.

The plan the Lavrenko commission released in 1957 could have been drafted by Kozhevnikov himself. At its core was the recommendation that representative *etalony* be chosen on the basis of the geobotanical maps of the USSR that Lavrenko's team at the Botanical Institute had compiled. Once again, the Soviet scientific elite would have its own "archipelago"—islands of natural diversity and research autonomy, embodying the broadest values of diversity and autonomy, a kind of "free territory of the intelligentsia," as radicals had once insightfully charged. Encouraging results followed endorsements by Academician V. A. Engel'gardt, the new secretary of the biological division, and by President Nesmeianov.[19] From a low of forty extant reserves of all systems in 1952 (with 1,465,668 hectares),[20] there were ninety-three by 1 January 1961 (embracing 6,360,000 hectares), led by the Russian Republic's Main Administration for Zapovedniki and Hunting (Glavokhota).[21]

The specter of another *zapovednik* war, not unlike those of the 1920s and early 1930s, increasingly loomed. Repeating history, the agriculture ministry's Glavpriroda, where utilitarians were still ensconced, sought to shift the newly created or restored *zapovedniki* of the Russian Republic's Glavokhota system to its own network. Eerily, the issues had hardly changed since the 1930s, for Glavpriroda's reserves were still pursuing the same income-maximizing goals as before,[22] while the Glavokhota reserves, heirs to the Kozhevnikov tradition, continued to reaffirm both the *etalon* mission of the *zapovedniki* as well as the inviolability of their regime. In this reprise of the debate, ecological questions became central.

Paradigms in Motion

Among conservationists themselves, voices of qualification, such as that of G. P. Dement'ev, the new president of the Academy of Science's Conservation Commission, warned that "it was time to renounce the view that *zapovedniki* are a 'higher' form of conservation." Instead, Dement'ev noted that *etalony* were only "conditionally natural" areas, and that the tasks of conservation transcended the preservation of natural areas and their denizens (no matter how worthy that cause).[23] His words, however, went largely unheeded by the restorationists.

Nonetheless, a conference called by the Academy's Conservation Commission and those of the republics, meeting at the Zoological Institute in Leningrad on 25 January 1956, revealed the incipient divergence within the ecologist-conservationist camp over research interests and other values. Professor V. B. Dubinin, microbiologist and vice president of the commission, emphasized not *zapovedniki* but resource problems in his keynote address. More important, for the first time there was a vigorous call for the study of the ecological impact of migrating radioactive compounds, now identified as a serious health threat to humans.[24] Articles on the ecological consequences of pollutants and pesticides also began to appear in the commission's journal, *Conservation and Zapovednik Affairs in the USSR.*

This represented a new direction for ecology on four counts. First, radiation and pollution were seen as human health issues, to be studied in the tainted earth and waters around nuclear test sites, reactor sites, farms, and factories, remote from the supposedly self-regulating *etalony* of virgin nature. Second, the analytic framework for studying the effects of radiation and pollution was the species *population,* and not the vaguely defined community. Third, this new current of ecological research was pervaded by the optimistic supposition that nature was fully knowable and would eventually be reducible to mathematical description. Finally, flowing from this was the belief held by adherents of the new ecology that each of these environmental problems confronting society was susceptible to a technical solution—in principle.

This trend was well reflected in the changing institutional arrangements. Exemplifying it was V. A. Kovda's Institute on Soil Science, and especially Stanislav Semenovich Shvarts's Institute for the Ecology of Plants and Animals (founded in 1955), attached to the Ural Scientific Center of the Academy in Sverdlovsk, which had been fortified by waves of physicists and mathematicians seeking to apply their latest theoretical models to the study of living systems.

S. S. Shvarts, a "new man" of Soviet biology not unlike the geneticist N. P. Dubinin, began developing his critique of the older school as early as the 1950s. A disciple of N. V. Timofeev-Resovskii (who, exiled to Shvarts's institute, studied ecological aspects of radioactive cycling) and P. V. Terent'ev, both of whom strongly championed the population approach and the use of mathematical methods, Shvarts revealingly was drawn to the problem of acclimatization, one of the pet programs of the nature transformers, although it must be said that he openly critiqued the neo-Lamarckian approach of P. A. Manteifel' on the basis of a modern, population-oriented standpoint.[25] Above all, Shvarts, traumatized by the arrest of his father, an Old Bolshevik, in 1937, resolved to follow a course of political acceptability, avoiding even the slightest whiff of dissidence.[26]

After having helped launch the journal *Ekologiia* (Ecology) in 1969, Shvarts began to speak out more and more emphatically on the relationship between ecology as a science and resource development. In a talk in 1973 at a special Academy-wide conference on conservation, he first underscored the sharp distinction that needed to be made between professional ecological science and conservation. Owing to a wrongheaded conflation of the two in the public mind, he said,

> broad circles of readers began to understand ecology as . . . a science with a social agenda, whose task boiled down to the protection of nature, the amelioration of the microclimate in urban areas, the development of various methods of detoxifying effluents, and so on. However, speaking about ecology, it is always essential to emphasize that ecology is a biological discipline with its own . . . specific research methods.[27]

Nevertheless, Shvarts did see a central role for ecology in addressing the environmental problems of the day.[28] But to play such a role, ecology needed to be unflinchingly "scientific," abandoning all traces of muddy, idealist thinking and values.

One year later, in a talk to party leaders in the Urals, Shvarts went further, deriding the ecological alarmists. "I am deeply convinced," he declared, "that their assertions are illegitimate." Discussions about the "exhaustion of nature," he continued, "sow doubt about the powers of man . . . There is a wise aphorism: 'A resource deficit is simply . . . a deficit of knowledge.'"[29]

The ultimate goal, he explained, was not some prehuman harmony but the ability "to direct natural processes." "We have no other alternative," he asserted, recommending the development of a general theory of ecological engineering.[30] While Shvarts's ecological engineering cannot be equated with I. I. Prezent's voluntaristic call in 1932 for Soviet biologists to become "engineers" in the great transformation of nature, there is at least one common thread: the notion that static natural harmonies do not exist in the real world. If, as Shvarts noted, ecology is "a science of the environment," then that environment has become increasingly transformed by humans. Consequently, "the most progressive ecologists see the main task of their science as developing a theory governing the creation of a transformed world." The world "could not remain untransformed," Shvarts declared, adding that such a process of transformation needs to be governed by considerations of human needs.[31] But—and this is where Shvarts radically diverged from his colleagues—he saw that process of the framing of economic and developmental strategies as the proper preserve of the political authorities, not of scientists with technocratic aspirations.

A year before his death Shvarts participated in a series of sharp debates with the writer and conservation activist Boris Stepanovich Riabinin, a member of the Central Council (*sovet*) of the All-Russian Society for Conservation.[32] Held during the spring of 1975 at the Academy's House of Scholars and the Ural Palace of Culture, both in Sverdlovsk, the debates marked the ultimate development of Shvarts's positions. It is useful to look at them, for they provide powerful ecological-scientific justifications for the prodevelopment point of view. Lest it be forgotten, Shvarts was perhaps the best-known ecologist in the Soviet Union at the time among the lay public.

Throughout the debates, Shvarts's main argument was that prehuman pristine wild nature no longer existed. Using the same example offered by Kozhevnikov nearly seventy years earlier, Shvarts noted that almost all of the forests of Western Europe were at least second growth. From this, however, Shvarts drew a diametrically opposite conclusion.

Kozhevnikov had conjured up the image of German forest plantations to serve as a warning to Russians to preserve what virgin nature remained to them. For him, the deceptive luxuriance of the human-altered vegetation concealed a less stable, less biologically diverse assemblage of organisms than the harmonious community that had been supplanted.

Reversing this idea, Shvarts asserted that there were no grounds to consider second growth inferior to original ecosystems. "In general," he noted, "it is not at all easy to determine how a bad or good ecosystem might be defined." (It is significant that Shvarts used the modern term *ecosystem* rather than the older and more platonic *biocenosis*.) The "luxu-

riant tropical forest," he pointed out, would be choked by industrial ef-
fluents in only a few years, while the relatively species-poor taiga was
able to withstand such abuse for centuries. The value of a given ecosystem had to be calculated in the context of its use-value to human society,
argued Shvarts, and not by some abstract principle of diversity or harmony.[33]

Another aspect of this problem cropped up during the discussion about
the place of predators in the modern world. Riabinin quoted from a newspaper article, "Nature Has No Stepchildren," to bolster his contention
that predators play a necessary role in the economy of nature and should
be preserved. He asked Shvarts to comment as an ecologist.

Shvarts addressed the fate of the wolf as exemplifying the problem of
large predators in the modern world. Through the mid-1950s the wolf
had been hunted down, even in the *zapovedniki*. With the triumph of the
etalon view in the 1960s, however, the campaigns ceased, and the wolf
population surged to over 100,000. Soon wolves once again became an
object of public concern.

In his response Shvarts distinguished between those few remaining
natural areas, such as the tundra, where the wolf still fulfilled a role of
sanitary predation, and elsewhere. In the vast, anthropogenetic majority
of Russia's modern agricultural landscapes, the wolf needed to be exter-
minated; there was no going back to the prehuman balance.[34] One practical conclusion that flowed from this argument was Shvarts's support for
active management within nature reserves, which he did not recognize
as incorporating self-regulating nature.[35]

In 1979 a round-table discussion was held in the pages of the journal
Hunting and Game Management. Not surprisingly, the head of the Ministry of Agriculture's Glavpriroda, A. Borodin, repeated Shvarts's argument that *zapovedniki* were only truncated islands of natural systems,
and therefore the ecological argument that wolves were necessary for
maintaining those systems' self-regulating properties was spurious.[36]
More revealing, however, was the polemic of O. Gusev, who accused
the bulk of Soviet biologists of "losing their objectivity" and "idealizing
nature" while wolves were destroying 30 million rubles' worth of agricultural stock each year. They were purveying a baseless ecological cate-
chism. Ridiculing the ecologists, Gusev suggested that they had fallen
into the teleological fallacy of believing that the wolf was created in order
to prey on ungulates, ungulates to eat grass, and "both, in order to tes-
tify to the glory of the wise Creator." Their "murky" theory of "natural

equilibrium" was the philosophical equivalent of a divine plan. Starry-eyed "idealization of nature" was to be contrasted to Gusev's hard-nosed realism: "The crux is this: that with the elimination of predators, their place will be taken by other factors of selection, including human beings, whom the entire course of evolution on earth prepared for a decisive role in the evolution of the biosphere."[37]

This belief (would the nature-transformers ever admit it?) in a fated role for humans as the new chiefs of evolution was also sketched out by Shvarts, who, like the Tomsk zoologist-acclimatizer Nikolai Feofanovich Kashchenko (and the Russian revolutionary intelligentsia, many of whom envisioned socialism as a time when humans would become "gods on earth") almost eighty years earlier, proclaimed the end of the wild. All species would come under the management and stewardship of humans. "But this is nothing to fear," he reassured Riabinin; nature in the future would be better suited to human aspirations and needs, at least according to Shvarts's material understanding of them.[38]

The nub of the matter was a conflict over values. Shvarts objected to Riabinin's contention that industrialization and urbanization were leading to the "impoverishment of nature" as simply an emotional reaction not deserving serious consideration. The only relevant understanding of impoverishment was in its quantifiable, "professional sense," namely a lowering of biological productivity. There was no room for aesthetic or emotional criteria. Riabinin, rhapsodizing about pristine nature, adhered to his critique of the urban "rat race," which "cut the heart out of life," warning that "blind faith in science is one of the modern varieties of ignorance." There were absolutely no grounds for technological optimism à la Shvarts: No, he warned, "there must not and cannot be easy and quick solutions." For his part, Shvarts declared that the "alarmists'" slogan "Back to Nature" was not only "reactionary" but also "antiscientific." "Man cannot return to the caves," he intoned.[39]

The Flight from the Biocenosis

One of the ironies of the history of Soviet nature reserves is that at its moment of resurrection, the ecological *etalon* concept fell into new contradictions. By the end of the 1960s the ecological program for reserves had swept the field. In 1967 the Ministry of Agriculture finally got around to banning acclimatization in its reserves and had even ceased its raids of

Glavokhota *zapovedniki*. Yet, in addition to the serious problems of poaching, recreational abuse, and the continuing harvesting of resources in the *zapovedniki* (particularly those of the Ministry of Agriculture), conceptual problems also remained.

Critiques such as Ramenskii's and Shvarts's had long pointed up ecologists' pervasive ignorance about the most basic problems of their science, beginning with the question of how to determine the boundaries of putative, integral, natural communities. A shift in thinking among ecologists, however, could occur only with the ripening of a number of other developments in their intellectual and social environment. These included the rise of biosphere studies, the emergence of island-biogeography theory, the growth of demand for leisure activities, a new attitude toward science, and the renewed legitimacy of aesthetic motives for protecting nature. By the mid-1970s leading conservation ecologists had begun their most daring intellectual journey, one that has not yet ended: the flight from the biocenosis. Its most notable consequence was a radical reconceptualization of the role and nature of protected territories, including *zapovedniki*.

Increasingly the *etalon* concept began to founder under the weight of its own theoretical and practical shortcomings. First of all, ecology continued to lack general agreement on the definition of the biocenosis (or, since 1944, *biogeocenosis*). Second, none of the numerous competing ecological approaches could satisfactorily resolve the nagging old and new problems that were seriously undermining the idea of the *etalon*. One major issue was that of the so-called downstream effect. Even if one accepted the possibility of encompassing a discrete biocenosis in a nature reserve, there was still no way to isolate such an area from potent, ambient in-migrating factors, such as air- and water-borne pollutants, feral dog packs, fertilizer runoff, and a drop in the water table owing to regional drainage. The most visible victims of this downstream effect were the *zapovedniki* at Berezina (in Belorussia), the Astrakhan' reserve (in the Volga delta), and the Khopër (in the Black Earth region). Irrigation water taken from the Khopër River, together with regional drainage, had dried up the pools in the reserve's floodplain, changing its entire natural character. Of the multiple biotic consequences the most dramatic was the further decline of the *vykhukhol'* (aquatic mole) in one of its last habitats on earth.

Another small reserve with a serious problem was the Central Black Earth *zapovednik*. This was a variation of the downstream problem, since the reserve was too small (4,500 hectares) to be a self-regulating system in any meaningful sense. Although the reserve's thoughtful director, the late A. M. Krasnitskii, believed generally that *zapovedniki* should incorporate only self-regulating systems, he was confronted with the reality of his own vulnerable territory. If he took no active hand in its management, the endangered species of steppe grasses might die out—owing to vigorous competition from ambient cultivars, from lack of grazing, or other reasons.

Ruminating about the problem, he posed a different question: Even if a steppe *zapovednik* were to exist that was a hundred or even a thousand times as large as the Kursk reserve, what sort of a model of nature would it be? To maintain a regime of strict protection, he argued, would lead to the accretion of litter and an almost complete end to the surface flow of water, as well as to changes in the soil-water balance and a more uniform soil temperature. In turn, species better able to exploit the now moist conditions of the ecosystem would oust those that had been more competitive under the previous arid conditions, leading to a succession of the steppe by meadow. Grass cover could probably not exist independently of some type of management, be it mowing, grazing, or natural foraging.

The problem remained, however, to determine *which* of those factors was responsible for the formation of the steppe. And for this there was no evidence. Since the steppe could easily have been created as a consequence of human economic activity (intensive grazing of livestock), attempting to restore the steppe in the tiny *zapovednik* would amount, ironically, not to the reconstruction of some type of natural *etalon* but to the recreation of a previous anthropogenetic, or human-caused, condition.[40] Reacting to this subversive thought, even Krasnitskii conceded that "the desire for vegetation in the *zapovednik* to have a preagricultural character seems antidialectical." Instead, he proposed that ecological communities in *zapovedniki* must meet at least two criteria: to be self-regulating, and to be maximally insulated from intrusive human factors. With these conditions met, the natural biota would be able to develop spontaneously, and the informational resource of the system would be saved. There was no need to make an unsupportable fetish of virgin, prehuman nature.[41] Here, for the first time, was a willingness to accord seminatural, human-transformed systems the same citizenship rights—as "communities" and even *etalony*—that virgin biocenoses had enjoyed. One result was that this opened up a whole new range of second-growth areas as candidates for status as protected territories—making a virtue of necessity.[42]

Moreover, since the 1960s a new sensitivity to a global level of ecological problems had given wide currency to the ideas of V. I. Vernadskii, who had long since pointed to the biosphere as a single system.[43] This led to new priorities for conservation. Because, in the words of Nikolai Fedorovich Reimers, one of the most influential of the new ecological-conservation theorists, "one can benefit through reshaping nature in some region of the biosphere only by losing in another area,"[44] it was no longer enough to plan land use policies even on the level of the local biocenosis; a nationwide, if not global, perspective was required. Something bigger than the *etalon* was needed.

Also on Reimers' mind were thoughts about how nature works. A creative yet sensible thinker who avoided extremes, Reimers found himself increasingly disturbed by the claims made by S. S. Shvarts and partisans of his school about the predictive and technological potency of mathematical ecology. In a series of popularized pamphlets for the Znanie (Knowledge) Society, he began to express serious doubts about ecological engineering. Why, he asked, have environmental disasters occurred? Sometimes, he responded, we are unable to come to grips with the facts at our disposal. Sometimes, he continued, we are not in possession of the facts. Sometimes unforeseen circumstances occur. And, finally, sometimes circumstances occur that are unpredictable in principle— "and such things exist!"[45] Given such epistemological limitations—a revolutionary conclusion for a Soviet scholar—he characterized the "fashionable 'prognosis' for the transformation of the biosphere into the technosphere" as folly. From an informational standpoint alone, description of some natural phenomena—let alone their simulation—looms as impossible. Just the variety of genetic combinations within one species, Reimers noted as an example, can range from 10^{50} to $10^{1,000}$ variations.[46] These warnings, in addition to the wide publicity given to ideas about limits to economic growth, have created a climate of opinion since the mid-1970s that encourages conservationists to think on a global scale.

Another new ingredient in *zapovednik* affairs had to do with the rise of a branch of ecology called island biogeography. Pioneered by Robert MacArthur and then adapted to conservation problems by Jared Diamond, Michael E. Soulé, and Bruce A. Wilcox, its focus is not on the dynamics of a putatively closed ecological community but rather on the study of those conditions that affect the viability of populations of individual species living in a particular area, considered as an "island." (Originally real islands were studied.) For those who were intellectually prepared to abandon the concept of the closed ecological community, here was an empirically oriented conservation program that studied more identifiable entities (populations) and whose success could be measured.

Sociodemographics also played a role in this story. By the mid-1970s the highly urbanized population of the USSR was taking to the country's back roads in increasing numbers in search of scenery and a small dollop of serenity. Where the prewar emphasis was on rest homes and sanatoriums, modern Soviet vacationers have increasingly sought more active forms of leisure.[47] Recreational geography became institutionalized,[48] and tourists streamed to *zapovedniki* to see what were frequently publicized as the "Soviet Yellowstones." True, the first reaction of the old-line ecologists to the *zapovednik* tourist invasion was alarm; *their* institutions were being subverted to a less rarefied purpose, for one thing, and the conditions of pristine inviolability were being violated as well. But the Soviet tourist was a potentially powerful ally in the fight to preserve natural amenities. Soon it became apparent that a new type of protected territory, the national park, could be a real boon in the struggle to save elements of natural diversity in the USSR.[49]

Enter the Philosophers

Finally, mention must be made of the renewed legitimacy of aesthetic rationales for conservation. Nowhere is this clearer than in a remarkable article, "The Philosophical Bases of Contemporary Ecology,"[50] by the influential philosophers Ivan Timofeevich Frolov, a key Gorbachev adviser and newly appointed editor of the Party's newspaper *Pravda,* and Viktor Aleksandrovich Los', of the Soviet Academy of Sciences' Institute of Philosophy.

The aesthetic attraction of nature, the authors assert, increases as society becomes more and more urbanized. Indeed, they explain, "it would be a mistake to conceive of the biosphere merely as a source of resources or a 'disposer' of wastes." Equally important is the need to reintegrate both aesthetics and values into our way of relating to the world and into our science. Did not Einstein, Bohr, and Heisenberg (no soft humanists or biologists, but *physicists*) invoke aesthetic criteria in their search for the "best" scientific explanation?[51]

Commenting specifically on the society-nature relationship, Frolov and Los' take pains to debunk a number of "myths." The first is that of the inexhaustibility of nature. Essential here is the authors' sensitivity to the meaningfully finite capacity of nature to assimilate wastes. Previously Soviet philosophical writing had stressed (as had Shvarts) the notion that resources are socially defined and not fixed entities, and that surrogates can be found for both resources and natural processes. The position of Frolov and Los' constitutes a reversal of decades of voluntarist thinking.

The second myth they take to task is that of the desirability or even possibility of "man's 'domination' over nature." Here they offer the revolutionary conclusion that

> under the influence of the crisis nature of the developing socioecological situation, man is gradually moving away from the illusion of anthropocentrism and rejecting the traditional hegemonistic relationship to nature. His thinking has ceased to limit itself to notions centering around needs and designs of him and him alone. His activity is acquiring an ever-broader biosphere orientation, and his thinking is drawn to "biocentrism" . . . Biospherocentrism assumes an orientation of human activity and thinking in directions that consider his interests both as subject and as object, as man and nature. [52]

Linked with this is the intriguing recommendation for a return to Marx's original monistic notion—subsequently elaborated by the great biogeochemist V. I. Vernadskii—namely, that human beings and our environment are both parts of a single, dialectically interactive whole. We act on nature both as subject and object; when we alter our environment, we often create dislocation and dangers for ourselves. [53] This point is especially crucial because we know that in the Soviet past, owing to a constrictingly narrow definition of the human being, not merely aesthetic but other psychological and, indeed, biological dimensions were disregarded in setting social and economic policies.

New Directions in Protected Territories

For decades Soviet conservationists had been in thrall to the biocenosis as *etalon,* whether in establishing the system in the 1920s, fighting to prevent its demise from the 1930s through the 1950s, or struggling to rehabilitate it in the 1950s and 1960s. Thus preoccupied, they did little to develop either ecological or conservation theory. By the 1970s, though, the general lack of practical utility of the biocenosis concept, combined with the appearance of new notions and social forces, began to generate a new agenda. Elements of this agenda came together most powerfully in the widely discussed monograph by Reimers and Feliks Robertovich Shtil'mark, *Protected Natural Territories,* published in 1978. Reimers was a biologist affiliated with the Central Mathematical-Economics Institute who had earlier studied the relationship between forest types (by age and species composition) and the number of game animals each could support, while Shtil'mark, trained as a game-management specialist, had long worked for the Glavokhota as a key planner of new *zapovedniki.* Their salient points were:

1. That the system of protected territories should be redefined as an integral, distinct branch of the economy—its stabilizing sector, enabling the rest of the economy to function. Popular perceptions of these territories as unproductive lands, reflected in their zoning status as "nonagricultural lands," should also be revised. Instead, the USSR should follow the lead of the Kirgiz SSR, which had established a republican state land fund as a special permanent category.

2. Principles of siting and determining the area of *zapovedniki* should be revised. Rather than *zapovedniki* being selected according to the old formula of one per biogeographical unit (the *etalon* principle), they should be created to provide enough healthy nature in the proper areas to ensure no breakdown of the socioecological equilibrium.

3. The socioecological equilibrium, defined as the balance between economic activity and the carrying capacity of the environment that permits a maximum level of production to be sustained, should be assessed from a broad nationwide, if not global, perspective.

4. To accommodate the various needs and levels of conservation, from the protection of rare species to recreation, a new, efficient multifunctional system of protected territories should be created. While appropriate minimum areas of individual reserves dedicated to preserving particular species complexes could be determined with the aid of island biogeography theory, for example, other types of protected territories might have more flexible requirements. These could be regulated so that all of the different protected territories taken together would then be integrated into an overall system of providing for the maintenance of the socioecological equilibrium.[54]

Addressing the constituency of Soviet tourists, Reimers and Shtil'mark warmly greeted the new national park movement while emphasizing that *natsional'nye parki* should not be established at the expense of or through the conversion of *zapovedniki.*

In 1979, shortly after the appearance of their book, Reimers and Shtil'-mark wrote a popular piece for *Man and Nature*. Although it was entitled "Etalony prirody" (Models of nature), there was little in the article about representative biocenoses. If the old-line biocenology was out of the picture, two new themes stood out: diversity and aesthetics. Arguments for diversity were reflected in the protest literature of the "Village School,"and especially by the bards of distinctive Siberia and the Far North (V. Rasputin, V. Astaf'ev, A. V. Skalon, and Shtil'mark himself). And while Shtil'mark and Reimers continued to make scientific arguments for diversity (especially for preserving genetic diversity), they increasingly propounded an unabashed, purely literary, nonscientific defense of aesthetic values and diversity. In their joint article they had resurrected an arresting quotation from the early Russian botanist and conservation leader Valerian Ivanovich Taliev:

> The virgin forest, the unplowed virgin steppe attract the contemporary mature individual not only with the prospect of clean air, wide open spaces, and freedom from the confines of everyday life. They are also sources of experiences of a higher order. They speak to us! . . . For our world view nature is not only something outside of us, but it forms together with us an integral whole; we ourselves are only a small unit within the one great organism of nature. To learn how to penetrate to this unity, to feel around oneself the beating of the unbroken pulse of life, means to create a positive foundation for spiritual development, to incorporate into the developing soul a powerful counterweight to the narrow, practical "I," and to develop the ability to apperceive the world in an artistic and aesthetic way.[55]

As most would doubtless agree, such ideas would hardly have been able to find an outlet in the days when ecology was dominated by Prezent, and would have been denounced as idealism as late as the mid-1960s. Bald-faced aesthetic argumentation had finally come out of hiding in the USSR.

Present Perspectives

While from the perspective of the 1960s it would appear that the old conflicts and positions of the 1920s were simply resurrected once again after the long night of Lysenko and Prezent almost without alteration (only the institutions' names had changed), from the perspective of the present the picture looks significantly different. I shall now try to explain why.

It seems to me that historically the main attraction of the paradigm of the closed biocenosis and of the *etalon* to Soviet ecologists has been (1) that it enabled them to present themselves as the only true guardians of the ecological health of the country on the basis of their expertise in these arcane matters; (2) that it gave them a network of research bases and a certain degree of professional autonomy; and (3) that it offered powerful scientific justifications for both an anti-industrial stance and the preservation of pockets of pristine nature—islands of diversity (including aesthetic diversity) in a sea of sameness (*odnorodnost'*). It is no accident that in the late 1920s ecological arguments were mobilized against collectivization and the construction of grandiose hydroelectric projects, and that then, as now, many conservationists were in the front lines of the struggle to preserve the cultural and architectural heritage of Russia.[56] *Zapovedniki* provided a different model of nature—and of life—from that of Soviet reality and constituted actual islands of inner emigration and professional autonomy.

As long as no better means were available to promote these values, the *etalon* concept retained its hold on Soviet biologists and their allies. It was even able to survive the Prezent-Lysenko episode, gaining renewed vitality during the thaw and the peaking of Soviet interest in the Biosphere Reserves Program, despite the powerful critiques of ecologists such as S. S. Shvarts.

New possibilities for expressing these values openly began to emerge slowly during the Brezhnev period and after. In particular, aesthetic discourse, legitimized by the leading philosophers of the land, reappeared in conservation and was nourished by (and nourished in return) the new movement for Soviet national parks (there are now twelve of these, with many more being planned).[57]

In addition, new scientific and ecological perspectives enabled key theorists in conservation ecology to revise their conceptions about the nature and function of protected territory. Island biogeography theory—currently promoted with great vigor by the conservation theorist and ecologist A. V. Iablokov of the Severtsov Institute of Evolutionary Animal Morphology and Ecology, who edited the new Soviet translation of Soulé and Wilcox's crucial anthology, and by the Academy's *Journal of General Biology*—has provided a more empirical way of both measuring and preserving diversity than did the vague notion of biocenosis. The field of biosphere studies—more relevant to the preoccupations of industrial society, where environmental problems arise over entire river basins, for example—has provided a more convincing justification for the protection of large natural and seminatural areas than had biocenotic theory. Indeed, biocenotic theory was unable to speak adequately to the problem of protecting agricultural and other anthropogenetic landscapes—themselves reservoirs of a certain natural diversity—because of its fixation with naturalness and purity. Like the Marxists of the 1890s, who had edged out their Populist rivals by arguing that socialism could not be built on the basis of the peasant commune because the latter was being swept off the historical stage by ascendant capitalism, island biogeographers and other more modern ecologists critiqued the *etalon* concept by noting that pristine natural systems, if they had ever existed, no longer did, owing to the ubiquity of pollution and other man-made effects.

In groping toward a less metaphysical view of the structure of nature, ecology was also helped along by a slow change in the broader intellectual climate. This development, in which such fields ancillary to ecology as philosophy and even economics have played an important part, may be described as a new middle-range realism. As the possibilities of real reform began to seem more and more urgent, and now, at last, within reach, styles of neoplatonic thinking, with their ideal types and utopian orientation, have been gradually supplanted by more empirical and yet more epistemologically tentative approaches. The overly ardent embrace of science—in as mathematized a form as possible—as an omnipotent key to Truth was a reaction against both Stalin's thirty-year orgy of irrationalism and political intrusion into science during that period. (Mathematics was seen as value-free and therefore invulnerable to interference by party philosophers.) The allure of scientism and the application of mathematics have waned in recent years, however. Even the mathematical modelers in ecology have recently admitted the validity of Timofeev-Resovskii's warning about the impossibility of attaining an exact fit between models and actual natural phenomena.[58] There is also the growing realization that science cannot be totally purged of aesthetics—or values.

For ecology itself, a final consequence of these transitions has been the forging of a new relationship between science and public policy, and with it the creation of a new subdiscipline: social ecology, or *noologiia*, to use the term offered by Frolov and Los'. Not a program based on the neoplatonic category of the biocenosis, social ecology equally rejects

Shvarts's "ecological engineering," based on the equally platonic faith in the absolute knowability of the world and on the omnipotence of technological solutions. Addressing as it does the desire to put a cap on open-ended growth, rejecting the idea that nature is inexhaustible in any meaningful sense, and emphasizing the importance of taking human aesthetic and psychological needs seriously—and, consequently, the need to preserve diversity—social ecology has been able to attract to its side many who previously saw the *etalon* as the scientific guarantor of those values. Yet, with its rejection of the fetishizing of nature and its focus on the need to establish a socioeconomic-ecological equilibrium, social ecology has retained its standing among those committed to a twentieth-century industrial society. The Gorbachev era, which opened with the election to the new Central Committee of such philosophers of the society-nature relationship as Frolov and I. D. Laptev (editor of *Izvestiia*), seems to be the optimal ideological milieu for the deepening and institutionalization of the trends I have described.

Where there is hope for reform and a readiness on the part of many in educated society to undertake it, where there is real opportunity for expert scientific input to be taken seriously in policy making, and where the prospects for intellectual freedom have grown increasingly bright, the likelihood is that nature reserves will lose their symbolic role as extraterritorial islands of freedom from the Stalinist state. They will take on other symbolic values, while their last-ditch partisans will consist of those who are unable to exploit the new opportunities presented by Gorbachev's political revolution and who nourish a profound and abiding fear of change and modernization, even if it is accompanied by democratization. In that they will resemble the extreme group of wilderness advocates in the West.[59]

Some Cross-Cultural Remarks

Those familiar with the fortunes of ecological science generally will find much that unites the Soviet experience with that of Europe and the United States. Everywhere, it seems, ecology has faced difficulties in defining its mission and gaining full academic and institutional accreditation. While the original motivating idea of ecology—the study of the interrelationship of life forms among themselves and with their inanimate environments—enjoyed great appeal and promise, ecologists could not agree on a coherent set of paradigms, methodologies, and techniques. Already established subfields in biology such as plant and animal physiology and ethology claimed that a separate new science was not necessary for the study of "ecological" problems.[60] One explanation for the widespread redefinition of ecology as the study of ecological communities is that it provides ecology with a *unique* object of study, safe from the claims of neighboring disciplines.

Nevertheless, there were other "externalist" or social factors that promoted the identification of ecology with the study of natural communities. Such study was intimately connected with conservation efforts, and ultimately with the pretension that ecologists, on the basis of their special scientific expertise, should wield wide influence over the larger society's decisions concerning the use of resources and economic development. Whether they themselves fostered this social mission for their science, as was the case in the United States before the Second World War,[61] or were tempted to capitalize on rising public concerns about the environment in a later period,[62] ecologists worldwide have frequently exhibited a tendency to promote their discipline as central to the fate of the earth. In the United States, however, to take one Western case, this must be seen in the light of continuing efforts to bolster ecology's evershaky disciplinary standing. Where the Soviet experience emerges as distinctive is that ecology there also became a standard-bearer for science and academe generally in their common struggle with an intrusive and repressive state. The prolonged appeal of the idea of the closed, self-regulating ecological system or biocenosis, and of the need to create a network of ecological nature reserves (*zapovedniki*) incorporating these models of healthy nature (*etalony*), can be understood only within the protracted struggle of the Soviet Union's creative intellectual elite for autonomy and a restoration of its rightful status in the running of the nation's affairs.

· 4 ·

The Soviet Nature-Nurture Debate

Mark B. Adams

During the past quarter-century the social import of human biology has been a subject of considerable interest and dispute in both professional and public settings.[1] In the United States and Great Britain much controversy has surrounded IQ tests, the issue of race and intelligence, sociobiology, genetic engineering, *in vitro* fertilization, cloning, and the relationship between crime and the XYY karyotype. The scientific validity and the ethical, legal, social, and political implications of this work have received a good deal of public attention. In English-speaking countries these discussions are often characterized as part of the ongoing debates about "nature and nurture"—the relative contributions to human physical and mental characteristics of hereditary (genetic) and environmental (social) factors.

During roughly the same period similar discussions of the relative roles of the "biological" versus the "social" in shaping human nature and behavior have been taking place in the Soviet Union. This debate exhibits some striking parallels with its contemporary Western counterpart. Both the Soviet and the Western discussions began in earnest in the late 1960s, heated up in the 1970s, and waned somewhat during the 1980s. Both involved geneticists, biologists, psychologists, educators, journalists, popularizers, and politicians. Both began in professional publications, spread into popular science journals, and soon became the subject of widespread public discussion. Both assumed political dimensions, with certain nurturists attacking the positions of certain naturists as racist and reactionary.

At first blush it might appear that the Soviet debate is simply a reflection of the Anglo-American one, or that both are manifestations of an

94

international scientific controversy, arising out of recent scientific research findings, that shows little national differentiation. As Loren Graham has noted, however, the Soviet discussion manifests some striking contrasts with its Western counterpart.[2] First, the positions advocated by some Soviet "hereditarians," on the whole, seem rather more extreme and categorical than those of most Western "hereditarians." A genetics textbook by M. E. Lobashev has called for the rebirth of eugenics.[3] A. A. Neifakh has called on Soviet authorities to breed superior people.[4] V. P. Efroimson has argued that altruism, heroism, self-sacrifice, conscience, respect for old people, parental love, monogamy, chivalry, and intellectual curiosity "have entered into the basic stock of man's inherited characteristics."[5] Iu. Ia. Kerkis has claimed that chronic criminality is hereditary.[6]

The Soviet debate has a second remarkable feature. In the West, liberals and leftists as a rule have been extreme "nurturists" or "environmentalists" and have often attacked hereditarians for deliberately or inadvertently supporting right-wing politics. By contrast, in the Soviet debate, democratic "liberals" who suffered under Stalinism and have called for greater civil liberties have been associated with the hereditarian position, whereas their nurturist opponents generally represent the "conservative" Soviet establishment and have been variously associated with Lysenkoism, ideological orthodoxy, and a political hard line.

What are we to make of these similarities and differences? Are the Western and Soviet debates really about the same things, or do they only appear so from our Western perspective? Four aspects of the Soviet debate are of particular interest:

1. *Timing.* The Soviet nature-nurture debate began to surface in 1966 and has continued into the 1980s. It is thus roughly concurrent with comparable discussions in the West. But if the Soviet debate is different from the Western one, it may also have different roots. How can we understand the timing in the context of Soviet developments?

2. *Participants.* Both the most outspoken naturist—V. P. Efroimson (1908–1989)—and the most outspoken nurturist—N. P. Dubinin (1907–)—are geneticists. Other important naturists in the debate include the geneticists B. L. Astaurov (1904–1974), V. V. Sakharov (1902–1969), Iu. Ia. Kerkis (1907–1977), M. E. Lobashov (1907–1971), A. A. Malinovskii (1909–), and D. K. Beliaev (1917–1985). By contrast, almost no prominent geneticists have sided with Dubinin, but he has been supported by a number of those formerly associated with Lysenko,

notably the philosopher G. V. Platonov. How can we account for these alignments, and in particular for Dubinin's special position, isolated from most of his fellow geneticists but supported by his old-time Lysenkoist rivals?

3. *Breadth.* The Soviet debate has involved not only geneticists but also psychologists, anthropologists, physicians, mathematicians, popular science writers, and Marxist philosophers. How can we account for the participation of representatives of these other disciplines and professions? In particular, what relation have they had in the past with Soviet genetics and with the nature-nurture issue?

4. *Soviet "naturism."* In comparison with the Western debate, the Soviet environmentalist position hardly needs explication: like so many Western social scientists and popular pundits, Soviet nurturists have maintained that humans are essentially social creatures whose cognitive and behavioral natures are shaped much more by their social environment than by any sort of biological determinism. But how can we account for the proclivity of Soviet liberals for naturism—and fairly extreme naturism at that?

Such public scientific debates present special analytical opportunities and challenges for the social historian of science. Philosophers and historians of ideas often take an interest in the substantive issues in such debates, seeking to determine the coherence, logic, evidential basis, and validity of the opposing viewpoints or to identify the world views or ideologies they represent. The social historian of science has the additional task of seeking to understand the debate in its historical and social context; to account for its shape, location, and timing; to explore the detailed biographies of its participants and their historical interconnections; and to assess the professional, disciplinary, and institutional interests that may be at stake. Such analysis of other scientific debates has proved instructive, revealing that they often involve underlying generational, institutional, and disciplinary conflicts that are not readily apparent from what is said.[7]

What follows is an attempt to analyze the recent Soviet nature-nurture debate in this way. I will argue that its special characteristics can be explained by a combination of historical and structural factors. First, I will argue that the linkage between certain scientific and ideological positions is not intrinsic but contingent: it has been shaped in particular ways by the unique historical experience of Soviet genetics. Second, I will argue that the recent nature-nurture debate is but one dimension of a broader conflict that attended the rebirth of genetics in the USSR after three decades of Lysenkoism, a conflict over which variant of the "old" Soviet genetics would be reborn. Finally, I will update my analysis and spell out some of its broader implications.

Historical Roots

The recent Soviet debate over the relative contributions of the biological and the social in shaping human mental and physical traits has its roots in pre-Stalinist and even prerevolutionary times. Widespread discussion of biosocial topics arose in many countries around the turn of the century in connection with attempts to establish the scientific basis and legitimacy of the social sciences (anthropology, psychology, and sociology) and as part of an efflorescence of interest in interdisciplinary research generally. Such trends formed part of the vigorous intellectual life of late tsarist Russia, and they found especially fertile soil in postrevolutionary conditions.

In the 1920s no traditional disciplinary boundary was sacred. Hybrid fields abounded, including A. N. Bach's "biochemistry," V. I. Vernadskii's "geochemistry," P. P. Lazarev's "biological physics," and N. N. Semenov's "chemical physics." The decade also saw an unparalleled profusion of interdisciplinary theories, including N. I. Vavilov's periodic table of genetic elements of 1920 (modeled after Mendeleev); A. G. Gurvich's theory of "mitogenetic rays" (1923), which held that a form of radiation given off by dividing cells triggers organic growth and development in a process analogous to radioactive decay; A. I. Oparin's chemical "colloidal gel" theory of the origin of life (1924); and V. I. Vernadskii's geochemical concept of the "biosphere" (1926).[8]

As we might expect, the interface of the biological and the social proved an especially active area after 1917. The psychiatrist V. M. Bekhterev published a textbook of "reflexology," a new "science of human personality studied from the strictly objective, bio-social standpoint" which constituted "a branch of biology . . . merging into sociology."[9] The Nobel prizewinning physiologist Ivan Pavlov expanded his earlier, prerevolutionary work on animal conditioning into a general account of higher mental function in humans. Nikolai Semashko and other physicians adapted German "social hygiene" into a distinctly Soviet variant.[10] The botanist V. N. Sukachev wrote a text for a field called "plant sociology."[11]

These trends can be seen in contemporary publishing. In Petrograd, the publishing house Obrazovanie had put together dozens of anthologies of "new ideas" in biology, psychology, sociology, physics, philosophy, and so forth. In Moscow a similar role was played by the Sabashnikov firm. We know from Soviet archives that M. V. Sabashnikov sought to remain in business in the early 1920s by commissioning books with a materialist approach to man and nature that he felt would appeal to political authorities. Many of the collections he tried to assemble dealt with the relationship of the biological and the social, including volumes on the evolutionary origins of human beings, the biological basis of behavior, and sociological works based on biology. One area he especially favored was eugenics.[12]

Soviet Eugenics

Probably the most successful Soviet attempt to integrate the biological and social during the 1920s was the Russian eugenics movement.[13] Interest in eugenics among Russian psychiatrists, neurologists, anthropologists, public health officials, and experimental biologists predated the Revolution, but in 1920 they were drawn together in a single organization with the founding of the Russian Eugenics Society. In succeeding years branches of the society met in Leningrad, Moscow, Khar'kov, Odessa, and Kazan', and two eugenics periodicals were published.

Russian eugenics research was institutionalized around 1920 by two experimental biologists with recent European training. One was Iurii A. Filipchenko (1882–1930).[14] In 1921 he founded the Bureau of Eugenics of KEPS, the Commission for the Study of Natural Productive Forces of the Russian Academy of Sciences. Based at Petrograd University, the bureau published an annual volume of reports and conducted extensive genealogical studies of the intelligentsia of the capital. In Moscow the leading organizer of eugenics was Nikolai K. Kol'tsov (1872–1940). In 1916, with philanthropic funding, he had created the interdisciplinary Institute of Experimental Biology. After losing its endowment during the Civil War, the institute became part of GINZ—the network of research institutes funded by the Commissariat of Public Health (Narkomzdrav). In 1919, in keeping with the new emphasis on applied research, Kol'tsov gave practical genetics increasing prominence at the institute with the addition of a breeding station, a genetics division, and a eugenics division. Serving as the base for the Russian Eugenics Society, this eugenics division also supported wide-ranging interdisciplinary eugenics research, including studies of genealogies and blood chemistry, as well as investigations of the inheritance of behavior in mice conducted by Kol'tsov's wife, Maria Sadovnikova. The profile of both the division and the institute as a whole fit the "synthetic" format so popular at the time.[15]

It is important to realize that in the 1920s there were no hard and fast distinctions between genetics and eugenics. University courses and textbooks in America and elsewhere often linked the two, and some of the leading monographs on heredity—ranging from Paul Kammerer's *Inheritance of Acquired Characteristics* (1924) to R. A. Fisher's *Genetical Theory of Natural Selection* (1930)—concluded with lengthy chapters on eugenics. This is especially significant when we consider the fact that Filipchenko and Kol'tsov headed two of the three institutions where Soviet genetics was created; the third was a plant-breeding institute headed by Nikolai Vavilov. Thus, almost all Soviet animal geneticists were trained by Filipchenko, Kol'tsov, or one of their students. Most of them became involved in eugenics; many published articles on human heredity; some chose the field for their degree work.

A good example is Vladimir V. Sakharov. He won a degree at Kol'tsov's Moscow institute in 1924 for two concurrent studies: one on pedigrees of Russian musicians, the other on mutations in fruit flies.[16] In the late 1920s he served as secretary of the Russian Eugenics Society. Then in the early 1930s he not only analyzed the incidence and distribution of human blood groups in Uzbekistan but also pioneered studies of chemical mutagenesis in drosophila. In the 1950s and 1960s, as Lysenkoism waned, Sakharov played a leading role both in establishing Soviet molecular biology and in reestablishing Soviet work in human genetics. Two other Kol'tsov students from the late 1920s, V. P. Efroimson and A. A. Malinovskii, followed a similar course, combining an interest in human heredity with experimental and mathematical work.

The pattern in Leningrad was similar. The interconnectedness of eugenics and genetics there is illustrated by the evolving name of Filipchenko's bureau: founded in 1921 as the Bureau of Eugenics, in 1925 it became the Bureau of Genetics and Eugenics and in 1927 the Bureau of Genetics. Among those from Filipchenko's group who maintained an interest in animal and human heredity throughout their careers are Iu. Ia.

Kerkis, N. N. Medvedev, and Theodosius Dobzhansky. Other Leningrad students who subsequently took up work in human heredity include M. E. Lobashev and A. A. Prokof'eva (-Bel'govskaia), until her recent death the leading Soviet expert on human cytogenetics.[17]

Anthropologists, psychologists, psychiatrists, neurologists, and physicians were also involved in the movement in the 1920s. The young anthropologist V. V. Bunak became interested in eugenics through his attempts to develop biometric techniques for measuring skulls; he met Kol'tsov through Moscow University and was employed in the eugenics division of Kol'tsov's institute. Psychiatrists were especially active in eugenics: one of the leading figures in the Russian eugenics society was T. I. Iudin, author of several books on eugenics, psychopathology, and constitution; another was psychiatrist P. I. Liublinskii, coeditor of the society's journal. Probably the most renowned psychologist associated with the movement was V. M. Bekhterev, founder (1908) and director of the Psychoneurological Institute in Petersburg, who supported Filipchenko's eugenics activities in the prerevolutionary years. Eventually even Pavlov got into the act. In Western circles, because of his work on conditioned reflexes, Pavlov is often seen as the archetypal behaviorist. This view is largely mistaken. In 1926, as a result of his contacts with Kol'tsov and his wife, Sadovnikova, Pavlov organized an experimental station at Koltushi, outside Leningrad, devoted to "the genetics of higher nervous activity." After Pavlov's death in 1936, the station was headed by his leading student, L. A. Orbeli.[18]

Soviet eugenics was dominated by political liberals of bourgeois backgrounds, and in the early 1920s they wrote articles on the dysgenic effects of the world war, the Revolution, the Civil War, and the famine that did not endear them to the authorities. Later they made more concerted efforts to accommodate their field to the Soviet context; these proved awkward and not wholly successful, leading some Marxists to criticize the movement for its preoccupation with middle-class virtues. But other Marxists found eugenics inspiring: M. V. Volotskoi championed the sort of Lamarckian socialist eugenics advocated by Paul Kammerer, for example, while a Kol'tsov protégé, Aleksandr Serebrovskii, propounded a Bolshevik eugenics based on Mendelism. In 1926 Serebrovskii defended genetics against the charge that it was politically reactionary by focusing on its eugenic implications for social betterment, declaring that the totality of human genes in a country constitutes "a national treasure, a *gene fund* from which the society draws its people."[19] Incidentally, his term *gene fund* (*genofond*) was translated into English by Dobzhansky in 1950 as "gene pool," from which the use of this term in modern population genetics originates—yet another illustration of the many ways eugenics and genetics were intertwined.[20]

In 1928, in anticipation of the First Five-Year Plan, the Party called for proposals. Carried away with visionary enthusiasm, Serebrovskii suggested "the widespread induction of conception by means of artificial insemination using recommended sperm," estimating that, given "the tremendous sperm-making capacity of men," and "with the current state of artificial insemination technology (now widely used in horse and cattle breeding), one talented and valuable producer could have up to 1,000 children . . . In these conditions, human selection would make gigantic leaps forward. And various women and whole communes would then be proud . . . of their successes and achievements in this undoubtedly most astonishing field—the creation of new forms of human beings."[21]

The Great Break

The period of the First Five-Year Plan is associated with a kind of sea change in Russian history encompassing the collectivization of agriculture, the liquidation of the kulaks, breakneck industrialization, the first show trials, a hardening of the Party line, and the beginnings of what has come to be known as Stalinism. Soviet scientific life experienced the Bolshevization of the Academy of Sciences, the reorganization of universities, the imposition of quotas and political controls, crackdowns on bourgeois specialists, antitechnocracy campaigns, and restrictions on international travel. The profound ideological, political, institutional, and cultural changes that occurred between 1929 and 1932 are often collectively referred to as the "Great Break."[22]

As it turned out, the Great Break ended an era: it terminated a half-century of vigorous and relatively free biosocial discourse, thereby transforming the Soviet nature-nurture debate forever. In keeping with its antitechnocratic thrust, the new Party line explicitly proscribed extrapolation from the biological to the social. The new biosocial taboo was immortalized in the creation of a new pejorative Russian word, *biologizirovat'*—literally, "to biologize"—which was understood as one of the several sins collectively referred to during the period by another newly minted phrase, "Menshevizing idealism."[23]

This new proscription on links from the biological to the social seems to have been quite clear-cut and its dangers perfectly apparent to Russian workers. We get a sense of this from a letter sent by the Russian ecologist G. F. Gause to Raymond Pearl, a distinguished American biologist, a pioneer in demography, and a faculty member in the school of Hygiene and Public Health at the Johns Hopkins University in Baltimore. Pearl was arranging the American publication of Gause's book *The Struggle for Existence,* which would soon become a classic of modern ecology. In a letter to Pearl dated 21 September 1933, Gause wrote: "I wish I could ask you a very great favor in regard to one particular point. I was trying to avoid in this book any mention of human competition and human populations . . . I may assure you that there are particular reasons for me to ask you to avoid any mention of human beings in your introduction as well, and I hope that you will kindly fulfill this wish. This favor is for me of the highest importance."[24] Pearl responded in a letter of 5 October that "I shall take pains in writing the introduction to your book to make no mention of human affairs/beings," and when the book appeared in 1934, Pearl was true to his word—although he could not resist allusions to Karl Pearson, whose eugenic interests were probably well known to most English-speaking readers.

By 1933 many of the new interdisciplinary fields that had grown up with such vigor in the previous decade were broken apart (experimental zoology), dissolved (biophysics), or renamed ("plant sociology" became "phytogeocoenology"). So far as I can determine, *no* field that linked the biological and the social survived the Great Break intact. As a biosocial, technocratic movement, eugenics was doubly damned. Shortly after the last issue of its journal appeared in 1930, the Russian Eugenics Society was disbanded. At roughly the same time the Eugenics Division of Kol'tsov's institute was abolished. Serebrovskii's eugenic proposal for human artificial insemination was roundly chastised and was even the subject of a lampoon in *Pravda* by Demian Bednyi, who asked whether Serebrovskii was proposing a new Marxist slogan, "Genes of the World, Unite!" Lest any lingering uncertainties remained, the 1931 article "Eugenics" in the *Great Soviet Encyclopedia* declared categorically that eugenics was "bourgeois," Kol'tsov's eugenic ideas were "fascist," and Serebrovskii's variant was "Menshevizing idealism." From the late 1920s Russian translations of Western genetics textbooks systematically excised the chapters on human heredity.

The Great Break had profound consequences for Soviet genetics and the institutions where it had flourished. In Leningrad, Filipchenko was relieved of his teaching responsibilities as of January 1930, and his Department of Experimental Zoology at the university was disbanded.[25] In 1930 the Bureau of Genetics at the Academy of Sciences became its Laboratory of Genetics. Following Filipchenko's sudden death from meningitis in May 1930, the laboratory was taken over by the plant geneticist N. I. Vavilov, who headed a cadre of Filipchenko's students that included Kerkis, Medvedev, and Prokof'eva. In 1933, with H. J. Muller's arrival in the USSR, the laboratory was elevated in status, becoming the Academy's Institute of Genetics (IGEN), headed by Vavilov. At its core was a genetics laboratory headed by Muller, whose staff included most of the animal geneticists trained by Filipchenko. The next year both the Academy and IGEN relocated to Moscow. In general, then, the Leningrad genetics group suffered the disruptions of the Great Break but was able to survive thanks to Vavilov's patronage.

The changes in Moscow were even greater. During the 1920s, genetic research in the Kol'tsov institute had been carried out by two groups headed by A. S. Serebrovskii and S. S. Chetverikov. In 1927 Serebrovskii had a falling out with Kol'tsov and moved his research base to the Moscow Zootechnical Institute and the Timiriazev Institute of the Communist Academy. In the spring of 1929 Chetverikov was arrested and sent into exile. Within a year the remarkable collection of young animal geneticists who had trained under Kol'tsov was dispersed. Three— N. K. Beliaev, B. L. Astaurov, and V. P. Efroimson—took up silkworm breeding in the Caucasus or in Central Asia. In 1930 Kol'tsov was removed from teaching at Moscow University and the Department of Experimental Biology was disbanded, to be replaced by five new ones, including a new Department of Genetics, chaired by Serebrovskii. Like the Leningrad group, the Moscow group lost its most outstanding young animal geneticist: as a result of the Great Break, both Theodosius Dobzhansky (from Leningrad) and Nikolai Timoféeff-Ressovsky (from Moscow) elected to stay abroad, where they soon gained international renown.[26]

Remarkably, by 1933 Kol'tsov had managed to recreate genetics at his institute by bringing in Nikolai P. Dubinin, an excellent and ambitious young geneticist who had studied in the late 1920s with Serebrovskii. In 1932 Kol'tsov invited him to head the genetics division, and in succeeding years, as a way of showing that his enterprise was in step with the times, Kol'tsov used public settings to highlight Dubinin's background as a *vydvizhenets*—someone promoted beyond his formal qualifications because of his political and class background. Nonetheless, there were many holdovers from the Chetverikov days, including V. V. Sakharov. In 1936 Kol'tsov managed to bring Astaurov back from Central Asia.[27] Understandably, because of differences in politics, background, and seniority, tensions existed between Dubinin and those who had worked with Kol'tsov and Chetverikov in the 1920s. These tensions were to be exacerbated in the late 1930s by the growth of Lysenkoism.

The 1930s and Stalinism

As we have seen, the eugenics movement in the 1920s involved most of Russia's animal geneticists and forged links between them and physicians, psychiatrists, and anthropologists. With the delegitimation of eugenics around 1930, the movement dissolved per se, but by 1933 it had begun to regroup in an institute under the direction of Solomon G. Levit.[28] A physician, Levit had joined the Party in 1919 and from 1926 through 1930 was an official of the science section of the Communist Academy. In the mid-1920s he had been a strident Lamarckian. Under Serebrovskii's influence, however, in 1927 Levit was won over to genetics and soon showed the enthusiasm of a recent convert. In 1928 Levit and Serebrovskii formed the Office of Human Heredity and Constitution under the auspices of the Biomedical Institute of the Commissariat of Public Health.

The office's original research program was shaped by Serebrovskii's ideas and involved genealogical studies and human population genetics. The publication of Serebrovskii's human breeding scheme as the lead article in the office's first research volume cast ideological doubt on the enterprise, however, and in 1930 he abruptly severed his connection with it. That year Levit was made director of the Biomedical Institute as a whole and expanded his human heredity office into the institute's Genetics Division. He spent 1931 on a Rockefeller grant studying genetics with Hermann J. Muller in Texas. Just prior to his return to Moscow in January 1932, Levit was suspended as institute director, but he resumed the post that fall.

Over the next three years Levit managed to assemble an impressive staff, launch widespread twin studies, and cultivate highly interdisciplinary approaches to human heredity. It was clear from the beginning, however, that what Levit and his group were doing could hardly go by the name "eugenics." In 1934 a conference jointly organized by the institute and the Commissariat of Public Health was able to settle on the right language: it called for the creation and widespread development of a "new" field called "medical genetics." In 1935 the institute was suitably renamed the Maxim Gorky Scientific Research Institute of Medical Genetics. Under this rubric it managed to absorb a number of younger Moscow researchers from various fields who had formerly been active in eugenics, including the anthropologist Bunak, the neurologist Davidenkov, and M. V. Volotskoi, a former secretary of the Russian Eugenics Society. In addition, the institute began an extensive investigation of the psychology of twins in a new division headed by A. R. Luria, whose subsequent studies established his international reputation as one of this century's leading psychologists. [29]

Levit's year in Texas apparently had some effect on his host. Newly acquainted with Serebrovskii's socialist eugenics, in June 1932 H. J. Muller broke with the American eugenics movement in the paper "The Dominance of Economics over Eugenics," which reiterated some of Serebrovskii's arguments and language. That fall Muller left for Europe, and, after a short time in Berlin, he accepted an invitation from Nikolai Vavilov to head a laboratory in the Academy's new Institute of Genetics. He was elected corresponding member of the USSR Academy of Sciences in 1933. [30]

There is strong evidence that Muller went to the Soviet Union in order to realize his eugenic program. Much of his time was devoted to drosophila experiments conducted with Filipchenko's former students Kerkis, Medvedev, and Prokof'eva, but he was also an active consultant to both the Levit and Kol'tsov institutes. In his many book reviews and articles for popular Soviet books and journals he rarely passed up an opportunity to argue for the implementation of a socialist eugenics program. In 1935 he published *Out of the Night: A Biologist's View of the Future,* which resurrected (without attribution) Serebrovskii's discredited human artificial insemination scheme. In May 1936, perhaps out of growing impatience, Muller took the fateful step of sending Stalin a copy of the book, together with a long letter arguing for a trial implementation of his plan.

Matters came to a head in December 1936 at the fourth session of the Lenin All-Union Academy of Agricultural Sciences. The main speakers were Vavilov, Lysenko, Serebrovskii, and Muller. All participants had been instructed to avoid mention of human heredity. Muller defied the political instructions, however, and, against the urgings of both Vavilov and Serebrovskii, he used eugenic language that had been politically unacceptable since 1930. Once Muller had broached the issue, Lysenko's supporters warmed to it. For the first time Lysenkoists drove home the argument that genetics, medical genetics, and eugenics were all ideological adjuncts of fascism. As evidence they recalled Serebrovskii's breeding plan; one commented ominously that the memory of that plan would live longer than its author. Toward the end of the meeting Serebrovskii arose on a point of personal privilege to declare that his 1929 article was so replete with the grossest ideological errors that it was painful for him to recall it.[31]

It is difficult to avoid the conclusion that, however inadvertently, Muller had played directly into the hands of the Lysenkoists. The integrity of both Levit's institute and the field it embodied depended on the fiction that medical genetics and eugenics were totally unrelated—a fiction unmasked by Muller's naive straight talk. By resurrecting Filipchenko's and Serebrovskii's eugenic arguments—so fitting in the 1920s, so inappropriate in 1936—he also resurrected the link between genetics and eugenics. In the process he unwittingly compromised the Soviet Union's leading geneticists: his senior colleague Kol'tsov, his patron Vavilov, his friend Serebrovskii, his protégé Levit. Only following the meeting did Muller learn that Stalin had been reading his book and did not like it.

Beginning in December 1936 the ax fell. Muller's student I. I. Agol was arrested as a Menshevizing idealist; he was shot in 1937. So too was N. K. Beliaev, a student of Chetverikov and Kol'tsov. The Institute of Medical Genetics was disbanded and its director Levit disappeared; he was probably shot in May 1938. Even the secretary who translated Muller's book for Stalin was reportedly arrested and shot. When word reached Muller that he was due to be arrested, Vavilov managed to arrange his passage out of Moscow through the quickest possible route—as a member of the International Brigade headed for the Spanish civil war.[32]

December 1936 also proved to be an ideological turning point. Before then, Lysenkoist rhetoric had been largely devoid of references to human heredity or fascism; afterwards, the charge that eugenics proved the fascist character of genetics became a prominent cliché of Lysenkoist propaganda. The charge helped to secure Lysenko's election to the USSR Academy of Sciences. In 1938 Kol'tsov and Lysenko were the two candidates for the same Academy slot in genetics. Kol'tsov had been a corresponding member since 1915 and was the clear favorite. On 11 January 1939, some two weeks before the election, an article in *Pravda* attacked him for his eugenic views and intimated that he had imported fascist ideas into the socialist motherland. Four days later Dubinin chaired a meeting at the Kol'tsov institute to "discuss" the eugenic mistakes of its founder, but, according to Dubinin, Kol'tsov failed to appreciate "the social and public scientific significance of the criticism of his eugenic errors by his students."[33] No doubt Kol'tsov felt betrayed by the very student whose career he had secured. On 29 January Lysenko was elected; in April, Kol'tsov was relieved as director of the institute he had created, its name was changed, and it was absorbed into the Academy of Sciences. He died of a heart attack in December 1940; the next day his wife committed suicide. Kol'tsov's students from the 1920s blamed his death on the political harassment he had recently suffered over eugenics, partially at the hands of Dubinin. On behalf of these loyal students, Astaurov published a thoroughly laudatory obituary in *Priroda,* a journal Kol'tsov had helped to create almost three decades earlier.

Lysenkoism

The strains among Soviet geneticists gave way to a united front as members of the discipline closed ranks in the face of the growing threat of Lysenkoism. During the quarter-century from 1939 until 1964, the Soviet state gave Lysenko strong support, and his version of "Michurinism" dominated official policy and rhetoric. The experience had a profound and lasting effect on all Soviet geneticists who survived the period and conditioned the mindset they brought to the task of rebuilding their discipline in the 1960s.

In 1937 Lysenko became assistant to the president of the council of the Supreme Soviet. In 1938 he became president of the Lenin All-Union Academy of Agricultural Sciences (VASKhNIL). In 1939 he was elected to full membership in the Academy of Sciences and made a member of its presidium. He used these positions to move against his archrival N. I. Vavilov and his associates. With the help of the secret police, Lysenko's supporters organized young Party workers to harass the Vavilovites within their own institutes. In 1940, apparently with Lysenko's complicity, Vavilov was arrested, and as the case against him was being prepared, several of his associates were also taken into custody, including G. A. Levitskii and G. D. Karpechenko, two of the world's most distinguished plant cytogeneticists. All three died in prison in 1942–43. Upon

Vavilov's arrest, Lysenko assumed direct control of IGEN as its new director. Some of its geneticists lost their jobs immediately; others managed to hang on for a time. Remarkably, only three or four young people there converted; even under the palpable threat of arrest, imprisonment, and death, most professional geneticists refused to go along with Lysenko's theories.

Despite these purges and institutional takeovers, Lysenko did not manage to uproot the discipline of genetics entirely. Although he came close to destroying most of the leading plant geneticists in the Soviet Union, many geneticists who had worked on animal material managed to get by. In particular, the Kol'tsov institute proved a strong center of resistance: in the postwar years its genetics staff was the best in the USSR. The universities also proved to be centers of resistance; at Moscow University, for example, Serebrovskii continued to head the department of genetics. Finally, Lysenko never did succeed in uprooting genetics from the affiliated disciplines where it had flourished. When Levit's Institute was abolished in 1937, Bunak moved his activities to anthropological institutions, where he continued his work on craniometry, heredity, and race. Davidenkov remained a professor at his Leningrad medical institute, became a full member of the Academy of Medical Sciences in 1945, and rose to become one of the elite corps of physicians treating Kremlin officials.

In the immediate postwar years Soviet genetics resurfaced with new confidence and energy. The physicist Sergei Vavilov, Nikolai's older brother, became president of the USSR Academy of Sciences in 1945 and, the next year, gave his support to plans for the creation of a new institute of genetics to be headed by Dubinin. An especially important role was played by Pavlov's scientific heir, Levon Abgarovich Orbeli: as head of the Pavlov institute (1936–1950), director of the Military Medical Academy (1943–1950), vice president of the USSR Academy of Sciences (1942–1946), and academician-secretary of its Biological Sciences Division (1939–1948), he did what he could to oppose Lysenkoism, giving some Leningrad geneticists a haven at the Koltushi station to study the inheritance of behavior. With the postwar resurgence of genetics came a predictable revival of the naturism to which it had become linked. In 1947 Davidenkov published a book on medical genetics, while A. A. Malinovskii, one of Kol'tsov's students, published two important works: a Russian edition of Erwin Schrödinger's *What Is Life*, which aroused the interest of Soviet physicists in genetics, and an article in *Priroda* on biological and social factors in the origin of human racial differences.[34]

In August 1948 the situation changed radically. At a meeting of the agriculture academy called on short notice, Lysenko announced that the Central Committee had read his report and approved it. Within weeks a special meeting of the presidium of the USSR Academy of Sciences brought its policy into line. One result of these meetings was a series of edicts that fired geneticists by name and reorganized or eliminated the laboratories and institutes where they had worked. Dubinin's genetics laboratory was abolished, and the Kol'tsov institute was amalgamated with another institute. Orbeli was replaced as head of the Biological Sciences Division by A. I. Oparin, a Lysenko ally who administered the liquidation of genetics. As a result of these actions Lysenkoists took control of the professional structure of Soviet biology, including its institutions, textbooks, and degree-granting commissions, and were more or less able to maintain that control until 1965.

These events were traumatic for Soviet geneticists. Two—Efroimson and Romashov—were arrested and sent to camps, where they joined Timoféeff-Ressovsky, who had been interned since 1945. Most geneticists lost their jobs, and some were unemployed for a year or more. About a dozen were able to find unrelated work as biologists in remote stations in Iakutsk or Central Asia (among them Sokolov, Sidorov, Kerkis); several were given pharmaceutical work by the Russian Ministry of Health (Alikhanian, Sakharov, Prokof'eva-Bel'govskaia); one (Nikoro) found work as a piano player in a beer hall. Dubinin was given haven by V. N. Sukachev in a forestry institute in the Urals, where he was subsequently joined by Romashov. Astaurov was able to maintain his post in the embryology division of the Kol'tsov institute but was forbidden to do genetics research.

During the most intense period of Lysenkoism (1948–1952), Michurinists had a field day at the geneticists' expense, and one of their principal themes was that genetics, eugenics, and fascism were of a piece. The flavor of their rhetoric is exemplified by two infamous broadsides of the period. In a pamphlet published in an edition of 110,000 copies, Minister of Education S. V. Kaftanov wrote that the propositions of genetics "led in our country, just as they did abroad, to eugenic ideas and the ideology of Fascism."[35] The next year A. N. Studitskii published an article in *Ogonek* that dwelt on both the uselessness of geneticists (because they worked on drosophila) and their fascist orientation, appropriately calling his article about them "Fly-Lovers and Man-Haters."[36]

The Khrushchev Period

With the death of Stalin in March 1953, criticisms of Lysenko reappeared and quickly increased in number and intensity. By 1955, with the release and rehabilitation of millions from camps, the atmosphere had changed sufficiently so that petitions were sent to the Central Committee requesting the dismissal of Lysenko and his supporters from their administrative control over Soviet biology, and in 1956 Lysenko resigned as president of the agriculture academy. Many regarded these events as setting the stage for the rebirth of Soviet genetics. These hopes proved premature. By late 1958 Lysenko had succeeded in consolidating a relationship with Nikita Khrushchev, who moved repeatedly on his behalf until his own ouster in October 1964. Nonetheless, from the mid-1950s genetics began to develop.

The most characteristic feature of underground Soviet genetics during the Khrushchev period was the extraordinary aid provided by some of the Soviet Union's leading physicists, chemists, and mathematicians.[37] The Watson-Crick model of DNA, published in 1953, brought the gene within their disciplinary reach, and their growing importance in providing the Soviet Union with nuclear weapons, missiles, and sputniks was mobilized on behalf of genetics. Although Lysenko was able to maintain his control of the biological establishment and the Academy's biology division, genetics grew up wherever physical scientists could become legitimately involved. At the Biophysics Institute, for example, they created a laboratory of radiation genetics in 1956, headed by Dubinin. At Leningrad University a key role was played by its rector, A. D. Aleksandrov— a Party member, a prominent mathematician, and a sometime member of the presidium of the Academy of Sciences—who brought Lobashev back to the university in 1957 to head the department of genetics. With the help of such atomic physicists as Igor Tamm and Ivan Kurchatov, and such chemists as I. L. Knuniants and Academy president A. N. Nesmeianov, genetics also began to reappear in Moscow, spearheaded by V. N. Sukachev, V. V. Sakharov, V. P. Efroimson, and A. A. Malinovskii.

Because they dominated the Academy's governing presidium, these physical scientists were able to engineer reorganizations of the USSR Academy of Sciences in 1957 and 1963 that greatly facilitated the development of genetics. Led by the mathematicians M. A. Lavrent'ev, S. L. Sobolev, and A. A. Liapunov, in 1957 the presidium created a Siberian Division of the Academy, based at a newly constructed Science City (Akademgorodok) near Novosibirsk, which included a new Institute of Cytology and Genetics isolated from Lysenko's control. Then, in the early 1960s, molecular genetics was given a great boost through the efforts of N. N. Semenov, chemical physicist and Nobel laureate, who mobilized support for an overall restructuring of the USSR Academy of Sciences that divided the biology division into three new divisions—one for physiology; a second for molecular biology, which included chemists; and a third, much-reduced division of general biology, which included Lysenko's institute. In 1963 the plan was approved. Semenov became a vice president of the Academy, in charge of all the chemistry and biology divisions and institutes (including Lysenko's), and appointed as his assistant the geneticist A. A. Malinovskii.

This 1963 reorganization allowed molecular biology to flourish in its own division, but the new structure also involved risks: in the smaller General Biology Division, devoid of molecular biologists, Lysenko now had more complete control. At Lysenko's urging, Khrushchev opened four new posts for genetics in that division, to be filled by Lysenko's nominees: N. I. Nuzhdin was to be elevated to full membership; the elderly breeders P. P. Luk'ianenko and V. S. Pustovoit were to be directly elected to full membership; and the polemicist V. S. Remeslo was to be elected corresponding member. The June 1964 elections, however, did not go as planned. Luk'ianenko, Pustovoit, and Nuzhdin were approved in the divisional election, but Remeslo was rejected. When the vote was brought to the Academy's general assembly, an almost unprecedented floor fight erupted, in which the physicist Andrei Sakharov and V. A. Engel'gardt (a world-renowned muscle biochemist and the leading Soviet advocate of molecular biology) spoke strongly against Lysenko and Nuzhdin. In the final vote Nuzhdin's candidacy was rejected. As if to make a point, the meeting approved the geneticist D. K. Beliaev's election to corresponding membership in the Siberian Division. These events reportedly enraged both Lysenko and Khrushchev, who threatened to replace the Academy with a state committee on science.

In the "little October revolution" of 1964 Khrushchev was removed from power. This led to a convergence of interests between a scientific leadership determined to see genetics reborn and a new political leadership eager to legitimate the ouster of a popular leader. Even before the

ouster was publicly announced, hurried calls went out to various geneticists to dictate articles for the press describing how Lysenko, with
Khrushchev's support, had harmed Soviet science and agriculture. In the
winter of 1965 the Academy was left free to handle the matter without
political instructions one way or another. The result was a certain amount
of understandable equivocation. An investigation of Lysenko's farm led
to a thoroughly condemnatory report. In official policy pronouncements,
however, Academy president Mstislav Keldysh made clear that although
molecular genetics could now develop untrammeled and Lysenko's "monopolistic position" had come to an end, "we should not indiscriminately
reject everything he has done . . . Attention must be focused on scientific and organizational questions, and all possibilities of administration by
fiat, of pressure and labeling, by no matter which side, must be ruled
out."[38]

It is not difficult to understand why Keldysh adopted a cautious attitude: Lysenkoists still dominated the Academy's General Biology Division and many of its institutes. The division's dozen full academicians
included at least three Lysenkoists but not a single geneticist; the dozen
and a half corresponding members included four outright Lysenkoists
and four or five others who had supported Lysenko. The only geneticists
in the division were corresponding members: N. P. Dubinin (elected in
1946), B. L. Astaurov (1958), and D. K. Beliaev (1964), the younger
brother of N. K. Beliaev, whose relation to the division was ambiguous
since officially he was a member of the separate Siberian Division. Then,
too, no political instructions had been received. With the political leadership of the country still unsettled and its ultimate sympathies uncertain,
the repudiation of thirty years of solemn pronouncements could not be
undertaken lightly. It was into this ambivalent structural situation that
genetics was to be reborn.

The Politics of Reconstruction

During the years 1965–1970 Soviet geneticists were busy reconstructing their science. Their immediate agenda was straightforward: the sol-
idification and expansion of their institutional research base, the creation
of new texts and curricula, the creation of new professional societies,
commissions, and journals—all the concerns of any new discipline. One
of the first items of business was to accord geneticists the degrees, cer-

tifications, and status they had been denied by Lysenko's monopolistic position, and to strengthen the profile of genetics within the Academy. These steps were taken in 1965–1967. On the basis of "earlier work," without the usual formalities of dissertation defense, dozens of geneticists were granted the higher degrees that they had been denied in earlier years. The Academy elections of 1966 promoted both Dubinin and Astaurov to full membership. This established two preeminent academicians in genetics in the midst of a heavily Lysenkoist division.

But the task of disciplinary construction was complicated by the fact that genetics was being not established but reestablished—after thirty years of Lysenkoism. This meant that genetics also had to reestablish its ideological legitimacy in the context of Soviet science: its public history had to be rewritten, its founders and achievements praised, its martyrs sanctified, and its long struggle vindicated. In the process some account had to be given of how so fine a science could have undergone such a troubled history in such a great country—an account that would not indict the Party or the government in unacceptable ways. Furthermore, this had to be done in a way that would protect genetics from political incursions and controversies, give geneticists control of their own field, and neutralize or abolish their entrenched Lysenkoist opposition.

Viewed in this way, it is clear that different approaches to the task of disciplinary regeneration were possible. Who would lead the recreated discipline, what shape it would take, and how Lysenkoists would be dealt with were open questions in 1965. In this context, ideological, disciplinary, professional, institutional, and personal conflicts arose that were to be reflected in the nature-nurture debate—conflicts over the shape the "new" Soviet genetics would take.

Emergent Conflicts

Dubinin was the natural choice of Academy pundits to lead Soviet genetics. He had spent twenty years as a corresponding member. He had an international reputation in population genetics. He was politically cooperative, but he had never wavered in his opposition to Lysenko's genetical theories. In 1946, 1956, and 1957 unsuccessful attempts had been made to establish an institute under his direction; he had had to settle for a laboratory at the Biophysics Institute. Accordingly, in its resolution of 25 December 1964, the Academy of Sciences presidium called on a committee of seven to plan a new Institute of General Genetics.[39] It was clear from the outset that this would be Dubinin's long-awaited institute. The newly appointed Scientific Council on Genetics and Selection was dominated by people from Dubinin's laboratory, but former Lysenkoists also figured prominently.[40] In 1966 these plans came to fruition. Not only was Dubinin elected a full member of the Academy of Sciences; he was also awarded a Lenin Prize. The new Institute of General Genetics was established, replacing Lysenko's old Institute of Genetics in the same building. Predictably, Dubinin was its new director.

In taking over Lysenko's institute and reforming it into the Institute of General Genetics, Dubinin brought in most of his colleagues from his laboratory of radiation genetics at the Biophysics Institute and gave them their own laboratories. But he also allowed two Lysenkoists to keep their laboratories. According to Dubinin's autobiography, he had kept them at their request because "a person can change his views, if he understands his errors and sees his place in the future."[41] Later, Dubinin justified his approach this way: "Some geneticists thought that it was necessary to get revenge for the events of 1948. Obviously, such an approach will yield nothing, besides a lowering of the moral standards of science . . . We had to break away from the psychology that controlled the hearts of the people at the 1948 session of VASKhNIL."[42] According to Dubinin, geneticists had been guilty of intellectual and ideological mistakes in the past, such as their belief in the immutable gene or their support of eugenics; Lysenkoists, of course, had been dead wrong in their genetics and had been intolerant and monopolistic. It was time to put these mistakes behind them and get on with the job of building the new Soviet genetics—an amalgam of the Lysenkoists and their victims, headed by Dubinin.

Given the structural situation that faced genetics in 1965, Dubinin's position is understandable. In the absence of political instructions to the Academy, his largesse toward his Lysenkoist rivals might minimize disruptive political conflicts, enlist their cooperation in the task ahead, and ensure their loyalty to Dubinin. With Lysenkoists so well entrenched in Academy structures, he may have felt that any attempts to purge, repudiate, or replace them would present a messy prospect at best. In addition, his strategic approach was also potentially self-serving: neither the Lysenkoists nor his fellow geneticists would be in a position to dispute his leadership of the field, since any attempt to do so would almost certainly provoke opposition from the enemy camp.

It is hardly surprising that Dubinin's position appalled many geneticists who had been repressed under Lysenkoism. For more than three decades they had seen their ideas and their teachers vilified. A dozen of their mentors and colleagues had been snatched up by the purges and had never been heard from again. They had spent their most productive scientific years harassed, unemployed, exiled, or imprisoned. They had watched their country's international preeminence in genetics research erode away while colleagues who had emigrated achieved international renown. Furthermore, they had no doubts whatever about who was to blame. They had endured the Great Break only to run into Lysenkoism and the purges; they had seen the beginnings of rebirth after the Second World War only to witness August 1948; they had anticipated its resurrection in the mid-1950s only to watch Khrushchev come to Lysenko's aid. Now their time had finally come. For these older geneticists, legitimating Lysenkoists for their ill-gotten gains was simply immoral and accommodation with them unthinkable.

Four outstanding geneticists from the earlier generation who headed labs at Dubinin's new institute were outraged by his approach to Lysenkoists and protested in an angry meeting. As Dubinin recounts the event in his autobiography, he chastised the four for their knee-jerk reaction to any position that Lysenko had ever advocated, whether it was right or wrong, and went on to comment that "Lysenko formulated several general propositions correctly," giving as an example the proposition that the organism and the environment form a dialectical unity. Hearing one of the clichés of Michurinism from the mouth of their director apparently led them to respond that this "principle" was so much claptrap. Dubinin in turn instructed them on their ideological errors, later commenting: "Past mistakes and incorrect ethical approaches to the tasks of science were ready to exert their morbid influence on the development of the social and philosophical principles of the new genetics."[43]

These tensions surfaced in the new N. I. Vavilov All-Union Society of Geneticists and Breeders. At its founding meeting, held 30–31 May 1966, Dubinin reportedly argued that eugenics and human genetics posed ideological dangers for genetics, and that accommodation with Lysenkoists was necessary; his views were not well received. With his seniority and a Lenin Prize in the offing, Dubinin must have expected to be elected president of the society. Instead, Boris L. Astaurov was elected president; Dubinin was not even among the four vice presidents.[44] This election set up a professional conflict between Dubinin as head of the scientific council on genetics and Astaurov as president of its professional society. At a joint meeting between the council and the society leadership in 1967, Dubinin reiterated his position, but, in his own words, he was opposed by "a small but monolithic group consisting of B. L. Astaurov, S. M. Gershenzon, S. I. Alikhanian, and D. K. Beliaev, who took mistaken positions on this question."[45] The conflict was institutionally reinforced in mid-1967, when Astaurov became the director of a reborn Kol'tsov institute. On 1 October 1967, as a result of their conflict with Dubinin, four of Russia's leading geneticists left his Institute of General Genetics to join Astaurov's Institute of Developmental Biology, and they took their laboratories with them.

Beginning in 1967 the conflict between Dubinin and Astaurov and what they represented became increasingly polarized. In general, Astaurov stood for the publication of candid and explicit accounts of the history of Soviet genetics, resistance to all vestiges of Lysenkoism, the resurrection of the intellectual legacy of Kol'tsov and the other founders, a science free of all political and ideological fetters, and the full and untrammeled rebirth of human and medical genetics. His organizational bases were the Kol'tsov institute, which he directed, and the new professional society of geneticists, which he chaired; the Moscow Society of Naturalists; and the popular science journal *Priroda*. Astaurov used his influence to support ideas, intellectuals, and dissidents who had been, or were being, repressed.

By contrast, Dubinin stood for a reconciliation with Lysenkoists, the condemnation of the ideological and political errors of the founders of Soviet genetics, close ties between genetics and Marxist philosophy, politically correct behavior, the open condemnation of eugenics as fascist, and a highly sanitized version of medical genetics. He was based at the Institute of General Genetics, which he directed, and the Scientific Council on Genetics and Selection, which he initially chaired by appointment. His principal support came from orthodox Marxist social scientists and philosophers, Lysenkoists, and certain Party circles.

Structurally, then, this polarization between the Astaurov and Dubinin camps reflected underlying institutional, disciplinary, and professional conflicts associated with the reconstruction of Soviet genetics. But these structural conflicts over the shape and character of the new Soviet genetics also had an ideological dimension. In the public arena this ideological polarization took the form of an increasingly outspoken debate over nature and nurture.

The Naturists

The attitude of most geneticists involved in the Soviet nature-nurture dispute was shaped by their historical experiences. In 1970 the average age of the dozen or so geneticists most centrally involved was sixty-three. Most had been born around 1907. These geneticists received their higher education in the Soviet Union in the 1920s and early 1930s, during the heyday of Soviet genetics, before the rise of Lysenkoism. Since Lysenkoism was especially hard on plant geneticists, they were almost entirely animal geneticists trained by Filipchenko, Kol'tsov, Serebrovskii, or Muller. They passed through their thirties, forties, and fifties during the Lysenkoist period (1936–1965), struggling to maintain their discredited discipline. They were entering their sixties when Soviet genetics was reborn and became, perforce, its leaders, seeking to vindicate their field and reestablish its institutional and intellectual base.

These geneticists seem to have envisioned the new Soviet genetics as a return to the traditions and spirit of its golden age during their student days, before the Great Break, and they wished to resurrect the genetics that Stalinism and Lysenkoism had destroyed. As we have seen, that genetics recognized no hard and fast distinctions between the genetics of animals and humans. Many members of this group had done work in what had variously been called "eugenics," "medical genetics," or "anthropogenetics," and their conception of the unity of genetics and eugenics had been reinforced by their own subsequent experience. They had seen plant genetics, animal genetics, medical genetics, and eugenics repressed at roughly the same time, by the same people, for the same reasons. It is hardly surprising, then, that with the end of Lysenkoism and an apparently free hand in rebuilding their discipline, they envisioned a field in which all that had been repressed would be born again in a more modern, molecular incarnation.

The period 1965–1971 saw the rebirth of Soviet naturism as part of the effort to reestablish Soviet genetics and its historical and ideological legitimacy. Predictably, in Moscow these efforts were spearheaded by people who had studied with Kol'tsov or were associated with the Kol'tsov institute, including Arsen'eva, Astaurov, Efroimson, Malinovskii, Rapoport, Rokitskii, Sakharov, Sidorov, Sokolov, and Timoféeff-Ressovsky. The legitimacy and prestige of these students depended, at least to some extent, on the legitimacy and prestige of their teacher, and

part of Kol'tsov's legacy was eugenics and medical genetics. In rehabili-
tating Kol'tsov, Astaurov and his colleagues worked closely with Kol'-
tsov's old journal *Priroda* and its managing editor, V. Polynin. Articles on
genetics and its history began to appear regularly in the journal. Astaurov wrote an almost poetic introduction to *Mama, Papa, and Me,* a popular 1966 book on human inheritance by Polynin, who followed it in 1969
with a heroic biography of Kol'tsov entitled *A Prophet in His Own
Land.*[46]

Efroimson and medical genetics. The most outspoken Soviet naturist
from the Moscow school was Vladimir Pavlovich Efroimson (1908–
1989), a geneticist whose career had embodied with poignant starkness
the historical struggles of his chosen field.[47] A student of Kol'tsov at
Moscow University, Efroimson first undertook studies of the genetics of
behavior under Kol'tsov's wife, Maria Sadovnikova. When he spoke out
against Chetverikov's arrest in 1929, Efroimson was dismissed from
Moscow University without a degree. Like Astaurov and N. K. Beliaev,
he then took up silkworm research, working at the Caucasian Sericulture
Institute (1930–1932). His method for determining human mutation
rates brought him to the attention of Levit, who invited him in mid-1932
to undertake work on human genetics in Moscow at the Biomedical Institute.

In December 1932, however, Efroimson was arrested, and he spent
the next four years in a prison camp in the far north. Upon his release in
1936, he replaced Astaurov at the Central Asian Institute of Sericulture
in Tashkent, then worked briefly in the Ukraine. During the war he
served as an epidemiologist in the Red Army, winning various awards for
his efforts. He completed work on his doctoral dissertation in 1947. At
about the same time, Efroimson wrote a letter to the Party Central Committee documenting the damage Lysenko's techniques had caused Soviet
agriculture. The letter led to his second arrest in May 1949, and this time
he spent six years in the camps.

Released in 1955 and rehabilitated in 1956, Efroimson returned to
Moscow, lecturing at the Moscow Society of Naturalists and taking an
active part in the temporary resurgence of genetics in the capital. From
1956 through 1961 he worked at the All-Union State Library of Foreign
Literature, where he became the best-read Soviet in foreign genetic literature and medical genetics. In the summers of 1961 and 1962 he participated in Timoféeff-Ressovsky's summer workshops at Miasovo in the
Urals. He was finally granted his doctoral degree in 1962. In 1963 Ef-

roimson completed work on a Soviet textbook of medical genetics of almost five hundred pages and submitted it for publication. His subject and background posed problems, and only one hundred copies of the book were printed, under the rules governing manuscripts. The next year, however, thanks to the intervention of the cyberneticist A. I. Berg, his text was finally published "on the recommendation and with the approval of the USSR Academy of Medical Sciences" in an edition of 9,500 copies. [48]

With the end of Lysenkoism, Efroimson quickly emerged as the leading spokesman for medical genetics. In 1965 he was appointed head of the anthropogenetics section of the Scientific Council on Genetics and Selection. At the founding meeting of the Vavilov Society in 1965, Efroimson gave a major address on the history of human genetics in the USSR that was explicit about the troubled history of the field. [49] In 1966, together with Prokof'eva-Bel'govskaia and E. F. Davidenkova (wife of S. N. Davidenkov, who had died in 1961), Efroimson wrote the survey article on genetics and medicine for the genetics society. [50] Two years later he joined the staff of the Moscow Institute of Psychiatry of the Russian public health ministry, also teaching genetics at the Second Moscow Medical Institute. As Efroimson's career demonstrates, the network of investigators in eugenics, anthropogenetics, and medical genetics from the 1920s and early 1930s was still active.

Efroimson had suffered for the right to speak his mind, and he was not going to pull any punches. In the 1960s his bibliographic expertise was mobilized in several dozen review articles that carefully scanned the literature for Lysenkoist vestiges and elaborated in detail the past and present misconceptions of their authors. In 1968 he published a second, expanded edition of his medical genetics text; he followed it in 1971 with an important book on immunogenetics. Finally, he expressed his broad naturist views in a provocative article entitled "Hereditary Altruism," which appeared, together with a supporting piece by Astaurov, in *Novyi mir.* [51]

Lobashev and the new eugenics. If the Moscow networks were still active, so too were those in Leningrad. [52] Indeed, the most remarkable expression of Soviet naturism came not from a Muscovite but from a Leningrader, Mikhail Efimovich Lobashev (1907–1971), who actually came from the background that Dubinin mimicked. [53] The son of a working-class family from the Volga, Lobashev was orphaned at an early age and received his early schooling at the Liebknecht Workers' Commune

School in Tashkent. He subsequently worked in Leningrad at a shipbuilding factory before entering the natural science faculty of Leningrad University in 1929 as a "promotee" (*vydvizhenets*), advanced because of his class background, despite his inadequate qualifications.

Lobashev went into genetics in the spring of 1930, after Filipchenko had left, but he soon became involved in the enterprises that Filipchenko had created. Upon his graduation in 1932, Lobashev took up graduate work on chemical mutagenesis in drosophila at the university's department of animal genetics, headed by V. P. Vladimirskii. Concurrently he worked on chemical mutagenesis in drosophila at the new Institute of Genetics, but when it moved to Moscow in 1934–35, he chose to remain behind. He became a docent at the department in 1938 and, after Vladimirskii's death the same year, took over the laboratory of animal genetics. In 1941, with the outbreak of war and the blockade of Leningrad, Lobashev joined the Party and served at the front in an artillery division, rising to the rank of captain. After the war he resumed work in the university genetics department, rising to become dean of the biological faculty in early 1948.

Following the events and edicts of August 1948, Lobashev was ousted from all posts and had to leave the university. In February 1949 Orbeli came to Lobashev's aid, inviting him to head a laboratory at the Koltushi station on the behavioral genetics of lower animals. There he joined the geneticists A. N. Promptov and R. A. Mazing, whom Orbeli had earlier given sanctuary. He also became acquainted with S. N. Davidenkov, who had close ties with the station. In 1957, on the initiative of the rector, A. D. Aleksandrov, Lobashev was called back to Leningrad University to head its reconstituted department of genetics and selection. He soon rose to his former post as dean of the biological faculty, and he again taught genetics, trained students, and organized conferences. With Aleksandrov's support, in 1963 the lectures for Lobashev's course were published as a textbook. This was the first Soviet genetics text since the late 1930s that was devoid of Michurinism; its single reference to Lysenko was negative.[54] Lysenko still enjoyed Khrushchev's support, however, so the text wisely steered clear of controversy: no mention whatever was made of human genetics.

After Lysenko's hegemony came to an end, Lobashev freely expressed his liberal vision of the new Soviet genetics based on the Filipchenko tradition of old. Although he had barely known Filipchenko, Lobashev inherited the department he had created—Russia's first university department of genetics—and he took pride in reconstructing the legacy of its founder. In 1967, in the second, completely reworked edition of his genetics textbook, he added a new final chapter on human genetics—and its final section was entitled "The Significance of Eugenics." There Lobashev wrote: "We regard it as essential that this branch of science be restored to full rights, shucked of its pseudoscientific husk."[55] He detailed his conception of the field in a four-page discussion that concluded the book. Furthermore, the inside covers of the text depicted the "tree" of genetics, with its various branches and their growing points, including animal genetics, tipped by "the genetics of behavior," and human genetics, tipped by "eugenics." Published by Leningrad University Press in an edition of 190,000 copies, the book was officially approved for use as a textbook by the Russian Ministry of Higher Education.

In 1968, together with two colleagues, Lobashev finished another genetics textbook designed for use in training biology teachers. It too depicted a disciplinary tree that presented "human genetics" as a major subject, subdivided (in ascending order) into demographic genetics, medical genetics, pharmacological genetics, "pedagogical genetics," and "eugenics." Once again, Lobashev concluded the chapter on human genetics by calling for the resurrection of eugenics: "Discarding all social perversions of scientific principles," he wrote, "eugenics as a branch of human genetics must exist and develop as a science based on exact biological and genetic knowledge, and must serve to improve the health of human society." "Thus," Lobashev concluded, "man's biological fate . . . like his social fate, is in his own hands."[56]

Timoféeff-Ressovsky. In the late 1960s moves were under way, spearheaded by Astaurov and Engel'gardt, to elect Timoféeff-Ressovsky to full membership in the USSR Academy of Sciences. From a scientific point of view there was no question that he deserved it: a founder of population genetics, modern evolutionary theory, radiation genetics, and molecular biology who had made Berlin-Buch a major world center of genetics during his two decades there (1925–1945), he was one of the world's most accomplished biologists and clearly outranked in quality and achievement not only all rivals but all sitting members of the division.[57]

Timoféeff had lived in Berlin throughout the entire Hitler period, including the war. Following his arrest by Soviet authorities in 1945, he had spent a decade in prison and in exile, but beginning in the mid-1950s he had played a key role in the rebirth of Soviet radiation genetics, organizing seminars and workshops and instructing an increasingly large scientific following. On the eve of the election a campaign was mounted to portray Timoféeff as a fascist traitor. Both Dubinin and the Lysenkoists were implicated. The attempt to elect Timoféeff may inadvertently have created suspicion in some Soviet circles that the Lysenkoist portrayal of the links between genetics, medical genetics, eugenics, and fascism may have had some validity after all.

Other doubts must have arisen in connection with the attempts to publish true accounts of the history of Soviet genetics. Survey articles listing achievements presented no problem, but biographies of the founders posed the difficulty of dealing with how they had died.[58] In the mid-1960s F. Bakhteev, Mark Popovsky, and Zhores Medvedev collected detailed information on Vavilov's last days in prison, and some of it was actually published.[59] There was even a move afoot, supported by Astaurov and some leading academicians, to publish Zhores Medvedev's book on Lysenko, which included high praise for Kol'tsov, Levit, Serebrovskii, and medical genetics. But the book's indictment of both the Lysenkoists and of the government and Party proved too much. The manuscript was sent to the United States, where it was translated and edited by I. M. Lerner and published by Columbia University Press in 1969 as *The Rise and Fall of T. D. Lysenko*. In addition, Medvedev circulated a manuscript on Timoféeff-Ressovsky, completed in 1968, that detailed the campaign to rehabilitate him and the difficulties it was encountering from Lysenkoists.[60]

The Nurturist Counterattack

The period 1970–1976 witnessed heightened conflicts that began to assume serious political dimensions. Having alienated much of the senior leadership of genetics, Dubinin turned increasingly outside his discipline to seek support elsewhere, especially among Lysenkoists, Marxist philosophers, and political circles. In January 1969 Dubinin became a member of the Communist Party. From his position at his Academy institute, and with new backing, he launched a nurturist counterattack.

The Medvedev affair. On 29 May 1970 Zhores Medvedev was confined at a mental hospital, and many believed that one of the principal reasons was his book on Lysenko. Medvedev's principal defender was Astaurov, whose efforts led to his release on June 17. Astaurov had been able to mobilize the leading physical scientists of the USSR Academy of Sciences in defense of Medvedev, including V. A. Engel'gardt, P. L. Kapitsa,

A. D. Sakharov, N. N. Semenov, and M. V. Keldysh (Academy president from 1961 to 1975); conspicuous by his absence was the politically correct Dubinin.

At first blush it may appear that the Soviet nature-nurture controversy had little to do with Medvedev's incarceration. On closer inspection, however, it becomes clear that it played a more central role than we might suppose. Medvedev's book had devoted one whole chapter to the Lysenkoists' repression of Soviet medical genetics (1937–1940) and the slandering of Kol'tsov. One of those most active in his release was A. A. Neifakh—a student of Astaurov who had worked at the Kol'tsov institute and was an open and outspoken advocate of the genetic determination of human behavior. When Medvedev was fired from his job in the late 1960s at the Institute of Medical Radiology in Obninsk, where he was a colleague of Timoféeff-Ressovsky, he applied for a job at the new Institute of Medical Genetics.[61]

The nature-nurture issue also entered into the core of the debate over Medvedev's diagnosis. In evaluating his sanity, psychiatrists questioned him about the views on Lysenkoism and the Party that he had expressed in his book. His "illness" was formulated as involving "paranoid delusions of reforming society," "poor adaptation to the social environment," and "split personality, expressed in the need to combine scientific work in his field with publicist activities." When the patient challenged the relevance of this to the medical diagnosis of his sanity, one psychiatrist responded that "psychiatrists are interested in all aspects of human activity," another that "psychiatry has always been a social science." By contrast, Roy Medvedev explained to a psychiatrist his right to know about his brother's medical evaluation in the following terms: "But my brother and I are identical twins and we have an identical heredity. If his mental state is unbalanced although he is unaware of it, then I must be in the same condition, and it is your duty as a doctor to warn me." And he later told a KGB official: "I shall regard any diagnosis of mental illness in my brother as a threat to myself as well—my brother and I are twins and we have an identical heredity."[62] The distinguished scientists who defended Zhores Medvedev against the psychiatric charges declared that the determination of his sanity was not a social question but a scientific, biological one, and that, as the nation's leading scientists, they were more qualified than psychiatrists to judge it.

The attack widens. In 1973 Dubinin's autobiography was published by Politizdat in an edition of 100,000 copies. The 444-page account of his life and career dealt with virtually all the major figures in the history of Soviet genetics. No prominent living Soviet geneticists escaped without some fairly serious direct or indirect tarring of their science, their politics, or their moral character. As Theodosius Dobzhansky remarked when he read the book in California in 1973, according to Dubinin the only geneticist who never made an intellectual or ideological blunder was Dubinin himself.[63]

The autobiography was especially hard on Kol'tsov. Dubinin claimed to have learned only recently, for example, that just after the Revolution, Kol'tsov was arrested as an active member of a counterrevolutionary organization. After several paragraphs detailing the reactionary nature of this organization and its crimes, Dubinin commented: "Seeing only his charm, and remaining ignorant of the complex twists in his life, those who surrounded him were not armed to resist its darker sides."[64] There followed a detailed discussion of the reactionary character of eugenic ideas generally and those of Kol'tsov, Filipchenko, and Serebrovskii in particular. Later in the book, he carried his critique of eugenics to Astaurov and his colleagues:

> Attempts by leaders of the genetics of the past to rehabilitate eugenic mistakes, and what is more, to put forth the content of the old mistaken eugenics as the ideal basis for the application to humans of the new achievements of genetics—such attempts have been made by B. L. Astaurov and A. A. Neifakh of the Institute of Developmental Biology, and M. D. Golubovskii of the Institute of Cytology and Genetics in Novosibirsk. V. P. Efroimson has set forth mistaken views on the purportedly genetic determination of ethical and social traits of the human personality. The writer V. V. Polynin has begun to propagandize the old eugenics. This already poses a serious ideological danger and is soil capable of nurturing a thirst for the uncritical glorification of the errors of the past. Once again a foreign ideology aimed at crushing the human personality, unjustified by the dictates of biology, has tried to enter and poison the pure wellsprings of our science.[65]

To add insult to injury, Dubinin dwelt on his own childhood as a *bezprizornik*—an orphan trained in special schools—and emphasized how much it had meant to him as a boy to see Lenin; indeed, the book includes a photograph that purportedly records the happy event. The book was entitled *Perpetual Motion* (*Vechnoe dvizhenie*).

Understandably, many Soviet geneticists referred to it as "perpetual self-promotion" (*vechnoe samovydvizhenie*), playing on Dubinin's actual history as a "promotee" (*vydvizhenets*). Most senior Soviet geneticists knew perfectly well, firsthand, that the portrayal of Dubinin's proletarian past had been carefully crafted to mimic Lobashev's by omitting certain telling facts. Dubinin had actually been born on Kronstadt into the privileged family of a naval officer, the captain of a minesweeper, and had grown up with German and English nannies. His father had died on the Volga in 1917, when Dubinin was ten years old. Although he had indeed subsequently been schooled in a special orphanage, it was because his mother was the orphanage nurse; far from being an orphan, Dubinin lived with his mother and brother in well-appointed quarters on the grounds.[66] Furthermore, Dubinin surely knew of Kol'tsov's arrest in the 1920s: it had been described in detail by Kol'tsov himself in a scientific article published in 1921.

To Astaurov and those around him, however, Dubinin's book—and the conditions of its publication—had ominous implications that went beyond its self-serving sins of omission. Dubinin's discussion of eugenics and its links with fascism were all too reminiscent of Lysenkoist rhetoric: it not only sullied the founders of Soviet genetics but also compromised the field's ongoing reconstruction. The pointed attacks on Kol'tsov cast a shadow over his students, since they were the ones around him who, according to Dubinin, had not been armed to resist his darker sides. In particular, Astaurov and other Kol'tsov students now sat in the Institute of Developmental Biology, which they were seeking to have named in honor of its founder. Even more ominous were Dubinin's personal attacks on leading Soviet geneticists for "once again" importing "a foreign ideology aimed at crushing the human personality" and "poisoning the pure wellsprings" of Soviet science: such phrases, suggesting that his opponents were fascists, smacked of Stalinism.

In 1974 Astaurov's worst fears seemed to be realized. While he was in the hospital with a heart ailment, a scientist on the staff of his institute, Il'ia M. Shapiro, failed to return home from a scientific conference in Italy. According to one account, Astaurov was summoned to the Academy and informed that "the unpatriotic act of his collaborator was not regarded as a chance event, and the very existence of his institute was threatened. After Astaurov returned home from that meeting, he suddenly died."[67] There was little doubt in anyone's mind that the threat to Astaurov's institute had been instigated by Dubinin, who did not attend the funeral. Many Soviet geneticists held him personally responsible for Astaurov's death.[68]

The following months saw a resurgence of Lysenkoism in the USSR

Academy of Sciences in which Dubinin was implicated. Astaurov's death opened a slot in the membership of its division of general biology. In the Academy elections of 26 November 1974, five months after Astaurov's death, the division directly elected to full membership in genetics and selection none other than V. S. Remeslo—the same unreconstructed Lysenkoist whose candidacy for corresponding membership had been voted down in 1964 over the protests of Lysenko and Khrushchev. In the early 1970s Remeslo had gained notoriety by claiming that he had used Lysenko's techniques to produce important new varieties of wheat. There is little doubt that Remeslo could not have been elected without Dubinin's backing. Understandably, Dubinin's rhetorical tactics, his attempt to use politics to take over the field, and his apparent alliance with Trofim Denisovich Lysenko and his followers led many Soviet geneticists to refer to him as "Trofim Denisovich Dubinin." A few weeks before Remeslo's election a second edition of Dubinin's autobiography went to the publishers and appeared in a printing of 100,000 copies.

Denouement. These events were felt at the Kol'tsov institute, but it proved highly resilient. Except for his six years in Central Asia in the early 1930s, Astaurov had been at the institute since 1922; many members of its staff lionized him and blamed Dubinin for his death. Thus the Kol'tsov institute continued to be a center of opposition to Dubinin, and the trends Astaurov had set in motion continued to prosper, albeit in less controversial ways. Before Astaurov's death, in an apparent response to Dubinin, Astaurov had written a scientific biography of Kol'tsov; it was published posthumously in 1975, and the following year the institute was officially renamed the N. K. Kol'tsov Institute of Developmental Biology.[69] The same year Efroimson joined its staff as a scientific consultant. V. A. Strunnikov (1914–) had been one of Astaurov's colleagues in Central Asia in the 1930s, and Astaurov had brought him into the institute in 1968; following Astaurov's death Strunnikov took over his laboratory, and in 1976 he was elected to the USSR Academy of Sciences as a corresponding member in genetics and selection. Perhaps the institute's most outspoken naturist was A. A. Neifakh (1926–), a protégé of Astaurov's who had been a graduate student at the Kol'tsov Institute (1946–1949). In the early 1950s, during the heyday of Lysenkoism, he worked as a cameraman; while making a popular science film on the "new cell theory" of Ol'ga Lepeshinskaia, an ally of Lysenko, he had had to fake the results to make the experiment come out as it was supposed to.[70] With the rebirth of genetics, Neifakh returned to the Kol'tsov institute.

He became director of its Laboratory of Biochemical Embryology in 1968 and has continued to head research in the field at the institute to this day, speaking out in various scientific and popular forums on the hereditary and biochemical basis of human mental qualities. [71]

Dmitrii K. Beliaev had been promoted to full membership in the Academy in 1972. Astaurov's death in 1974 left Dubinin and Beliaev as the two ranking geneticists. Beliaev owed much to Dubinin, who had appointed him assistant director of the Novosibirsk Institute of Cytology and Genetics in 1958 (he replaced Dubinin as director in 1959). Beliaev was the younger brother of N. K. Beliaev, who had died in the purges of 1937, and Kol'tsov and his students had helped find him a post. Beliaev must have felt torn between the two; for example, he had sided with Dubinin in opposing Timoféeff's election. [72] But ultimately Dubinin's tactics of political denunciation drove Beliaev into the Astaurov camp: Dubinin's 1973 autobiography had impugned his activities as director of the Novosibirsk institute, explicitly attacking him as a member of the naturist clique that was trying to "rehabilitate eugenic mistakes" and whose "foreign ideology" would "poison the pure wellsprings of our science." [73]

In 1976 Beliaev criticized Dubinin in an article in *Priroda* for his unreasonable and ideological stance on the nature-nurture issue. At Astaurov's initiative, in 1973–74 the journal had published an article by Ernst Mayr discussing mankind as a biological species, invoking the internationally renowned evolutionary biologist on the side of the naturists. [74] In an article in 1975 Dubinin had written that Mayr's position "leads to a series of reactionary conclusions," including racism. [75] In cool and reasonable language Beliaev discussed the issues and cited Dubinin's extreme ideological statements. Denying that the article had given any grounds for declaring Mayr to be a reactionary racist, Beliaev declared that Dubinin's ideas on human mental characteristics "are mistaken from a biological point of view, and are even more doubtful from the philosophical and methodological perspective." [76] Meanwhile, Dubinin remained undaunted, teaming up with a Moscow anthropologist in 1976 to publish a book rooting his nurturist views in the writings of Marx, Engels, and Lenin. [77]

In November 1976 Lysenko died—and the event went almost unnoticed in the Soviet press. Since the early 1970s there had been a virtual proscription on published discussions dealing with Lysenko and the troubles he had caused Soviet biology. Those who were setting about the task of reconstructing the legitimacy of Soviet genetics, free of ideological controversy and nagging historical references, must have regarded Dubinin's highly public polemics as unhelpful. In particular, Soviet genetics was preparing to reemerge onto the world stage. The Seventh International Congress of Genetics had been scheduled for Moscow in 1937, only to be postponed (and eventually held in Edinburgh) because of Lysenkoism; now, Soviet scientific authorities wished to show that Lysenkoism was past history, and the Fourteenth International Genetics Congress was scheduled for Moscow in August 1978. The highly publicized difficulties with the dissident Andrei Sakharov were creating enough international problems for Soviet science, and a Western boycott of the congress was threatened: what purpose was to be served by Dubinin stirring up old history and fresh dissension? It also seems likely that his ideological campaign was annoying some Party circles. After all, Dubinin was not the only geneticist who was a member of the Party: Alikhanian, Lobashev, and Kerkis—all of whom Dubinin attacked in his book—had been Party members since the 1940s, well before Dubinin. So too were many leading physical scientists in the Academy who had supported genetics, and they may have been perturbed by his strident ideological rhetoric.

The denouement of the controversy came at a meeting of the general assembly of the USSR Academy of Sciences, held on 21 November 1980, at which A. D. Aleksandrov upbraided Dubinin for his behavior. An important Party member and the former rector of Leningrad University, Aleksandrov had been a principal patron of Lobashev and the development of genetics at the university since the mid-1950s—indeed, in the Lobashev text that had called for the re-creation of eugenics, Aleksandrov had been singled out for special thanks. He was also one of the most prominent scientists to move to the Novosibirsk Science City in the early 1960s, where he had supported the development of its genetics institute, headed by Beliaev, which Dubinin had attacked. Aleksandrov spoke on the role of biological factors in the development of humans, taking a naturist position, and criticized Dubinin pointedly and at length for his mistaken and unreasonable position. In a rare move that could only be seen as a punishment, Aleksandrov's criticism of Dubinin was published, together with his response, and soon thereafter it was announced that "the presidium of the USSR Academy of Sciences has relieved Academician N. P. Dubinin of his responsibilities as director of the Institute of General Genetics AN USSR in connection with the expiration of his term, retaining for him his directorship of the laboratory of mutagenesis of this institute." An agricultural biochemical geneticist and vice president of the agriculture academy, A. A. Sozinov, was appointed his temporary replacement and was soon confirmed as director.[78]

Analysis of Recent Trends

An era in Soviet life ended with the death of Leonid Brezhnev on 11 November 1982. The interregnum that followed (presided over first by Iurii Andropov, who died on 9 February 1984, and then Konstantin Chernenko, who died on 3 March 1985) came to a close only when Mikhail Gorbachev consolidated his leadership in late 1986. During this period the Academy of Sciences was headed by the physicist A. P. Aleksandrov, who served as president from 1975; he was replaced by Gurii Marchuk in 1986.

With shifting political leadership at the highest levels, there was a kind of uneasy truce in the Soviet nature-nurture debate in the early 1980s. Medical genetics continued its slow, conservative growth under the leadership of Nikolai Pavlovich Bochkov (1931–). Trained as a pediatrician, Bochkov studied radiation genetics in Dubinin's laboratory until 1961, joined the Party in 1962, then took a job at the Institute of Medical Radiology in Obninsk, where he worked with Timoféeff-Ressovsky. In 1969, when the USSR Academy of Medical Sciences opened a new Institute of Medical Genetics, Bochkov became its director and has continued in that post to this day, publishing a series of solid professional books in his field. Although he has occasionally discussed ethical questions related to medical genetics, he has stayed away from the nature-nurture dispute in all of his writings, apparently seeking to insulate his medical specialty from ideological controversy.[79] Recent textbooks in general genetics have taken a similar tack, describing developments in human and medical genetics but steering clear of the nature-nurture dispute.[80]

The abiding ambivalence is reflected in the way Soviet philosophers have treated the nature-nurture debate. Since the 1960s they have had little difficulty reconciling dialectical materialism and molecular genetics. Indeed, a 1968 book by Ivan T. Frolov, *Genetics and Dialectics,* argued that the positions of Soviet geneticists in the 1920s and 1930s were much more consistent with dialectical materialism than those set forth by Lysenko. This book played a major role in uncoupling Lysenkoism and official Marxist ideology. Yet, perhaps unable to break utterly free from

Marxist social determinism, Frolov and other Soviet philosophers have approached the nature-nurture issue gingerly. In his many writings on the topic over the years, Frolov has tried to formulate some reasonable middle ground that defends the genetic basis of human nature and at the same time reasserts its essentially social character; it is not clear, however, whether he seeks to advance the discussion or to end it.[81]

Under the surface, of course, the dispute continued to simmer. In the years following his dismissal from the directorship of the Institute of General Genetics in 1981, Dubinin produced no fewer that three books stridently arguing for his nurturist views and castigating his naturist opponents.[82] At times it appeared that nurturism might be resurgent: as Loren Graham has pointed out, Chernenko's daughter, Elena, had stridently advocated nurturism in a 1979 book, *The Social Determination of Human Biology*, that made heavy-handed references to the classics of Marxism.[83] It is difficult to judge what influence the Party chief's daughter may have had on her father's views; for example, Khrushchev's daughter was a biologist who is known to have vexed him for his support of Lysenko. Whatever Elena's influence, her father was in power for only thirteen months. In any case, the natural leader of the nurturist cause, Dubinin, had clearly alienated much of the scientific establishment, which held him at bay throughout A. P. Aleksandrov's tenure as Academy president. In these years Dubinin's nurturist writings appeared principally under the imprimatur of Marxist social science and the Party.

Glasnost

Since Mikhail Gorbachev's consolidation of power in 1986 and 1987, the Soviet Union has experienced a remarkable rebirth of open expression that the world has come to know as glasnost. Such periodicals as *Ogonek* and *Moskovskie novosti* have published uncensored articles on Lysenkoism and the repression of Soviet genetics which have reopened old wounds. For example, a long essay in *Ogonek* by Valery Soyfer documented Lysenko's relationship with Stalin and criticized Soviet geneticists for collaborating with Lysenkoism; its publication led to letters strongly attacking Lysenko, and other letters strongly defending him, notably from his son.[84] This exchange illustrates what is probably the general pattern. Under glasnost various individuals and groups have been freer to express their opinions in print, professional disputes have become more public, and old controversies have gained new intensity and candor. This is certainly true of the nature-nurture debate, but in that debate—as in others—quantitative change has not become qualitative. This is illustrated by the way longstanding issues have resurfaced.

Although Timoféeff-Ressovsky died in 1981, still unrehabilitated, the controversy surrounding him has intensified. In early 1987, he was the subject of a heroic biography, *Zubr* (Bison), first published in *Novyi mir* by the popular Soviet author Daniil Granin. By year's end the book had appeared in a hardbound edition of 200,000 copies and a paperback edition of 300,000—both of which soon became bibliographic rarities. [85] The novel has been the subject of widespread discussions at professional clubs and institutes, including the Institute of Medical Genetics. A documentary film of Timoféeff's life has been made. But the new publicity surrounding Timoféeff has reanimated the campaign against him, and letters have been published alleging his complicity with Nazi atrocities, written largely by former Lysenkoists, one of whom reputedly headed the "sharashka" prison camp where Timoféeff was incarcerated. In late 1987 Timoféeff's case was judicially reexamined, and the government petitioned the USSR Supreme Court to rehabilitate him. A date was set for the announcement of the decision (usually a formality in such cases), but because of the renewed controversy the date has been repeatedly postponed. In 1988 there appeared another novel on the history of Lysenkoism, by Vladimir Dudintsev—*Belye odezhdy* (White robes)—which is provoking controversy. [86]

Meanwhile, at the Institute of General Genetics, Dubinin's earlier politicking had become more spirited. With the call for discussions of the coming Party Conference in June 1988, Dubinin sent a letter to Academy president Marchuk with political complaints against his replacement as director: Sozinov, he claimed, was heading a Zionist cabal at the institute and supporting ideologically dangerous naturist research. At a public meeting at the institute he repeated his charges. In response Sozinov reminded Dubinin that in earlier years such a denunciation could have resulted in his death, but that, fortunately, times had changed. As a result Dubinin resigned from the institute, or, more accurately, seceded; with Marchuk's support, Dubinin's laboratory, which occupies the building's fourth floor, has become an independent administrative unit. In the fall of 1988 Dubinin published yet another four-hundred-page history of Soviet genetics. In keeping with glasnost it was as candid as ever, renewing attacks on his late nemesis Astaurov and his "clique" (Beliaev, Efroimson, Lobashov, Polynin, and Rokitskii) for their eugenic and ideological errors.[87] Sozinov subsequently resigned, and the process of electing his successor has reportedly shown all of the confusion, excitement, and controversy that have characterized Soviet democratization elsewhere.

Even today the nature-nurture issue remains among the touchiest of subjects. Soviet medical genetics has only recently been reestablished as a specialty, and its administrators can hardly welcome a controversy that may once again throw its ideological legitimacy into question. Millions of people trained in biology during Lysenkoist days still occupy prominent posts throughout the Soviet system. In the meantime, Frolov's compromise position on the nature-nurture issue has gained new standing as his fortunes have risen: editor of *Voprosy filosofii* for a time, he has become a member of the Central Committee and a Gorbachev adviser. With so many other matters on the agenda, those close to the leadership may not want to encourage the resurrection of a bitter and divisive controversy.

As regards the nature-nurture issue, then, glasnost is still incomplete. This is most clearly shown by four examples of what has *not* thus far been published. On 12 September 1986, A. A. Neifakh expressed his naturist views in a heavily attended public lecture, "Genes and People," sponsored by the popular science society Znanie; yet he is still encountering difficulties in getting his views published.[88] In 1988 there appeared a new edition of *The History and Development of Genetics* by A. E. Gaissinovitch, Russia's premier historian of the field; even though it was issued by the Kol'tsov institute, Academy editors refused to permit the inclusion of archival material dealing with Lysenkoism that had not already been published.[89] V. P. Efroimson's crowning work, a two-volume study of the biosocial basis of genius, completed in 1982, has never been published; the manuscript remains on closed reserve at the All-Union Institute of Scientific and Technical Information.[90] Finally, consider the two-volume work by the Leningrader I. I. Kanaev, now deceased, entitled *A History of Human Genetics in the USSR:* it was advertised in the "Nauka" publishing house catalogue in 1976, but it never appeared.[91] On a recent trip to Leningrad I inquired after the manuscript and found it— locked in a cabinet and forbidden to be seen. Will it ever be published? If it is, it may prove to be a significant contribution to the history of the Soviet nature-nurture controversy—and its appearance would be an indication of the willingness of a new Soviet generation to face that history and come to terms with it.

Understanding the Debate

The Soviet nature-nurture debate arose in the 1960s in the context of the rebirth of Soviet genetics after three decades of Lysenkoism. The recreation of the field required the reestablishment of its institutional foundations and disciplinary legitimacy in organizations with a strong Lysenkoist presence and in the absence of political instructions. In this context, conflicts arose over the shape of the new genetics, its relationship to Lysenkoists, its ideological profile, and its history. The nature-nurture issue was a particularly sensitive one in light of the virulent nurturism of Lysenkoist ideology.

Lysenkoism had begun its attacks on Soviet genetics in the mid-1930s, and since 1948, with few exceptions, Michurinism had been the politically sanctioned view in biological textbooks and curricula. As a result, in 1965 many of the figures central to rebuilding genetics were in their sixties, and, since Lysenkoism had exercised an especially devastating effect on plant geneticists, most were animal geneticists. Their view of the character of the new genetics was conditioned by their historical experience. Almost all had trained under Kol'tsov, Filipchenko, Chetverikov, Serebrovskii, or H. J. Muller at a time when there were no hard and fast distinctions between genetics and eugenics, and many had actually been employed in bureaus, institutes, and laboratories where research on human genetics had been pursued. They saw genetics, eugenics, and medical genetics repressed at roughly the same time, by the same people, for the same reasons. Meanwhile, Dubinin's social background, his ideological and political history, and his special status brought him into conflict with many of his senior colleagues over the question of eugenics and human genetics.

The increasing polarization of the conflict between Astaurov and Dubinin grew out of different approaches to a shared history. The two sides were able to use networks and alliances to support their views of the new genetics. Dubinin's most natural allies proved to be the more conservative, ideological, and politically correct figures in Soviet science, who welcomed his accommodation with Lysenkoists and the minimal ideological and structural disruptions his approach to the reconstruction of genetics would entail. Astaurov's most natural allies were those who remembered the golden days of Soviet biology before Lysenkoism and who sought to renew them in the modern context. Soon, however, the debate took on a dynamic of its own. Dubinin's growing ideological stridency alienated many of his potential allies, leading him to rely increasingly on ideological support. As the debate expanded, the nature-nurture issue became linked to other Soviet issues that had little to do with biology per se, and "united fronts" began to form encompassing disparate conceptions, agendas, and experiences. True, Western discussions of the nature-nurture issue provided a contemporary counterpoint, and both sides in the Soviet debate were able to draw upon them in attempting to legitimate their positions. But the two sides mobilized the Western discussions only selectively, in ways they found useful to their cause. For the Soviet debate, then, the concurrent Western discussions served less as an inspiration than as a resource.

Although my account of the Soviet nature-nurture debate has concentrated on the group that "drove" the debate—the geneticists—it helps to explain the participation of those from other disciplines. For example, we have seen that the Russian eugenics movement involved anthropologists such as V. V. Bunak and Ia. Ia. Roginskii, who were able to continue their research during Stalinist times within Russian anthropology; the current anthropological participants in the debate may well have historical and institutional connections with these figures. Likewise, the involvement of Bekhterev, Iudin, Liublinskii, Pavlov, Orbeli, Davidenkov, and Luria in eugenics, medical genetics, and the inheritance of behavior helps us understand the naturist strand in current Soviet psychology and psychiatry. Do some nurturists who come from philosophy, anthropology, psychology, psychiatry, or medicine have historical links with Lysenkoism, Stalinist ideology, or repression? Of course, members of different disciplines may have experienced the Revolution, the Great Break, and Stalinism differently, but we will know *how* differently only by making a comparable analysis of their careers, personal and professional networks, and institutional histories. The remarkable patterns revealed by the detailed analysis of one Soviet discipline—genetics—suggest fruitful ways of analyzing others.

Broader Implications

The Soviet nature-nurture debate is only one of many in Soviet scientific life, but, as we have seen, it has had implications for broader questions. Similarly, the foregoing analysis of that specific debate has methodological implications that go well beyond studies of the troubled history of Soviet genetics.

The role of history. Because of the way historians have divided their turf, Russian history is usually punctuated at the 1917 Revolution, with those focusing on the postrevolutionary, Soviet period forming a distinguishable subgroup with its own problems and agendas. As we have seen in the history of Russian science, and especially the "biosocial" disciplines, however, the most pronounced discontinuity occurred not around 1917 but around 1930. The Great Break brought to an end a half-century of vigorous, continuous, and relatively free biosocial discourse; it did not begin to reappear until the 1960s, and has only reached the pre-1930 level of vigor under Gorbachev. If the same pattern is evident (as it appears to be) in other areas of Russian intellectual and social life—if much of what occurred in the 1920s was a continuation of trends and agendas originating in the late tsarist period—then we must extend our analysis of Soviet developments into tsarist times. This perspective may also help us to understand current events in the USSR. As the rehabilitation of Bukharin, Zinoviev, and Kamenev suggests (as well as the almost heroic current status of such scientific figures as Nikolai Vavilov and V. I. Vernadskii), resurrecting the intellectual legacy of the 1920s has become a major preoccupation of glasnost. As in the Soviet nature-nurture debate, so too, more generally, we may profit by seeing current public arguments as a competition between different interpretations of a common history, with divergent groups vying over agendas for reconstruction modeled on different times.

Institutions, disciplines, and networks. One of the most striking features of the Soviet nature-nurture debate is its institutional continuity. To a remarkable degree it has been a tale of two institutes: the two centers of eugenics in the early 1920s reemerged in the late 1960s as the two centers most embroiled in the nature-nurture debate, and geneticists trained at those centers became its chief protagonists. The department Filipchenko founded at Petrograd University in 1919 would have a chairman in 1967 who would call for the rebirth of eugenics. Filipchenko's Bureau of Eugenics, after four name changes, would serve as the base in the 1930s for H. J. Muller's campaign; seized from Vavilov by Lysenko in 1940, it would reemerge (after being renamed yet again) as an ideological base for Dubinin's nurturism in the 1970s—and the institute's dual historical personality continues to show itself to this very day. Kol'tsov's institute, home of the Russian Eugenics Society in the 1920s, has re-emerged (after three name changes) as a center of naturism, and has maintained close ties with *Priroda,* the popular journal that Kol'tsov helped to found in 1912. These continuities, spanning radical social and ideological change, show the remarkable momentum of institutionalized scientific traditions.

If the intellectual styles embodied in Soviet institutions have proved remarkably enduring, the same cannot be said of disciplinary rubrics. The scientific studies first launched as "eugenics" became "anthropoge-netics," "medical genetics," and still later "human genetics." Not that these names were unimportant; to the contrary, each presented a slightly different way of dividing knowledge, defining boundaries, and packaging problems, and each acquired a quite particular ideological res-onance. Rather, the history of the Soviet nature-nurture debate suggests that such disciplinary rubrics are communal constructions created and deployed for their intellectual, ideological, and political utility. A change in disciplinary nomenclature, together with the new classification of knowledge it entails, is a significant event in the social history of science, but does not have a simple relation to its intellectual history: the abolition of "eugenics" did not entail the end of the research agenda that had sprung up under that name.

This research was maintained, of course, not by physical structures or linguistic inventions but by people. Historians know of the extraordi-nary intellectual breadth and personal interconnectedness of the Russian intelligentsia; current events suggest that it is alive and well. The career patterns of individuals involved in the Soviet nature-nurture debate dem-onstrate that the Petrograd and Moscow networks of Filipchenko, Kol'-tsov, and their students have endured, tested and tempered by war and repression. Such personal human networks—formed by old school ties, extended families, work settings, and common interests—span diverse institutions and disciplines, connecting them in ways that cannot be seen in organizational charts or the published paper trail. The analysis of the dialectical play between institutions, disciplines, and networks—the three great organizing principles of the modern scientific enterprise that structure its interactions with society—is currently reshaping science studies in the West. When Soviet developments are studied in the same way, we may come to see more clearly both the unity and the diversity of world science.

Science and values. Perhaps the most abiding philosophical issue presented by the Soviet nature-nurture debate concerns the relationship between science, ideology, and values. In the past, both sides in the dispute have made claims about the inexorable logic linking certain scientific positions with particular social values; such claims, however, can find little justification in the history of the Soviet debate. Rather, we are struck by how very contingent have been the links between science, ideology, and values. As contemporary Lamarckian eugenics demonstrates, the link between Russian genetics and eugenics in the 1920s was not an inevitability: it was an artifact of the strategies and ideas of the founders of the Russian movement, the foreign models that inspired them, and the institutions and networks they created. Nor was there any necessary relationship between Lysenkoism and human nurturism: in his pronouncements on inheritance, Lysenko wrote much about plants but almost nothing about humans. These linkages were not necessary, but contingent—not logical, but ideological—and each triumphed over alternative ideological readings that embodied alternative associations. Had the Nazis not come to power in Germany, would Stalinists have felt the same need to assert a converse ideology? In any case, under Stalinism these particular contingent linkages were cemented and reified, only to reemerge when repression eased.

Finally, a few comments on the peculiarities of Soviet naturism are in order. As should be apparent, I view both the "extreme" character of Soviet naturism and its liberal ties as by-products of the historical and structural development of Soviet science. In the West, the eugenics movement lost support among many geneticists around the time of the Great Break, but it was never repressed; its repudiation occurred only after postwar revelations of Nazi atrocities, and shortly thereafter "medical genetics" and "human genetics" arose as disciplinary alternatives. The gradual dissociation of eugenics from human genetics was so successful that today many medical geneticists are unaware of, and sometimes deny outright, any historical link between their field and eugenics.

By contrast, no comparable Soviet evolution could occur in Stalinist times. As a result of its repression, Soviet genetics—with all of its eugenic and human aspects—went into a kind of deep freeze. When it thawed in the 1960s, what emerged was a concept of the discipline, its agenda, and its mission closely resembling what had existed on the eve of Lysenkoism. Forbidden links had been cemented by harsh experience: such are the peculiar dialectics of repression. By now, of course, the senior geneticists who rebuilt Soviet genetics are being replaced by a new generation of their own students, trained after 1965. Those older geneticists have labored to make this new generation rightly proud of its scientific heritage. Have they also passed on their legacy of anger, courage, and pain?

I began by noting that despite their apparent similarities, the Soviet and Western nature-nurture debates are really very different. It is appropriate to conclude by pointing out that they may in fact have something important in common. I have argued that the character of Soviet naturism, as well as the character of Soviet nurturism, is the product of particular historical circumstances—and so too are their political associations. If Soviet naturist positions appear extreme or "conservative" from an Anglo-American perspective, might this not be because *our own* perspective is also a cultural product, molded by our own scientific, institutional, social, and political history and experience? And if the social "implications" of biology are in fact contingent associations, broken or cemented by historical experience, and shaped differently in different institutional, cultural, and national settings, must we not devote more attention to comparative historical analysis in setting policy and shaping our biomedical future?

· PART THREE ·

Engineering and Big Technology

· 5 ·

Engineers: The Rise and Decline of a Social Myth

Harley Balzer

Half the engineers in the world work in the Soviet Union. This fact is a source of pride for Soviet leaders, a source of alarm for some American observers, and a source of perplexity for economists and historians. Why, with all that skilled personnel, does the Soviet economic system perform so poorly? Is the problem with the system or the people? For the social scientist the question is somewhat different: Can the tremendous number of engineers help explain why engineering has been the path to political leadership in the USSR? If engineers play the sort of role in Soviet politics that lawyers assume in America, what does this tell us about the engineers and the society?

The tremendous number of engineers is due in part to the fact that technical training has been a major mechanism of social mobility in Russian and Soviet society. Beginning in the second half of the nineteenth century, and accelerating rapidly in the 1930s, engineering served as a path for individuals from the lower social classes to move first into the intelligentsia and then into the political elite. Social mobility was accompanied by a perception of engineering as a high-status profession, with engineers portrayed accordingly in Soviet literature. But in the past few decades the status and prestige of engineering have declined, and the quality of Soviet engineering is now under sharp attack from internal critics. Engineers are referred to as "the gray people," and prestigious technical institutes experience difficulty filling their freshman classes. To understand fully the rise of engineering as both a mechanism of mobility and a source of prestige, we must also account for its perceived decline.

This chapter sketches the social history of engineers in Russian and Soviet society. I have attempted to trace the same key topics in each

141

period: engineering education, the work environment, professional identity and activity, social mobility, and the "cultural" image of engineers in literature (and, in a few instances, film). While basically adhering to this outline, I have diverged to stress the key aspects of each individual era. This approach should compensate for the inevitable blurring of historical specificity that results from covering an extended sweep of time.

Russian Engineers

Lenin was correct when he stated that Russia occupied the last place in Europe in terms of technical cadres.[1] But the most serious problem was not simply the shortage of engineers. It was that to compete with other nations, Russia had to catch up with a moving target, a situation with obvious parallels to the era of the "scientific-technical revolution" and Mikhail Gorbachev's reforms. In 1890 perhaps four hundred engineers graduated from the half-dozen higher technical schools in the Russian Empire.[2] During the period 1896–1902, mainly though the efforts of Finance Minister Sergei Witte, the number of technical institutes was doubled and the number of students more than tripled. This growth must be compared, however, to a tenfold increase in the number of American engineering students during the 1890s.[3]

Prior to the 1890s, advanced technical training in Russia predominantly followed a "ministerial" pattern.[4] Each department of the tsarist government sought to train its own specialists for the specific activities under its purview, and most technical personnel spent their entire career within one ministry. This was true of the mining engineers, transport engineers, and military engineers, all of whom were recruited mainly from the privileged classes. Exceptions were graduates of institutes under the control of the ministries of education and finance. It is testimony to the tenacity of the ministerial ethos that new, more flexible institutions were considered "second rank" even after they began to provide an education qualitatively superior to that at the older institutes.

With the development of industry and capitalist economic activity, engineers increasingly played entrepreneurial and other roles. The economic changes cut two ways. New economic organizations provided opportunities for the graduates of the "practical" industrial institutes who were interested in production and entrepreneurship. At the same time, experience in a ministry proved of immense value for subsequent commercial activity, and foreign firms found it essential to employ Russian citizens with government experience to deal with the bureaucracy.[5]

On a small scale beginning in the 1870s, and much more decisively in the 1890s, we see the emergence of engineering generalists. Trained at the Petersburg Technological Institute or in a few cases at other schools, and with experience in Europe, these engineers sought to break the constraints of the narrow ministerial pattern. Prototypes here were I. A. Vyshnegradskii and V. L. Kirpichev, two of the major figures who assisted Witte in the expansion of engineering training.[6]

The anvil for forging the new engineering cadres was the polytechnical institute. Witte established three of these institutions, as well as new mining and transport institutes offering shorter courses and less encyclopedic curriculums than in the existing schools.[7] In 1895 only seven technical institutes conferred engineering degrees; by 1902 there were thirteen. The number of students increased from four thousand in 1895 to thirteen thousand in 1904 and over twenty thousand in 1914. On the eve of the First World War more students were attending the Petersburg Polytechnical Institute than had been enrolled at all Russian higher technical schools in 1900. Although the graduation rate declined even from the disappointing one in ten of the 1890s, the geometric growth in numbers of students produced some increase in the number of graduate engineers.[8]

While hardly on the scale it would assume later, engineering education provided a means of social mobility in tsarist Russia. Russian higher education was more "democratic" in student enrollments than other European systems, and much of the non-noble enrollment was at institutions other than the universities.[9]

With so much emphasis placed on training engineers to staff new railroads and industrial enterprises, less attention was devoted to producing technicians. Support personnel were usually drop-outs from higher education or *praktiki* (practicals) with no formal training. Secondary specialized education was the weakest link in the system of technical training. Through a complex political compromise in the 1880s, technicums wound up occupying an intermediate place in the educational system. By the 1890s they were widely regarded as a dumping ground for poor students, since anyone aspiring to higher education chose a more direct route.

Engineers' professional activity grew with expanded numbers and diversified economic activity, but never broke out of constraints imposed by the tsarist government. Membership in Russian technical societies was narrowly circumscribed. Most were virtual alumni associations of individual schools, since organization on broader principles was prohibited. A few groups sought to become local or regional organizations, but even these—for example, the South Russian Society of Technologists— were based primarily on a single institute. The only organization asserting a claim to universality was the Imperial Russian Technical Society, which was too diverse in membership and, as its name implies, too closely tied to the state to express effectively the professional interests of Russian engineers.[10]

Engineers made sporadic attempts to establish professional associations based on geography or specialization rather than ties to a state-run school. When the journal *Inzhener* (Engineer) began publication in 1882, it articulated the need for a nongovernment professional engineers' organization, a call it made repeatedly until 1917.[11] Other groups' efforts to establish independent professional associations met resistance from the government.

Much of the impetus for growth in organization membership and activity was economic, reflecting increased numbers of engineers and employment difficulties during the economic slump after 1900. In the events surrounding the Revolution of 1905, however, many engineering associations began to pay more attention to broader political questions. In 1904 a group of politically aggressive engineers organized one of the first professional unions, and in 1905 the Union of Engineers played a leading role in seeking to establish an all-Russian professional organization.[12]

Despite a strong desire for unity, neither the Union of Engineers nor any other organization could speak effectively for all Russian engineers, much less all technical personnel. In the highly charged atmosphere of 1905, such an organization was impossible. Groups formed both to the left and the right of the Union of Engineers. Not all engineers were in sympathy with even the mildest forms of political activity. L. N. Liubimov, one of the elite transport engineers, recounted having to run for his life after angering a mob by refusing to doff his cap to the "new freedoms."[13] A reaction typical of many engineers was that of E. O. Paton, who sought to immerse himself in technical problems while remaining neutral politically.[14] For politically active engineers, a key issue was to define their relationship to workers and employers, a question that became even more pressing in 1917.

Electrical engineers played a leading role in organizational activity, reflecting the importance of electrical technology as Russia joined the industrial revolution as well as the influence of foreign colleagues. Their activism presaged significant contributions in the Soviet era.[15] Electrical engineers were among the first to discuss engineering ethics.[16]

Ethical dilemmas reflected new career opportunities as capitalism developed. Rather than remaining subject to the whims of bureaucratic superiors or corporate employers, engineers could work for multiple employers and even engage in that penultimate activity of the professional middle class, consulting. The chemist V. N. Ipatieff recalls in his memoirs the lucrative consulting fees he received.[17] By 1905 engineering societies were drafting model contracts for consulting arrangements with Russian and foreign firms, and Russian engineers were participating in meetings to create an international society of consulting engineers.[18]

The range of engineers' legal incomes, however, varied tremendously. For every talented and savvy expert like Ipatieff there were dozens of engineers barely scraping by. If the rewards of success in private enterprise were enormous, the risks were at least as great, causing many engineers to opt for government positions that guaranteed salaries and pensions.[19]

Young engineers frequently reported difficulty finding "suitable" positions, yet Russian industry remained woefully short of trained personnel. The geometric increase in the numbers of engineering students produced a surplus of engineers in some locations and in particular specialties but did not begin to meet Russia's overall need for specialists. Those with higher education often took jobs outside their field in order to remain in urban areas, while there were persistent shortages of personnel in rural areas, and especially in Siberia.[20]

Many positions that should have been filled by engineers went by default to people who had to learn their skills on the job. These *praktiki,* who were operating going concerns, frequently reacted with hostility to the appearance of institute graduates lacking applied skills. The metallurgical engineer M. A. Pavlov recounted how a lab technician at his first workplace, fearing competition from a school-educated engineer, put sand in Pavlov's test samples to distort the results.[21] The problem of the *praktiki* took on a particular edge in the Stalin era, when political leaders sought to reassert the virtues of uneducated specialists.[22] But these conflicts had a very long history.

If engineers themselves managed to develop a strong sense of identity, this was not reflected in the literature of tsarist Russia. Engineers appear infrequently in novels and short stories, and when they do appear, their identity as engineers rarely carries special significance. Hermann in Pushkin's *Queen of Spades* is a young officer; his engineering specialty is incidental. Dostoyevskii did initially refer to Kirilov as "The Engineer" in his notebooks for *Devils,* but one searches in vain for some trait that would make his being an engineer rather than his membership in the radical intelligentsia his distinguishing characteristic.

Only with industrialization do characters emerge whose engineering identity is integral to the literary work. N. G. Garin-Mikhailovskii's *Studenty* (1895) and *Inzhenery* (1906) are unique in featuring the milieu of engineering education and professional activity.[23] Tema Kartashev is not really plausible as anything but an engineer, and Garin chose his environment carefully to show the general problems of the intelligentsia as well as specific concerns of engineers. Another major "engineer novel" of the prerevolutionary period is Bogdanov's *Engineer Menni,* but here the Martian engineer is a foil for Bogdanov's utopian visions rather than an example of any existing figures in Russian life. Mikhail Slonimskii's *Inzhenery,* a vivid portrait of prerevolutionary engineers, was published after the Second World War and is really a historical novel reflecting the issues of the later period.

What, then, was the tsarist era's legacy to Soviet engineering? There was an inadequate but growing education system; and while technical training was a path to social advancement, especially for individuals from the lower classes, and also for impoverished noblemen, it was one path among many. Patterns of poor training for support personnel and uneven distribution of specialists were well entrenched. In literature, engineering identity was only beginning to take on social and professional significance. There did exist a highly competent core of engineering professionals and educators, striving to liberate their economic system from foreign control and their professional life from government restrictions so that they might become full-fledged international professionals.

Revolution and Reconstruction

The first decade of Soviet power witnessed a civil war followed by an attempt to repair the damage resulting from seven years of conflict. The Civil War and New Economic Policy established two extremes of a policy pendulum that has characterized Soviet life ever since.

By the outbreak of the First World War, Russian engineers had developed a sophisticated professional consciousness. The main barrier to constituting themselves as a profession on the model of their Western colleagues remained opposition from the tsarist government. While wartime conditions gave the government a rationale for preventing convocation of an all-Russian congress of engineers, the government's handling of the wartime economy also convinced many engineers that the autocracy was a disaster in economic and technical as well as political terms. Few engineers mourned the passing of the Romanov regime.

Following the February Revolution, engineers returned to the professional agenda they had established before the war. They convened an all-Russian congress, sought to establish a unified engineering society, and devoted particular attention to establishing an identity distinct from both labor and management. Although Western experience would lead us to expect engineers to be staunch supporters of a bourgeois regime, many Russian engineers were quite susceptible to the Bolshevik appeal. What there was of a technocratic movement in Russia found significant allure in an ideology promising rapid economic growth and technical transformation. Some saw the Bolsheviks as the only group capable of protecting their property and even their lives in an era of increasing anarchy.[24] Still others believed the Bolsheviks represented the best chance to defeat Russia's enemies.[25] If they did not welcome the Bolsheviks, many engineers were at least willing to give them a chance.

We may never know precisely how much of a toll the war, revolution, and civil war exacted from the technical intelligentsia. Battle casualties, disease, and emigration reduced the number of trained engineers by perhaps as much as half. Aftereffects continued well beyond the Civil War. The opportunity costs of students not trained, skills not shared, and professional communities not perpetuated made the damage far more extensive. The Bolsheviks' task was not merely to take up where the tsarist system had stood in 1913 but to repair massive losses.

Some of the damage after 1917 was self-inflicted. Revolutionary educational policies adopted in 1918 abolished entrance requirements, grades, and standards in general.[26] Widespread public demand for higher education resulted in many secondary schools becoming higher educational institutions with no significant alteration other than to their title. A report by the Ministry of Education noted that schools "developed for the most part by anarchy, according to local initiative."[27]

The educational process itself was dubious. In a situation of civil war and general crisis, students were more likely to be at one of the various military or political fronts than in a classroom.[28] A revolution carried out in the name of the working class inevitably gave rise to pressures to reward workers. Preferential treatment for proletarians (and often, but not always, peasants) and discrimination against children from the middle and upper classes were considered to be logical consequences of the Revolution. Discrimination proved much easier to enforce than affirmative action.[29]

There were two major obstacles to preferential treatment for workers in education. One involved the knowledge base. Prospective students lacked both the secondary education and the cultural background needed for advanced study. Even when the higher school curriculum was reduced to the essentials of technical training, most workers found it beyond their capabilities. Workers' faculties were created to provide remedial education to prepare workers and peasants for admission to institutions of higher education. But it was not possible to compensate overnight for decades of cultural deprivation or a lack of basic education.[30]

Overcoming the lack of education among workers was child's play compared to the task of ascertaining just who was a worker. In a situation of revolutionary social flux, in which proletarian credentials could mean the difference between life and death (during the famine in 1920–21 some students continued to receive front-line rations),[31] the individuals able to secure proof of proletarian origins were often precisely those who had the least claim to that heritage.[32] After 1921 the New Economic Policy (NEP) brought a period of relative stability. Yet, as in most instances, a closer examination reveals important fluctuations within a supposedly coherent period. Debates among the leadership resulted in vacillating social policy in the schools, including a purge of the student body in 1924.[33]

Administrative conflict between education and industrial interests carried over from the tsarist era. During the NEP, technical schools remained under the education administration. The emphasis continued to be on higher education for engineers, with short-term courses the norm for workers. Little was accomplished in the area of specialized secondary training, and after 1921 expansion of the education system all but ceased. The new schools that had been established during the Civil War either vanished or continued a precarious existence with inadequate funding and minimal infrastructure. At the best schools the program reverted to the five-year encyclopedic curriculum prevalent before the war.

By the late 1920s the education system appeared to be restored to normal operation. The need for technical cadres was being met to a limited extent by a combination of old specialists, new Red Directors, and a large number of *praktiki*. But the supply of specialists was not adequate, especially if the tempo of economic development was to be increased.

An emphasis on proletarian leadership infused Bolshevik policy toward engineers' professional organizations. Despite the official preference for specialty-based unions uniting engineers, technicians, and workers, however, existing professional organizations continued to function with remarkable continuity during the Civil War. Government conferences in 1918 and 1919 resolved to establish Engineering-Technical Sections (ITS) under the mass-membership trade unions that would be responsible for "material" issues. But the All-Russian Association of Engineers was permitted to continue its activities directed at economic reconstruction and technological development,[34] and the Russian Technical Society continued a precarious existence.[35] Even during some of the darkest moments of the Civil War the government provided financial assistance for these organizations, which suggests that they retained some patrons.[36]

The ITS played a more active role in the mid-1920s. In addition to continuing to support engineers' material interests as economic relationships changed, the sections took on responsibilities in education and cultural work, including raising qualifications and providing training for workers. But the direct interests of members continued to take precedence. In 1923 the central bureaus of the construction and railroad workers' unions were able to set aside a specific number of places at technical institutes for children of ITS members.[37]

Members of the prerevolutionary Russian Technical Society sought to continue their corporate existence and play an independent role in economic and technical policy. Most striking in this regard is evidence that the influence of international technocratic ideas resonated strongly. P. I. Pal'chinskii considered the engineers to be the only "reliable" group in Soviet society, and the sole group capable of negotiating successfully with foreign governments and corporations.[38] These engineers and their technocratic musings hardly represented a significant threat to the Soviet government. Yet their discussions and group activities were more extensive than has previously been believed, and it is obvious what could be made of these activities in a supercharged political environment.

Influential members of the leadership, including Lenin himself, intervened repeatedly to assist engineers and scientists experiencing political troubles and problems with living conditions.[39] These initiatives were expanded during the NEP. But central policy frequently encountered obstacles from local officials, from workers, and from Party committees. Specialist baiting, while only a temporary government policy under War Communism, proved very difficult to curtail. Harassment, persecution, and even murder of specialists was not uncommon. One specialist recalled the mid-1920s: "In one plant specialists were threatened, cursed with the vilest language as bourgeois and "former people"; in another plant unexpectedly doused with water; in a third trundled out of the factory in a wheelbarrow; in a fourth their apartment windows were broken; in a fifth an engineer was struck in the face."[40]

Measures taken under War Communism and the NEP did little to alleviate the shortage of specialists. Those with education continued to prefer administrative positions, while most of the jobs in production were filled by *praktiki*. The majority of graduates did not actually work in their specialties.[41] The major difference between the mid-1920s and late 1920s in this respect was the degree to which the situation was regarded as a crisis.

Little artistic literature was generated in the desperate conditions of the Civil War. The strongest themes in this period were utopian yearning and antiutopian critique. Representative of the numerous utopian fantasies was Bogdanov's *Engineer Menni*, frequently republished during the early 1920s.[42] The most famous, and best, of the antiutopian critiques was Evgeny Zamiatin's *We*, published in 1922. The designer of the spacecraft in *We* is never described as an engineer, yet his experience mirrors the inner conflicts between technical opportunities and spiritual qualms experienced by many technical specialists in this period. The chaos and devastation of their surroundings drove writers to seek sanctuary in idealized fictional worlds.

Katerina Clark has noted that during the NEP, literature demonstrated an "almost perverse" preoccupation with the war and War Communism. Yet we must remember that novels require time, and often are written after reflection on personal experiences. There is an inevitable delay between events and their recording in fiction. To write about the present moment entails the risk of being overtaken by events—a particularly dangerous situation in the Soviet context.

Electricity is the hero of much NEP fiction, not the engineer. There were few models of Red engineers available, and bourgeois specialists, even if temporarily acceptable in the economy, were not permissible role models. This is perhaps additional evidence for the view that the compromise with bourgeois specialists was never more than a temporary accommodation. Kleist, the old engineer in Gladkov's *Cement,* is hardly a role model. And *Cement* was perhaps the preeminent novel of the NEP, establishing conventions for the entire genre of production novels which occupies such a prominent place in Soviet literature. [43]

Cultural Revolution: The Great Divide

The Cultural Revolution of 1928–1931 was a major turning point. Whatever the reasons for the shifts in policy, an evaluation of the consequences must regard the events of these years as a catastrophe for education and engineering professionalism. Successes in social engineering have to be weighed against long-term costs in the quality of training, the nature of institutions, and the character of the profession.

New technical institutes and mass production of engineering cadres were hardly the primary reasons for or consequences of the Cultural Revolution. But generating new communist specialists was both a chief motive and a major result. While drafting a program of rapid industrialization for the First Five-Year Plan, Soviet planners recognized that the already severe shortage of personnel would be exacerbated. Initial proposals for expansion of the higher education system took account of available resources. In 1928 the government proposed creating forty-seven new institutes. While not meeting all the needs for personnel, this probably represented the extent of what was realistically achievable. [44] Typically the plans provided for 100 percent of the projected need for engineers but only one-third of the required technicians. [45] Secondary technical training remained an orphan.

It quickly became apparent that the specialists to be trained were scarcely adequate to staff the old industrial structure, much less the vast number of new facilities established by the industrialization program. At the November 1929 Plenum, Stalin insisted on accelerating tempos across the board. The number of schools and students was vastly increased; the term of study was cut to four years; and the curriculum was reduced to bare essentials, with narrow specializations suited to specific needs. The main method of creating new schools was to split off faculties from existing institutions. [46]

The student body was increased largely by lowering standards and adopting a blatant policy of class discrimination favoring party nominees—*vydvizhenie* (promotion). This period has been described at length by others.[47] But it is important to note that achievements in social mobility were accompanied by negative consequences in the quality of education and character of educational institutions that have persisted into the 1980s.

Faced with an impossible task, Stalin and his colleagues typically opted for a solution emphasizing quantity rather than quality. Educational institutions suffered from rapid expansion and competition for scarce resources. Most of the expansion of the school system came in a period of about eighteen months in 1930–31. Under these conditions, faculty, equipment, classrooms, and dormitories were in scarce supply at virtually all schools. By 1935–36 Soviet educational administrators were categorizing the existing institutions as "strong" (*moshnye*), "average" (*srednye*), and "dwarf" (*karlikovye*).[48] The number in the "dwarf" category was far from inconsequential.

In a situation of weak institutions and uneven student quality, education officials gave up any hope of broad training. The goal became to impart basic skills to poorly prepared students in a minimal amount of time. The directors of one leading Moscow institute were typical in concluding that while broad knowledge of science and technology might be desirable and even useful in subsequent work, the limited amount of time available for study made it necessary to "sacrifice knowledge of general value and replace it with deeper specialized education." At its current stage of development the nation's industry demanded "precisely narrowly specialized engineers."[49] A subsequent report noted: "It has reached the point where they train narrow engineers, *on the level of technicians.*"[50]

Each commissariat sought to train its own staff in specialties so limited that they bordered on the absurd. There was an engineering position (*dolzhnost'*) for each specific aspect of production. The Commissariat of Light Industry included engineering specialties for the compressors in each type of machinery. The Commissariat of Heavy Industry insisted on separate engineers for oil-based paints and non–oil-based paints. The Commissariat of Agriculture trained agronomists for individual crops and veterinarians for each type of animal. Each commissariat was afraid to trust specialists trained by another.[51]

Enterprises responded by using students in menial jobs, so that they never learned about production in general but merely how to perform the narrowly specialized tasks for which their education prepared them.[52] Since "direct production practice" was a fundamental aspect of higher education in these years, assigning students to workers' jobs was not only convenient and profitable but also ideologically correct.

Mass production of poorly trained individuals with engineering credentials did more than debase the currency of higher education and undermine professional identity. It exacerbated the already inefficient allocation of personnel. Even before the Cultural Revolution many individuals with engineering educations preferred positions in administration and other forms of nonindustrial employment. Now this flight became an epidemic. Completing secondary specialized or higher education was a path for escaping the pressures of work in production.

At the same time that the engineering profession was flooded with poorly trained *vydvizhentsy,* the atmosphere created by the show trials of engineers in 1928 and 1930 led to the elimination of the older professional organizations. The All-Union Association of Engineers and the Russian Technical Society could not survive the atmosphere of the First Five-Year Plan. Yet the supposed disloyalty of bourgeois specialists is open to serious question. At the April 1928 plenum the Party had approved documents stating that "the great majority of the technical intelligentsia has come over to genuine cooperation with Soviet power."[53] The precise motives for Stalin's assault on the technical intelligentsia still require elucidation.

Despite the declining status of engineers' professional organizations, the suspicion of specialists, and the extolling of practical proletarians, this was a period in which the prestige of an engineering degree soared. But prestige did not automatically accrue to *all* engineers. It is necessary to refine our understanding of the social processes at work in the first five-year plans. Whereas old specialists owed their prestige to their knowledge and the quality of their work, most of the new *vydvizhentsy* derived their prestige from being among the elect: they had been selected for their services to the Party and were destined to occupy important positions after completing their specialized training.

The quality of *vydvizhentsy* education was questionable at best. Despite their inadequate preparation for advanced study, they were expected to take on tremendous amounts of political, agitational, and social activity at their schools.[54] The attrition rate was tremendous, but the survivors formed a fraternity of leaders with special traits—political acumen, blue-collar credentials, and superb networks.

The prestige of engineering resulted from its role in permitting social mobility and political advancement, not its economic significance. And this social sea change carried over through the Brezhnev era. In 1980, 80 percent of the members of the Politburo and 65 percent of Central Committee members had received an engineering education. In the case of the Politburo, most of them were *vydvizhentsy*.[55]

The fiction of the First Five-Year Plan in many respects takes up where *Cement* leaves off. Rapid transformations, utopian aspirations, and accelerated tempos dominated the era. This was not a backdrop against which an engineer could feel comfortable. Like Kleist in *Cement*, and Nalbandov in *Time Forward*, engineers are dubious heroes at best. They manifest an outmoded inability to believe in the great leap. The positive character is the practical man who overcomes the limits imposed by technical rationality. Katerina Clark has noted that critics were harsh on writers who got their technology wrong in this period.[56] At the same time, there was an emphasis on overcoming the limits imposed by old and rational technology.

Stalinism

The era of Stalinism following the Cultural Revolution consists of three discrete periods: the second and (abbreviated) third five-year plans, the war, and postwar reconstruction. Although the period should be treated as a whole to emphasize commonalities, we must recognize that there were also important differences.

By 1931–32 even Stalin recognized the damage that had been wrought by the Cultural Revolution. A new administrative body, the All-Union Committee for Higher Technical Education (VKVTO), headed by Gleb Krzhizhanovskii, sought to restore a sense of standards.[57] It provided a central administration but did not have the power to force industrial ministries to comply with its suggestions. Problems of quality persisted, as indicated by repetition of most of the 1932 criticisms in another decree reorganizing the committee in 1936. The narrow specialties developed during the Cultural Revolution manifested tenacious staying power. The specialty list was cut from nine hundred to three hundred categories, and in technical fields it was reduced even further. But within three years many of the narrow specialties had reappeared.[58]

On the eve of the Second World War, *Pravda* noted that in the Soviet

Union there were five times as many higher school students per one thousand population as in Europe. The quantitative problem had been solved. Quality was another matter. Recognition of the need for a smaller number of better-qualified specialists led to reduced admissions beginning in 1939, along with cuts in student stipends.[59] But the war changed everyone's orientation. Once again a crash campaign for more specialists was instituted. In part this was a response to perceived wartime needs, but it was also what the system did best.

Many of the emergency measures adopted during the first two years of the war resembled policies of the Cultural Revolution. Graduations were speeded up, courses were shortened, education was combined with regular work in production, and students were required to add political and military studies while also participating in extensive "social labor." The rapidity with which these measures were abandoned—in most cases beginning as early as 1942—provides a trenchant commentary on how they were evaluated. By 1944 the emphasis on quality was genuine and widespread.[60]

The war also accelerated geographical change in the school network. Educational opportunities were expanded in Siberia and Central Asia, laying the foundations for later development of new scientific centers.[61]

During postwar reconstruction there was again a serious shortage of cadres, and quantitative growth again took precedence. In the first postwar decade the number of higher schools more than doubled—in part a restoration of the prewar institutions, but also reflecting the rise of new institutes and continued geographical expansion. The number of higher schools in the east quadrupled. But once again quantitative expansion took place without adequate provision for teaching cadres, equipment, and housing. Local officials did not always attach the same importance to education as central authorities. The infrastructure of educational institutions was frequently appropriated for other purposes, and even several years after the war, schools were struggling to reacquire scarce building space occupied by local administrative bodies.[62]

Social mobility was a less pressing issue after the First Five-Year Plan. While workers' faculties and preparatory divisions continued to exist, most students came from secondary schools, where academic merit usually counted for more than proletarian origins. Yet one crucial episode of social mobility deserves mention. During and after the war an influx of demobilized veterans into technical schools had a major influence on the student body and subsequent engineering cadres.

Working conditions improved only marginally through the Stalin era, and massive turnover of personnel persisted. During the early five-year plans, engineers frequently "volunteered" to sign pledges that they would remain in their jobs for a fixed period, usually until the end of the current five-year plan. Technical institutes could have a new director four or more times in the course of a year, a pattern that might be repeated year after year.[63]

Technical competence was scarce in the 1930s. Old specialists were under a cloud, and they had acquired their skills on outmoded technology. Many of the new Communist managers lacked adequate training. Commissar of Heavy Industry Sergo Ordzhonikidze complained that the factories had outgrown the managers.[64] One gets the sense that no one was competent, while all were being pushed to the wall by demands for an increased tempo of growth.

The Second Five-Year Plan was supposed to be the plan of "quality," but by the fall of 1935 this began to be undermined by a new campaign for a faster pace. The Stakhanovite movement placed engineers in an almost impossible position, not very different from what they had experienced during the First Five-Year Plan. The pressure was soon compounded by more intrusive police activity, as the inevitable errors and accidents of speed-up were attributed to espionage and sabotage.[65]

The engineer-manager in theory benefited from one-man management (*edinonachalie*). But in reality he was under constant pressure from bureaucrats above and workers below to revise plans upward. Managers lost control over wages, rations, housing, and other incentives, and even a loyal apparatchik who supported the increased tempo was not in a position to be a dictator. Some shock workers (participants in accelerated work) elected their own managers. Engineers were able to participate only tangentially in the informal shop-floor networks that became crucial to success in production.[66] While these factory-floor systems allowed workers to feel a sense of group identity and solidarity, and to protect their interests, engineers were always at the edges. And no efforts to unite engineering or supervisory personnel were tolerated.

After the assault on bourgeois specialists during the Cultural Revolution, old specialists no longer figured prominently in campaigns directed from the center.[67] At the local levels, however, they often suffered from specialist baiting and a general climate of mistrust. At some technical institutes old professors were dismissed in the late 1930s.[68]

Employment patterns reflected the pressures on specialists and their inadequate preparation. Engineers with diplomas continued to work in administration, while production was overwhelmingly the province of *praktiki* and *vydvizhentsy* without higher education. Soviet enterprises continued to be heavily overstaffed with administrators in comparison to their European and American counterparts.[69]

On the eve of the war in 1941 there were 214,000 specialists with higher educations in the industrial commissariats, of whom only 68,000 (less than 32 percent) worked in production.[70] The situation in regard to technicians was slightly better, with perhaps 50 percent working in production. At individual enterprises the situation varied widely. For example, at the Kiselevsk mechanical plant all twelve engineers occupied administrative posts. But this may have been an exception. The Trud plant had seven of twelve engineers and twelve of seventeen technicians working in the shops; but the plant also had thirty-eight *praktiki* occupying engineering positions.[71]

The character of professional life is demonstrated by the fate of the new All-Union Council of Scientific-Technical Societies. Established in 1933, this organization was not able to convene its first congress until 1959. The "transmission belt" image of organizational activity reached its fullest development in this period.[72]

Despite—or perhaps because of—crash education initiatives, the major source of new cadres for production during the war was the transfer of specialists from administrative positions.[73] A second major influx was of females, whose proportion in the student body increased significantly under wartime conditions. In the Soviet cultural context feminization has had a long-term negative impact on the prestige of engineering.[74]

During the war attention was devoted to rationalization and invention, and to improving the organization and productivity of labor.[75] Under wartime conditions genuine scientific research and innovation were even more difficult than previously, and were focused on activities directly connected with military products. Methods used to solve logistical and production problems often required little in the way of research and development. Army quartermaster A. V. Khrulev described the improvements in transport derived from using horses, reindeer, and camels—all better suited to the unpaved roads of the hinterland than modern vehicles, which were in any case unavailable.[76]

It is not surprising that the war encouraged extensive militarization and the expansion of political involvement in science and technology. Political involvement in Soviet science has always been two-edged, constituting both a disturbing intrusion into the creative process and a means to expedite the realization of high-priority goals.[77] The apparent lack of administrative control before the war was due as much to "undergovernment" as to assertions of independence. The Party simply did not have sufficient trained personnel to exercise close control over scientists and engineers engaged in research and development. To some extent this can also be attributable to a basic respect for natural science in Russian-Soviet culture. Political problems in the Academy of Sciences in the 1920s affected mostly the social sciences, and it is likely that a detailed history of the purge era would show a similar picture.[78]

The political treatment of scientists has generally differed from the treatment of technical specialists. The specialists appear to have been regarded as a greater threat, particularly at the end of the 1920s. It also seems, however, that engineers were sufficiently intimidated by the measures taken against them in 1928–1930 that there was little need for special purges among them in the late 1930s. But their docility was purchased at an enormous cost to the economy. Disincentives to innovation and to work in production contributed to serious technological backwardness in Soviet industry, a condition that has still not been alleviated.[79]

We still know far too little about the identity and collective biography of the victims of the purges, and these questions will remain unanswered until further archive materials become available (if they exist). We do know that virtually the entire group of activists in the engineering-technical section of the metallists' union was purged, suggesting that the terror frequently struck the most active and professionally conscious among the technical cadres.[80]

The capacity to focus talent and resources on high-priority areas of research and development, as opposed to specific construction projects, was mainly a product of the postwar era. Following the war, however, the Party remained desperately short of technically competent members. A solution to both shortages was sought in co-opting the scientific-technical intelligentsia into the Party and fostering a greater Party role in local-level R&D organizations.[81]

The wartime experience helped to change the attitudes of a large portion of the scientific-technical intelligentsia. Many who would not even have considered joining the Communist Party before 1941 became members under wartime conditions.[82] The proportion of Party members with higher and specialized secondary educations increased during the war years from 39.8 percent to 57.4 percent. By 1945 over one-third of the 1.2 million specialists with a professional education were Party members.[83]

The postwar decade is the least-studied period in Soviet history, and one that requires much more investigation. It was not a time for new initiatives in professional organizations. But, as Vera Dunham has demonstrated in her book *In Stalin's Time*, it was a period when the professional ethos and reliance on scientific professionalism that had emerged before and during the war became part of the Soviet social contract. Dunham has described this as the "Big Deal" between the regime and the new, largely professional middle class. Its wartime origins are seen clearly in Korneichuk's novel *The Front*. Following the war a plethora of popular novels conveyed the same message. The emphasis was on life style and ethos rather than on engineering, but even the production novels of the period feature an acceptance of the engineering professional.

Where the regime had declared war on neutrality and apoliticism during the First Five-Year Plan, it now accepted a much more modest goal. It was more important to say the right things in public than to live the credo to the fullest. Inner emigration might even be acceptable, provided one denied it publicly. The basis for the pervasive apathy of the Brezhnev era was already being laid.

But the significant changes initiated during the postwar period took root only gradually. While Azhaev's *Daleko ot Moskvy* expresses a certain positive attitude toward knowledge, there are still echoes of specialist baiting as the pipeline is built in one-third the time projected by experts.[84]

Literature is also our major source regarding the work environment during this period. Despite the demands of reconstruction, one gets the impression of a much less overwrought tone in industry. Serious, stable, and consistent work are valued, rather than the increased tempos and stunts of the early five-year plans. An engineer who meets the plan and fulfills basic social norms can expect to enjoy the rewards in relative peace. And fulfilling a plan is itself a relatively peaceful process, in comparison to the prewar environment.

But this was still the Stalin era. Even a decorated military engineer might find it impossible to live where he wished, and many continued to labor in the netherworld of special camps.[85]

De-Stalinization and Stagnation

The Khrushchev era witnessed an attempt to escape from the departmentalism that dominated Soviet administration, but the leadership never found the means to end this condition. In education the Big Deal came under attack, as Khrushchev's policies made it clear that the middle-class status acquired by those with education was not necessarily hereditary. New rules for higher education required a period of employment before admission to higher school, and called for education along with full-time work in production. The result was something of a new *vydvizhenie*, with admission preferences given to those with at least two years' work experience and those willing to continue working while in school. Fraud was once again rampant. Since that time class-based education policies have not been abandoned, although they have generally received less emphasis since 1960.[86]

Repeated reforms resulted in fluctuations in the number of schools and students, but overly rapid changes frequently had little or no impact. By the time local education authorities geared up for reforms, they had been superseded. After 1960 consistent expansion resulted in a significant increase in the number of graduates, along with a continuing tendency for engineering graduates to find employment in both administrative and blue-collar positions. Many still avoided production.[87] Growth continued at gradually declining rates until 1983. And the preference was still for higher education, not technician's training.

Expansion strained resources, demoralized the teaching staff, and increasingly undermined the credibility of the education system. Yet the plan had to be fulfilled. This resulted in the pervasive practice of mechanically awarding passing grades to students merely for attending classes. Faculty whose jobs depended on the number of students enrolled had little desire to jeopardize their positions by giving failing grades, and the students understood this game only too well. The social myth of access to higher education for all was maintained by providing evening and correspondence programs. Today these part-time programs continue to account for about 40 percent of Soviet engineering graduates and are widely criticized for their poor quality.

Despite persistent problems with cadre allocation, the belief in planning was thoroughly institutionalized. It became impossible to conceive of a solution to personnel problems involving anything other than more or better planning. Planning had to be more scientific, based on better data, or conducted according to new formulas. It was unthinkable to question the idea of planning itself.[88]

The Khrushchev period did witness the beginning of two trends with long-term import for technical specialists. One was a reorientation in specialty preferences and planning that increased the proportion of students in automation and computing. This was the sole significant statistical shift in specialty distribution in the post-Stalin period.[89]

The other major change was a shift in the ratio of wages for engineers and workers. In contrast to the rapid change in specialty distribution, wage leveling was a gradual but consistent process over more than two decades. By the 1970s it had reached absurd proportions. In construction the average wage for engineers actually fell below the rate for workers. Wage leveling reflected the regime's social policy, as well as the fact that many engineers were trained in narrow specialties more suitable for technicians.

As was the case through much of the period under study, the retarded technical base of Soviet industry meant that workers moving from village to city were simply exchanging one form of physical labor for another.[90] For specialists, technical backwardness had a similar impact on prestige. "Rationalizers and automators" had little to do in an industrial environment where physical labor predominated. As recently as the 1980s production lines with robots used the modern equipment for only limited aspects of the production process, while workers hand-carried products from ordinary lines to the automated sections and back again.[91]

Employment trends reflected the continued fragmentation of specialties, and their debasement. Not only did most institutes, following instructions from their ministerial patrons, continue to train specialists in accord with precise, narrow needs. They also spewed forth an absurd plethora of pseudotechnical specialists such as "engineers for wage and norm setting" and (my favorite) "engineers for socialist competition."

It is hardly surprising that under these conditions the prestige of engineers declined sharply.[92] In part this was a process seen in all industrial societies in the postwar decades. Technocracy and rationalization were not bruited about after the Second World War as they had been in the 1920s. It was also a function of changes in the structure and character of scientific and technical activity. The decline in prestige of engineers and scientists has been a nearly universal phenomenon in the age of the industrial research lab. Technical improvements have always been the product of multiple input, but now this phenomenon has often effaced the individual inventor completely.

In the Soviet Union the global trend has been accentuated by the economic system's abysmal performance in technological innovation.[93] Incentives for scientific research and invention are weak, but creativity has not been stamped out. Incentives to diffuse innovations, however, are almost nonexistent. Of the new technological processes introduced in the Soviet Union, 80 percent are introduced at a single enterprise. Of the remainder, almost all are introduced at fewer than five enterprises. An appalling .6 percent of new processes reach more than five enterprises.[94]

The decline in engineers' prestige and status has combined with changes in the Soviet social structure to make engineering a less desirable career. Competition to enter technical institutes began to decrease in the 1970s and dropped precipitously in the 1980s, despite an overall increase in the number of applicants to institutions of higher education. Not only are salary and employment possibilities in engineering considered to be bleak, but it is not the career of choice for second-generation students. The applicants to technical institutes come mainly from worker and peasant families, or else desperate students for whom *any* higher school is preferable to the potential loss of intelligentsia or middle-class status. As the Soviet Union becomes increasingly a middle-class society, the difficulty of filling the engineering schools will increase.[95]

Conditions in professional life have reinforced the decline in prestige. Shortly after Stalin's death the government introduced a new form of engineering organization. Once again engineers and workers were combined in a single union. The inevitable effect was to reduce the status of engineers.[96] Without a professional society, it remained very difficult for Soviet engineers to develop group cohesiveness or links with the international community.

Emigrés speak of a palpable increase in the Party's role in science and technology after roughly 1967 or 1968 (the exact timing varies in different geographical areas and specialties, and even among individual institutes). Increased Party activity manifested itself in vetting of personnel, declaring the Party's right of control over research and development institutions, and in the growing number of specialists and managers holding Party membership.[97] Once again political involvement was double-edged. It could result in priority for particular institutions and guarantee stable funding for significant projects like the space program. But personnel decisions were often made on political rather than scientific grounds.

Ironically, it was precisely when engineering began to lose status among the public that it achieved full acceptance in literature. By the late 1950s the Big Deal was consummated, and expert knowledge became a mark of the hero. In this context scientists and engineers could finally become positive protagonists. The temptation persisted to portray them as proletarians or gadflies, as somehow not typical, but specialized knowledge itself was no longer a stigma.

Cultural images of the engineer in the Brezhnev period came predominantly from two genres: village prose and production novels. In the Brezhnev-era production novel there is some ambivalence about the role and status of engineers. The genuine hero is an engineer who rises from the work force rather than one who is a second-generation member of the intelligentsia. In Kolesnikov's *School for Ministers* trilogy, the academic engineer Karzanov and the upwardly mobile worker Altunin are able to talk frankly about the greater chances for a new invention's being accepted if the worker is involved in its development. And it is the worker Altunin who moves on to the heights of power after struggling through night school. On the very day he receives his engineering degree, Altunin is shifted from production work to management, demonstrating both the promise of education and the shortage of cadres.

In contrast to the production novel, village prose reflects the decline in status of the entire scientific-technical ethos. In works of the village school technical specialists appear infrequently, and when they do they are harbingers of destruction, corrupters of values, or at best providers of Hobson's choices, as in Rasputin's "Farewell to Matera," or the tremendously popular film *Siberiade*. Films provide a significant source of characterizations of engineers in the Brezhnev era. *Pena* is a farce dealing with the threat posed to old-style managers by the scientific-technical revolution. The hero of *Moscow Does Not Believe in Tears* is a practical chap who hand-rigs the apparatus many of his institute's engineers use for their dissertations. While still retaining older ideological elements, the literature and film of the Brezhnev era provide a basis for full integration of the middle class.

Gorbachevshchina

Mikhail Gorbachev inherited a situation in which the quantitative aspect of the technical cadre problem had already been solved. In 1975 the USSR employed three times as many engineers as the United States, and the number was continuing to increase at double the American rate. Not until 1983 was there some indication that the expansion had ceased.[98] Inertia and continuing excess demand for higher education, however, make it difficult to curtail enrollments.

The new leadership has demonstrated a willingness to confront the plethora of difficulties in engineers' training and employment. The Basic Directions for Restructuring of Higher and Specialized Secondary Education, approved in 1987, cite a long list of problems, most of which are familiar from the 1932 decree on the same subject. The State Committee on Education has stated its intention to deal with uninterrupted and unwarranted increases in numbers; a level of education that does not meet contemporary demands; fragmentation of specialties; overloaded curriculums; overcrowded classrooms; poor knowledge of new technology; failure to appreciate the role and value of secondary specialized education; lack of teaching cadres and basic infrastructure; low pay for teaching; serious shortcomings in employment of specialists; and anarchy in plans for the number of specialists developed by departments with no accountability for the figures.[99] Boris Eltsin even noted the constant repetition of the same problems over several decades (although he did not look back to 1932).[100]

Thus far the major accomplishment has been a Confucian one: things are finally being called by their proper names. This is a crucial first step, but the next moves will be more difficult. The February 1988 plenum of the Central Committee adopted a resolution consolidating the several strands of education reform and placing all education under a single bureaucratic agency for the first time since the 1920s. It remains to be seen, however, how the new State Committee for Education will handle relations with the industrial ministries which are being called on to finance much of the projected improvement in educational facilities.

Aside from the education reforms, there have been other initiatives, including measures to increase salaries for some engineers. This will not affect all engineers but rather will be limited to those with important jobs in productive enterprises or sectors. It reflects the new regime's acceptance of major income differentials, and has already had a positive impact in a limited number of instances.

A significant development in a number of professions has been public appeal for new professional organizations, including inventors, designers, and education workers. There has even been a call to establish a new union of engineers and scientists—quite a claim in a society that has experienced purges of supposed technocrats. It is striking that Soviet professionals frequently tend to pick up the threads of the professional programs articulated by their prerevolutionary predecessors.[101]

It will be interesting to see how these changes are manifested in literature. Thus far we have not really seen the full cultural results of perestroika, only those of glasnost. The novels, plays, and films released during the first few years of Gorbachev's term of office represent the cultural product of earlier periods that had been denied public release. The literature of perestroika is only beginning to appear and is still limited to memoirs and journalism.[102] Perhaps in the 1990s we will see literature reflecting the changes of the 1980s.

The ambiguities surrounding engineers' status, numbers, and varied roles in Soviet society become less challenging when we separate out the component parts of the scientific-technical intelligentsia and refine our understanding of the term *engineer*. A great many of the individuals called engineers in the Soviet Union would not have that title in Europe or America. The criteria for membership in the engineering elite are of at least two types: technical and social-political. Since the 1920s professional leadership has not been in the hands of those selected by the engineers themselves, or those who have the greatest claim to international professional recognition.

Soviet higher education perpetuated many attributes remaining from the nineteenth century, including departmental orientation, encyclopedic programs of study, and poor success rates. Neither tsarist nor Soviet administrators were able to resolve whether technical training should be the province of educational or industrial administrators. The problem of supply was resolved not so much by providing adequate numbers of cadres as by debasing the meaning of an engineering degree. Formal rationality pervaded the system from 1929 to the mid-1980s, and the poor quality of many of the cadres graduated during that half-century was the result.

A corollary of the problems in engineering training is the persistent orphan status of secondary technical training, with the related shortage of technicians. In this respect Soviet education and industry reflect characteristics of their Russian precursors, exacerbated by the emphases of the Soviet educational and economic systems and individual preferences. A shortage of auxiliary personnel, the tendency of educated specialists to avoid production, and an excess of supervisory personnel have persisted for over a century.

Providing second-rate higher education to a large number of people and calling them engineers might not in itself constitute a major problem. But the practice brought with it a decline in professional standards, wages, and identity that has been devastating. Recent measures to raise wages and status for *some* engineers constitute a plausible corrective. It does not matter if everyone is called an engineer, as long as the "real" engineers are given genuine incentives and the opportunity to perform.

There are, of course, superb specialists in the Soviet Union. Even during the Cultural Revolution skilled engineers graduated from the top institutes. But there have never been enough of them. To this day all but a few of the elite institutions can trace their lineage to prerevolutionary higher schools. Resources are scarce, and engineering has consistently been called on to play multiple roles, providing technical specialists and managers as well as the political elite.

Throughout the period under study engineering has served as a means of social mobility. Before the Revolution this was a gradual process. During the Civil War, the Cultural Revolution, and again under Nikita Khrushchev, efforts were made to force the pace. Since the mid-1960s the principle of advancing workers and peasants has remained, although it has not been applied on a mass scale. The constant in this story is engineering as a first step up the social ladder: it continues to represent social mobility for individuals from peasant and proletarian backgrounds. In the 1920s and again in the 1930s, engineers sought to provide privileged access to technical institutes for their children. Only scattered efforts of this sort have been evident since the Second World War.

Thus technical training emerges as the transitional stage between worker and intelligentsia, or between working-class and middle-class status. As we have seen, as the Soviet Union becomes a largely middle-class society, the difficulty of finding qualified candidates to enter engineering schools will increase.[103] This is another modern problem hardly unique to the Soviet Union. But it is more alarming in a society that prides itself on planning.

Here again the Soviets appear trapped by their own formal rationality. For a century *praktiki* have occupied important positions in the shops, while a majority of technical specialists have been employed in administration or other positions outside direct production. Many other engineers have been employed as technicians and production workers, and so there continues to be a very poor correlation between individuals' training and status. The real issue here is that the Soviet authorities insist on making it an issue. Planners and bureaucrats become upset when their charts of positions and responsibilities are contravened, and when scientific planning of people's education and career patterns fails to correspond to actual behavior. Yet the genuine problems in education and industry lie elsewhere. The solution has always been cast in terms of more or better planning, rather than a sense that the labor market might need to operate with a degree of freedom similar to what some have proposed for the entire economy.[104] It is time to consider the returns on general investment in human capital, but new financing mechanisms based on contractual relationships point in the opposite direction.

Finally, a return to the tantalizing question of the political role of engineers is appropriate. Technical training in the USSR has led more often to political leadership than to political dissidence. The scientific rather than the technical intelligentsia has been the more vocal force in dissident activity. Despite a shift in preferred elite career patterns from engineering to economics, law, and international studies, and young people's stated preferences for careers in science and the humanities, technical training remains a major route to high party positions. Yet the myth is in decline, and we may be witnessing a struggle between engineers and economists for the soul of the Soviet system.

· 6 ·

Rockets, Reactors, and Soviet Culture

Paul R. Josephson

Within two decades after the Second World War the Soviet Union had achieved parity with the United States and exceeded the capacities of most other nations in the development of such-large scale technologies and systems as rockets, reactors, jet airplanes, particle accelerators, electrical power grids, and, to a certain extent, automobiles. These achievements captivated a large international audience. Until quite recently, however, most Western observers have focused on the internal aspects of those technologies—parameters and specifications, speed, volume, and accuracy—ignoring the social and cultural context of their development. This focus was in part the result of the effort to demonstrate which country was first, or had faster, bigger, or more powerful hardware. Scholars are now turning toward a consideration of such external factors as the political and the ideological as well as the cultural and social environment for research, development, and construction of large-scale Soviet technologies.

In this chapter I will attempt to consolidate some of the more recent trends in the history of Soviet technology, exploring what Soviet advances in nuclear power engineering and rocketry between the end of the Second World War and 1970 tell us about Soviet culture. I will examine how scientists and Party members had to move beyond scientific considerations to create the social and cultural foundation for massive expenditures on forms of technology that promised only limited immediate social benefits, but that offered constructivist visions of applicability for communist society. The rise of a cult of science and technology, atomic and space culture, and heavily publicized visions of uses for rockets and reactors was central to this process. Using popular scientific, literary, and political journals (and some newspapers), I will explore what Soviet officials and scientists perceived as the proper public understanding of these technologies, and will show how the public responded with patriotism, prose, poetry—and incredulity. My focus will not be on the views of the Moscow and Leningrad literary and scientific intelligentsia (although only some members of this intelligentsia were critical of what they viewed as an unquestioned faith in large-scale technologies; most wholeheartedly embraced Soviet development programs). Rather, I will describe atomic and space culture as manifestations of an official but publicly embraced cult of science which viewed technology as a panacea for social and economic problems. I will examine how both the atomic power and space programs encountered opposition in their early stages, although both were seen as indispensable to the construction of communist society. My conclusions will focus on what we can learn about Soviet culture from the study of large-scale technologies.

Since the 1920s large-scale technologies have played a major role in Soviet hopes, plans, and expectations for the future. The State Plan for Electrification (GOELRO), the introduction of the five-year plans, and the Dneprostroi hydroelectric project are some of the main examples of the Soviets' reliance on technology as a solution for social and economic problems. In ideological and philosophical pronouncements, leaders from Lenin and Bukharin to Stalin and Khrushchev have presented technology as the highest form of culture and the key to the Soviet future. In the postwar years, even in the face of economic devastation, and despite the need for massive reinvestment to rebuild the economy, the state poured capital into military research and development, in particular nuclear power and rocketry, trebling the salaries of scientific workers. But scientists soon recognized the need to create a broad base of support for programs developing peaceful applications of nuclear energy and space research. Taking advantage of the Stalinist gigantomania and the postwar cult of science, scientists, engineers, and technologists quickly moved to build a general acceptance and scientific understanding among the Soviet population.

The Cult of Science

Stalinist policies toward science were relaxed as the cult of science gained currency in the postwar USSR. The cult was based on a reductionist view of physics as the key to solving scientific, technological, and socioeconomic problems, initiated by postwar successes in nuclear weapons research and development, cemented by achievements in the peaceful uses of atomic energy and space exploration, and nourished by the general environment of de-Stalinization. Soviet scientists, and in particular physicists, began to reclaim control over the scientific enterprise. As physics and technology showed success after success, the Party came to recognize physics as the "leading science." This contributed to the development of the cult, and physicists benefited from it, especially after the death of Stalin.

During the last years of Stalin's rule the cult was characterized by a fascination with grandiose and gigantic architecture;[1] large-scale construction projects,[2] especially those favoring prefabricated concrete forms;[3] faster, more modern elevators;[4] bigger and better machine tools; hydropower stations and other forms of geological engineering,[5] telemechanics and automation; and the Moscow subway system. By the early 1950s, many scholars had adopted a more sophisticated view of rapid technological progress, the growing interconnectedness of different branches of science, and man's increasing control over nature as part of an ongoing scientific-technological revolution.[6] Of all aspects of this revolution, atomic energy (through the mid-1950s) and space[7] (through the mid-1960s) stood at the center of Soviet development efforts and contributed to the growth of a cult of science.

This cult embraced the attitude that such large-scale, expensive, and highly visible projects as nuclear fission and fusion reactors, high-voltage long-distance power lines, particle accelerators, and spaceships should be at the center of national programs. Throughout the postwar period such Academy functionaries as S. I. Vavilov and A. N. Nesmeianov and other Party officials stressed the central place of physics in the economic development plans of the USSR, especially in the areas of automation, telemechanics, and atomic energy.[8] In one of the most striking reductionist statements about the importance of physics, L. L. Miasnikov drew a direct link between atomic energy and the anticipated utopian achievements of Soviet society. Miasnikov, calling physics the leader of science, argued that physics played a central role in the development of philosophy and dialectical materialism, the other natural sciences, and technology. All branches of science "absorbed the richer physical material," and such newer branches of science as acoustics, electrotechnology, heat engineering, and atomic energy grew primarily on the foundation of physics. Finally, Miasnikov concluded, "The physics of the atomic nucleus, not long ago 'hatched from the egg,' will lead to the kind of results which will call forth a true revolutionary transformation of the productive forces of society."[9] As the "leading science" within this cult, physics, and physicists, found the strength and political and social basis with which to reestablish control over the institutions of the discipline, to offer refuge to geneticists from Lysenkoism within these institutes,[10] and to advance more and more fantastic visions of a future communist society based on the achievements of science and technology, in particular atomic energy and space research.[11]

One more aspect of the cult of science must be mentioned before we turn to a discussion of space and atomic culture: the role of Nikita Khrushchev in the development of large-scale technologies. Under Khrushchev, scientists pressed to regain control of research and development in such areas as the philosophy of science and fundamental research. This process began in earnest in 1956 at the Twentieth Party Congress, when I. V. Kurchatov, the father of the Soviet atomic bomb project, gave a short speech noteworthy for its brazen support of a nuclear energy program that bordered on scientific fantasy.[12] Kurchatov's appearance at the congress is significant in and of itself because it signaled that the Party had recognized scientists as political actors, something for which Kurchatov thanked Khrushchev indirectly.[13] What is more, a number of popular science writers immediately recognized the significance of the Twentieth Congress for atomic power fission and fusion.[14]

In addition to unleashing the forces of de-Stalinization, Khrushchev supported the efforts of scientists both directly and indirectly. For example, he played a major role in the development of nuclear power, recognizing its value for both the Soviet economy and his own purposes. Khrushchev supported reactor research and development with much the same vigor he gave to large-scale technological development programs in agriculture (chemicalization, the planting of corn, the sowing of the Black Earth regions) and space. He took a personal interest in nuclear energy, accompanying Kurchatov to Harwell, England, in April 1956 to visit the central British nuclear research facility,[15] and participating in the Geneva Conference on the Peaceful Uses of Atomic Energy. It is clear that Khrushchev hoped to use Soviet scientific and technological successes to bolster his own position within the Party.[16]

Because of the contributions of scientists to the ultimate victory of the

USSR in the war, and to the development of nuclear weapons, the construction of atomic reactors, and the launching of the first artificial satellites, scientists gained tremendous prestige and political power in the postwar Soviet Union. Their continued success in these areas enhanced their efforts to regain control of the scientific enterprise. But in order to proceed any further, they also needed to develop popular support for their programs, especially in view of the fact that to the average Soviet citizen, potential applications seemed long in coming to fruition. As a result, scientists and Party officials alike set out to create a culture of atomic power and space exploration as a manifestation of the cult of science.

Atomic Culture

Atomic power was incorporated into Soviet visions for the construction of communism in two stages. The first stage involved the development of the cultural and social environment for the advance of nuclear energy in the USSR. During the immediate postwar years, in the face of significant technical problems and public concern over the atomic bomb, nuclear weapons, and war with the West, Soviet physicists began to develop an atomic culture.[17] Explaining the physics of the atom to anyone who would listen, scientists, science writers, and ideologues described in grandiose terms the potential for the application of nuclear energy throughout the economy. Acknowledging the need to build bombs because of "hostile capitalist encirclement," they nonetheless stressed the peaceful uses to which this new form of energy could be applied.

Soviet atomic culture was based on several visions—some reasonable, some farfetched—of nuclear energy's potential. All of these schemes were based on the assumption that atomic energy would accelerate the construction of communism. During the second stage, beginning in the mid-1950s and lasting until the early 1970s, nuclear scientists and engineers struggled to put nuclear power on the political agenda and to overcome the opposition that had arisen as a result of concerns about both safety and economics.

In an effort to make atomic energy more palatable to a Soviet population with strong recent memories of war, the popularizers of atomic energy proclaimed its potential peaceful applications in industry, agriculture, biology, medicine, and transportation (nuclear automobiles, locomotives, jets, and rockets), and developed a peculiar iconography of the constructive potential of atomic energy in such areas as the Soviet version of Project Plowshares—"geographical engineering."[18] They also predicted that, while accelerating the construction of communism in the USSR, atomic energy would exacerbate the contradictions inherent in the capitalist system. By the late 1950s these arguments had taken the standardized form that under socialism, as opposed to capitalism, energy research and development served the masses, not corporate monopolies, and peaceful, not military, purposes, and was therefore safer than in the West, and thus did not require the same degree of regulation. One writer made the claim that only under socialism was the rapid commercialization of atomic energy possible.[19] V. S. Emel'ianov, then director of the Main Administration for the Utilization of Atomic Energy, dismissed the notion of some Western physicists that civilization was not ready for peaceful uses of nuclear power: in the socialist USSR, where "the progress of science and technology are tied to the building of communism . . . atomic energy in its turn opens new, grandiose perspectives in the matter of the creation of the material-technological basis of communism."[20]

In his book on American culture at the dawn of the atomic age, Paul Boyer discusses the attempts to recast atomic power in the United States in terms of its potential benefits for society.[21] In the USSR, as part of the cultural preparations for atomic energy, Soviet scientists presented similar utopian visions. While biological and medical uses stood at the forefront of Soviet visions of the future, no branch of the economy was untouched by the magic wand of this modern alchemy. Wide applications were foreseen in agriculture, where the use of phosphorus and carbon isotopes facilitated the study of plant physiology—the process of photosynthesis, the action of fertilizers, the role of viruses and bacteria, and so on. Isotopes could be used as radioactive tracers for the study of bees and fish.[22] Room-temperature irradiation would prolong the shelf life of meat and other agricultural products such as potatoes.[23] (No one seems to have thought of the potential danger of radioactivity moving up the food chain into humans.) In mining and metallurgy Soviet engineers spoke hopefully of applications in prospecting, oil surveying, and the monitoring and weighing of coal, with especially great expectations for the Sixth Five-Year Plan.[24]

But of all the areas of application, atomic engines for transportation excited the highest hopes. What were the technical reasons for faith in atomic motors for automobiles, boats, locomotives, submarines, airplanes and rocket ships?[25] First, atomic energy had limitless potential, a major asset since oil and gas reserves were limited and someday would run out. Second, atomic energy did not pollute as did conventional internal combustion engines. Volumetric considerations also favored the atom: a few grams of uranium went as far as several tankersful of gasoline. This would permit lengthy trips or flights without refueling.[26] Many scholars acknowledged the imposing technical problems that stood in the way of realizing the use of atomic energy for transportation purposes; these included the need for very thick containment on reactors for safety and the resulting increase in weight of the engine. But concerns about nuclear-powered airplanes crashing to the earth, for example, never entered into the discussions.

Finally, just as in the United States, Soviet scientists entertained the notion of using peaceful nuclear explosions (PNEs) "to increase man's control over the environment,"[27] to fill deserts with water, to strip-mine, to build canals,[28] and even to correct the "mistakes" of nature.[29] One engineer explained how "atomic energy permits us to dream about the alteration of the climate of whole countries," the use of explosions (as in Project Plowshares) to create a "winter sun," melt the arctic snow, and turn deserts into lush gardens.[30] In short, atomic energy would permit man to alter nature.

Knowledge of atomic energy moved quickly into the public domain. In an attempt to bring it to the people, the Ministry of Culture introduced a number of courses at the elementary and middle-school levels, entitled "Atomic Energy and Perspectives for Its Utilization in the Economy." The courses covered theoretical concerns such as the structure of the nucleus and applications of nuclear power in the economy, including fusion and atomic-powered transport. They stressed the peaceful uses for atomic energy being developed in the USSR, as opposed to the military uses being planned in the United States.[31] The public became so fascinated with atomic power that one science writer bemoaned the fact that a national search for uranium had not been organized, calling for bands of citizens to be equipped with Geiger counters to locate the ore and speed up applications.[32]

The effort to convince the Soviet population of the benefits of atomic energy was only half the task. Once atomic energy had become part of the iconography of Soviet culture, Soviet nuclear engineers had to fight for its acceptance in the political and economic arenas. Until the Twentieth Party Congress in 1956, however, discussions of atomic energy in newspapers such as *Pravda* and *Izvestiia* were for the most part technical rather than political. But after I. V. Kurchatov's appearance at the congress, discussion of the potential of nuclear energy was elevated to a policy debate. By the end of the 1950s Soviet scientists had succeeded in making nuclear power an issue for consideration by all Party congresses.

In order to combat the argument that nuclear energy was not economically feasible, the engineering community made a number of fateful decisions to make it more competitive with conventional sources of generating power. The standard argument was that eventually nuclear energy would be a rational choice, given the geographical imbalance between fossil fuel resources and energy demand in the USSR, the limited nature of oil, gas, and coal reserves, and the safety of atomic energy. To advance the efficiency of nuclear power, engineers pressed for early "standardization and industrial assimilation."[33] The industry suggested several ways to cut down on capital and generating costs associated with nuclear reactors: they adopted several schemes to use less steel and reinforced concrete;[34] they sought to commercialize nuclear power in ever-larger reactor units, forecast on the order of 1,000 to 2,000 MW (a size that would never be licensed in the West); they sought to lay to rest concerns regarding fuel-cycle costs by pushing for the creation of a network of breeder reactors; and they adopted a plan calling for nuclear reactors to produce heat for industrial and domestic purposes at a significantly lower cost than conventional plants.[35] All of these plans reflect a view of large-scale technologies as panaceas.

As an indication of the degree to which technology was seen as a cure for economic problems such as the geographic imbalance between fuel resources and population, Soviet engineers approved several projects to build "mobile" and even "floating" nuclear power stations for application in regions where, because of climate or absence of fuel, it is difficult to produce electrical energy from diesel fuel or coal. The Arbus, a 1,500-kilowatt reactor for use in the far north, Siberia, and far east, was built in two forms in the early 1960s, one using an organic coolant and moderator, the other a water-water model.[36] Engineers have also advanced a model of a floating 6,000-kilowatt nuclear power station to provide electricity for oil and gas exploration in the far north and east.[37] The construction of these reactors reflects more than faith in the infallibility of technology. It suggests that technological determinism—the fact that technologies are built because of the momentum provided by the institutions that develop them, in part simply because they can be—also plays a role in the Soviet nuclear energy program.

Space Culture

The culture of space was also based on technicist utopian and constructivist visions. These included the predictions that space research would soon contribute to revolutions in radio communications, weather forecasting, and elementary particle research. In a few decades cosmonauts would have a base on the moon, producing billions of kilowatts of energy for transmission to the earth by laser, and would be traveling to Mars to mine its resources. In terms of ideology, advances in space, like those in atomic energy, demonstrated the advantages of the Soviet socialist system. Here, however, Soviet primacy and American failures were apparent to all. Successes in space exploration beginning with *Sputnik* in 1957 (and in some respects continuing to this day) were the capstone of the increasing power and prestige of scientists under Khrushchev. A series of "firsts"—the first satellite, man in space, two-man shot, woman in space, space walk, soft landing, and so on—convinced the Soviet populace, if not the majority of Party officials, of the superiority of Soviet science. While Khrushchev and other officials knew of the technological failings and even backwardness of their program, and must have realized that the United States would respond to Soviet achievements with redoubled efforts in its space program, they still embraced *Sputnik* as confirmation that a policy which gave scientists increasing autonomy in the formulation of research programs and administration of research and development was in the best interests of the country.[38] In much the same way that physicists had benefited from the atomic culture, they now found increased prestige and influence in the national euphoria over *Sputnik, Luna, Voskhod,* and other space shots.

From the beginning scientists and ideologues alike were drawn to the inescapable conclusion that successes in space technology confirmed the advantages of the Soviet social order and the planning of scientific research and development under socialism. A. P. Aleksandrov, later president of the Academy of Sciences, saw *Sputnik* as an indication that Soviet science and technology was no longer backward; as something valuable to all who believe in progress; and as a manifestation of the advantages of the Soviet system. Sputnik's significance was not only technological but also social. As Aleksandrov wrote:

> The initial decisive step of a new [scientific-technological revolution] was taken in the first country of socialism. In this region of technology socialism has surpassed capitalism. The scientific-technological superiority of the new, more progressive social order is clear. And if we still haven't passed the most progressive capitalist countries in all regions of science and technology, then in any case their superiority is a thing of the past. [39]

The achievements were also an indicator of high scientific-technological culture and organization and "a real demonstration of the advantages of the socialist order [of] . . . our Soviet system." [40]

A. V. Topchiev, vice president of the Academy of Sciences, echoed Aleksandrov's views in an article written for the first anniversary of *Sputnik* in October 1958. He called *Sputnik* a victory of Soviet technology and a symbol of the achievements of socialism. Apologists for the American way of life tried to characterize *Sputnik* as a chance success, he said, but as of 1958 no one had succeeded in duplicating it. *Sputnik* also demonstrated that the Communist Party had "wisely and purposefully" harnessed the creative forces of the masses "for the realization of the glorious dream of humanity: the construction of communism." According to Topchiev, successes in the space program foreshadowed the way in which science and technology would continue to play a major role in creating "the material bases of communism" during the 1959–1965 Seven-Year Plan, especially in such regions as nuclear physics and rocketry. Finally, "scientific-technological victories strengthen our positions in the struggle for communism, for the progress of humanity." [41] In a companion article Academician L. Sedov emphasized Topchiev's basic contention that Soviet successes in space, as in other regions of knowledge, were possible owing to the socialist social order and the "constant care" of the Party and government for the development of science and technology. [42]

Articles in such literary journals of the national and republican Union of Writers as *Raduga* (Rainbow), *Smena* (Change), *Don, Prostor* (Space), and *Sibirskie ogni* (Siberian fires) also described how achievements in science and technology confirmed the advantages of the Soviet system. According to one, the launching of the dogs Belka and Strelka indicated that in socialist society "people of free labor and truly free flight of thought will never be stopped by any kind of barrier. There was a time when we didn't have enough strength to unfold our wings and we only dreamed of great things."[43] But the Soviets were more than visionaries when they thought of the future. Because of the leadership of Lenin, the role of the Party in arming scientists and engineers with the best techniques and research and development apparatus in the world, and the idea of communism, these achievements were preordained. The flowers that had gone into space were a symbol of the peaceful assimilation of the cosmos. Now there was the possibility that man would also go into space: "And when the first man—our Soviet man—flies to the farthest planets, we will say modestly, but not hiding our pride, 'That's the way it should be!'"[44]

In stressing the peaceful aspects of cooperation with the West and of the Soviet space program in general, Soviet writers pointedly castigated the military design of the U.S. space effort, as with nuclear power. Two philosophers, V. Petrov and I. Ovchinnikov, described two major tendencies in space research. In the USSR space flight was used for research, and for peaceful "material and spiritual" purposes; in the United States it served military interests, and reflected an effort to put a military base on the moon and spy satellites in space as "the younger brother of the U-2." The program was, furthermore, run solely by the Pentagon.[45]

Newspaper headlines often blazed, "American Sputniks Are Spies!" As Petrov wrote,

> Soviet sputniks and . . . rockets first provided humanity with valuable scientific and practical information about cosmic space, first helped to conduct important biological research. With the help of our rockets the unseen side of the moon was first photographed. But the American militarists pursue different goals: reconnaissance of our territory from the cosmos, wild ideas of cosmic war, and the utilization of the moon, sun, and interplanetary space for military purposes.

This was in the minds of the "statesmen" at the Pentagon.[46] NASA's goals were to produce intelligence from territories where one could not go by legal means.[47] Another writer described NASA's program as two-faced, referring to its official, peaceful visage and its real, military mission.[48] Another depicted the Gemini program as civilian in name but military in practice.[49]

This is not to say that Soviet writers failed to appreciate the military significance of their feat. Academician Aleksandrov saw *Sputnik* as a harbinger of change in the cold war for those who insisted on dealing from a position of strength. *Sputnik* meant that the Soviets would soon be able to deliver nuclear weapons on U.S. territory with ICBMs, and those American militarists who saw it as nothing special, a "chunk of metal which anyone could launch," were merely deluding themselves.[50] More to the point, it was the West, not the Soviets, who had "kindled the arms race" and "dropped the atomic bomb on the defenseless inhabitants of Hiroshima and Nagasaki."[51]

Many writers took offense at the fact that the West, and especially the United States, tried to minimize the scientific importance of *Sputnik,* their beloved "little moon." In fact, the Soviet Union had achieved greater heights, distances, and weights. Indeed, most authors took every available opportunity to point out how little America's satellites weighed: *Sputnik* weighed over 80 kilograms and *Sputnik II* over 500 kilograms, but the first U.S. satellite, which had yet to be launched, would weigh less than 10 kilograms.[52] One author admitted that the weight of a sputnik was not a goal in itself, but it did indicate payload and the potential to hold more equipment and to conduct more exact experiments. Still, the Americans simply could not match the Russians. You could not even see their sputnik![53]

One journalist who was in New York when *Sputnik* was launched and heard its magnificent "beep-beep" on American television was surprised that such a peaceful, unthreatening device nonetheless sent the capitalist world into a tailspin, and that a look of embarrassment had replaced yesterday's haughty and self-satisfied physiognomy.[54] When John Glenn orbited the earth three times, *Izvestiia* reported, "Ten months and eight days after Iurii Gagarin, and six and a half months after German Titov, the first American was sent into cosmic orbit and safely returned to the earth."[55] The fact that the United States had finally launched rockets from Cape Canaveral would always be obscured by the fact that the socialist world and the men of communism had first "crossed the threshold of the cosmos . . . on the path to the stars."[56] A number of other writers also reminded the United States that there were no easy paths into the cosmos, and that attempts to paint failures as successes were common in the West.[57] What were the reasons for Americans' backwardness? Was it Harry Truman's advisers, or the fact that the USSR had begun its program earlier—both common explanations offered by the American press? It was not, Khrushchev explained, the fact that they had shortages of dollars, materials, or scholars but that they lacked the social system of the USSR.[58] All agreed that the October Revolution made Soviet success inevitable. While the USSR could launch a half ton into space, the United States struggled "with an unripe lemon"![59]

Popular Perceptions

Soviet perceptions of the potential for nuclear power and space travel went beyond such slogans and catchphrases as the "scientific-technological revolution" and "constructing the material basis for communism." The public responded, much like the scientists and ideologues, with enthusiasm for the potential benefits that large-scale technologies would provide Soviet society. Popular conceptions reflected an interest in the personalities of the cosmonauts, concerns over safety, questions as to whether man was needed in space at all or whether automatic, telemetrically controlled satellites would suffice, and a general outpouring of emotion largely in support of the space effort. There was, however, some evidence of opposition to massive expenditures on space.

From the beginning Soviet citizens embraced the success of the sputnik program and Soviet superiority in space with literature, cartoons, popular films, and state-sponsored courses in schools. They responded more vividly and emotionally than they had to nuclear power, primarily for two reasons. First, nuclear power had yet to be put into general use. Its applications were limited to creating isotopes for industry and agriculture and some modest electrical production. Granted, applications from space science in weather forecasting and communications were long in coming, but Soviet predominance in space over the United States dispelled any doubts that applications would eventually be realized. Second, and similarly, the United States was the pioneer in harnessing the energy of the atom, the construction of a small nuclear reactor in the USSR in 1954 notwithstanding. With *Sputnik*, however, the Soviet Union was first—again and again!

From the earliest days of manned space flight, Soviet journals from *Ogonek* (Little flame) and *Smena* to *Partiinaia zhizn'* (Party life) and *Kommunist, Priroda* (Nature) and *Nauka i zhizn'* (Science and life) were filled with articles presenting a common view of the cosmonaut as someone simple yet wise, possessed of a steely disposition yet kind, working in the spirit of the collective, and being, above all else, an ordinary Soviet citizen. Journals frequently published interviews illustrating how even a common worker with the right attitude and Soviet upbringing could become a cosmonaut.[60] The role of willpower and training, of sport, of the collective spirit, the need to act deliberately, cold-bloodedly, and quickly yet to be simple, calm, humanitarian, and thoughtful—all this was conveyed to the reader.

Journals with a more literary bent, such as *Prostor, Smena,* and *Neva,* depicted the more personal side of the cosmonaut,[61] while journals for schoolteachers attempted to convey the impression that Iurii Gagarin, German Titov, and the rest were "people of labor—sons of the working masses," lovers of music and reading.[62] Letters from the cosmonaut Vladimir Komarov to his wife, Tamara, written just after the Second World War, before "cosmonautics" even existed, were said to demonstrate that he was a courageous and heroic individual who "all his life thought about how to be and to remain a real man [*chelovek*]."[63] Still another journal contributed to the effort to make Komarov more personable, since much of the literature spoke of his "strong-willed, steadfast personality" and "iron bearing": *Neva* presented him as calm, kind, attentive, and polite.[64]

Several literary journals provided insight into the daily life of cosmonauts at their launch base in Kazakhstan. One article consisted of interviews with Pavel Popovich, Valerii Fedorovich Bykovskii, and Valentina Tereshkova, the first woman in space, showing them to be ordinary people. Popovich had thought of his mother first, not his wife or daughter, when he heard the ignition of his rocket on the launch pad. But takeoff caused him no fear: all cosmonauts are certain they will return to see the earth and the flowers.[65] Bykovskii spoke of his certainty that his mission would be a success since he had trained hard under the watchful eye of Soviet scientists and the Party.[66] As with the male cosmonauts, Tereshkova first thought of "Mama" while considering the significance of her feat,[67] but unlike them she was a bit worried before her flight.[68] The beauty of Soviet life was the fact that anyone might be the next hero. This did not mean that the cosmonauts were merely lucky. "Success came to them through hard work, difficult study, through special tempering of will and self-possession . . . They were lucky in one way only: like you and me they were born and live in our Soviet country."[69]

The comfort of the cosmonauts also attracted the attention of readers. A. G. Nikolaev and P. R. Popovich, the pilots of *Vostok III* and *IV,* respectively, commented on the comfortable nature of their ships, created by "our talented scholars and constructors," which enabled them "to solve still more complex problems."[70] The air conditioning permitted longer and longer flights. Working diligently on a series of experiments, the astronauts "did not suffer from loss of appetite." Meals prepared especially for the cosmos, packed in tubes and polyethylene packets, catered to the tastes of the individual: "The quality of taste was excellent . . . I love meatballs, fried chicken, and kolbasa. And I ate these in space. You won't waste away on this ration on the moon."[71]

Many writers responded to *Sputnik* with poetry and prose. While not of high quality but rather naive and immature, these works nonetheless reflect many of the key themes of Party and scientific tracts on the social and cultural significance of *Sputnik:* the advantages of the Soviet socialist social system, the concern of the Party, the meaning of the scientific-technological revolution, and the inevitability of the victory of communism over capitalism in peaceful competition with the United States. Such well-known writers as Il'ia Ehrenburg, Semen Kirsanov,[72] and Mikhail Sholokhov, and many lesser-known poets such as Mikhail Sokolov, N. Skrebov, Sergei Vasil'ev,[73] Aleksandr Farber, Nikolai Aseev,[74] A. Rogachev,[75] Vasilii Zhuravlev, Anatolii Radygin,[76] and Iurii Iakovlev, graced the pages of *Literaturnaia gazeta* (The literary gazette), *Don, Novyi mir* (New world), and *Oktiabr'* (October). Sholokhov responded with excitement and pride: "There you have it! And you really can't say any more, having been struck dumb from rapture and pride in the face of the fantastic success of our dear native science."[77]

After *Sputnik III* flew by the moon taking photographs, Vasilii Zhuravlev offered this praise in a poem entitled "Unfailingly":

> The third rocket flies in the depths of the universe
> Under the flag of my country.
> The craters of the moon freeze over under her.
> She penetrates into their secret of secrets unfailingly.
>
> So it is, day after day,
> Stubbornly, inspired,
> Loyally, with great aspirations,
> We storm the heavenly virgin lands
> And we master the cosmos unfailingly.
>
> For the world our affairs are concealed,
> As the clear day of spring is concealed.
> We are building communism,
> Full of resoluteness.
> And we will build communism unfailingly.[78]

After Gagarin's spectacular journey N. Skrebov called for the Soviets to go to the stars:

The radiant tomorrow has arrived!
Rapture has stirred the entire planet.
The East and not the West visited the cosmos
With a courageous cosmonaut!
Carrying a valuable cargo,
Having developed unprecedented speed,
Time and space having been subordinated to a citizen of our *Soiuz* [Union]!
He is radiant with unparalleled glory,
Filled with heroic strength.
Iurii Alekseevich Gagarin blazed a trail to the stars.[79]

Some were more circumspect, concerned that the achievements of Soviet science and technology had obscured the need for art and literature, or even doomed it in modern Soviet society.[80] For example, Il'ia Ehrenburg wrote:

I said that in the rejection of art by several young people, in the approach to any artistic work . . . weak aesthetic training is guilty; consequently, a part of the guilt falls to us writers. Leaders of the party, economists, engineers, agronomists quickly raised the standard of living of the population. Astrophysicists sound out their way to the moon. But people judge our work according to existence, laws, the relationship of man to his comrades, to the family, according to his responsiveness and indifference, sensitivity or rudeness. The writer to a significant degree answers not only for his readers, but also for those whom he cannot make his readers. Is it really closer to the moon than to the heart of a man who lives in the neighborhood?[81]

Yet most writers were concerned with the glorification of Soviet achievements, not questioning whether they had turned attention away from the more fundamental issues of day-to-day life. Schoolchildren were overcome with joy over sputniks and spaceships. One composer wrote a song for children, to be played "cheerfully";[82] another wrote "heavenly" tunes (*chastushki*).[83] Indeed, the monthly journal of the Academy of Pedagogical Sciences soon published short stories and lectures to be read aloud to schoolchildren on the history of man's space exploration,[84] current Soviet efforts,[85] the state of the art,[86] and Gagarin's character and career in school.[87]

Cartoons and drawings also carried the message of the Soviet Union's unparalleled success—and the weak position of the United States. Drawings on New Year's Day conveyed the least subtle messages. On January 1, 1958, the cover of *Komsomol'skaia pravda* carried a drawing of a Christmas tree covering half of the front page. At the base are a cow offering a full milk pail and two *kolkhozniks* (collective farmers) putting a coin, symbolizing increased agricultural production, into a Komsomol piggy bank. All of the objects on the tree are symbols of technological achievement: a railroad, a coal mine, the Kuibyshev Hydropower Station, the atomic icebreaker *Lenin,* several new airplanes, a rocket ship, and a sputnik at the top; electrical power lines and the Stalinsk-Abakan railway line decorate the length of the *elka* (fir tree). [88] The most striking drawing I encountered shows a Russian peasant in a troika, labeled "1959," driving toward "1960," pulled along by three satellites. [89] Other cartoons belittle America's efforts and draw attention to its failures. [90]

Not only in newspapers and journals but also in public, citizens greeted *Sputnik.* From the start the Soviet masses were involved in the celebration of its achievement. Hundreds of thousands of people gathered in Moscow near the Square of the Uprising at five o'clock one evening to applaud the appearance of the satellite in the northwest sky. As a lecturer at the Moscow Planetarium wrote, "This applause belonged to all Soviet people who had guaranteed through their heroic work the success of the first and most difficult step into space." [91] Soon, long lines made it impossible for people to get into the planetarium. [92] Nonetheless, thousands visited regularly to become familiar with models of sputniks and hear lectures on the solar system. [93]

Just as curious citizens had become involved in the search for uranium, so thousands of university and pedagogical institute students and scientific workers participated in regular observation of the first artificial satellites. Seventy observation points equipped with AT-1 telescopes in such cities as Pulkovo, Moscow, Uzhgorod, Arkhangel'sk, Riga, Kiev, Iakutsk, Vilnius, Tartu, and Frunze were organized by the Astronomical Council of the Academy of Sciences in cooperation with the Ministry of Higher Education. While they did not provide exact measurements, they did help ensure that the sputniks were not lost, and they offered a first approximation of the orbits. Bad weather plagued observation, but sightings ranged from ten to forty on any given day. [94] According to one source, Soviet citizens also used chance meetings in such places as the *elektrichka* (the electric train) to learn about the most recent successes from well-informed fellow travelers, gaining knowledge that with each success Soviet science surely had paved the way toward a new world.[95] The Academy of Sciences contributed to the public dissemination of information about Soviet space science, publishing collections of articles in *Iskusstvennye sputniki zemli* (Artificial satellites of the earth); by the fifth anniversary of the first sputnik, thirteen volumes had already appeared.[96]

In an attempt to bring the achievements of Soviet science closer to the masses, the state film industry produced feature films providing answers to the many questions the average citizen had about space travel concerning weightlessness, radiation, spaceships and space stations, and eventual travel to the moon. Such films as *The Road to the Stars*[97] and *To the Stars Again!*[98] were released shortly after the most recent achievements.

Just as in the United States, however, there was opposition to the extensive Soviet space effort.[99] V. Liapunov noted that some skeptics questioned the program: "In perhaps another year . . . people will understand why the force of thousands of minds was directed toward this great goal."[100] One of the leading space scientists also explained the need to study space in terms of building the material bases of communism. He granted that huge resources were being allocated on rockets and satellites, but argued that the expenditure would not hold back the solution of more pressing domestic economic problems. More important was the fact that the United States had embarked on space research. To maintain Soviet superiority, money had to be spent.[101]

Occasionally the editors of a daily newspaper would publish a letter from a citizen who questioned the efficacy of the space effort. The most striking came from one Aleksei N., who asked the editors of *Komsomol'-skaia pravda*, "What's in it for me?" What had these sputniks and rockets given to the simple mortal such as himself? "I, for example, on the eve of the launch of a rocket, received 300 rubles salary, and this is what I still receive, in spite of the successful launch. Doesn't it seem to you that the enthusiasm for these sputniks and the cosmos in general is inopportune and, more precisely, premature?" He continued:

> Now this rocket, I don't doubt, devours so much that everyone would surely gasp if they knew its price. Tell any worker, "Look, you, Ivan, if we didn't launch this rocket, . . . you won't be able to buy an electrical iron in the store"—then, I am sure, he would say, "Thank god, launch these rockets." Rocket, rocket, rocket—what's it needed for now? To hell with it now, and with the moon, but give me something better for my table. After that, then it will really be possible to flirt with the moon.[102]

Through the mid-1960s resistance persisted. The popular science writer (and critic of relativity theory from the 1930s) Vladimir L'vov discussed rather farfetched visions of the Soviet space effort to quiet skeptics who questioned the need to go further. He described how billion-hectare fields of solar cells on the moon would permit the transmission of trillions of kilowatts of energy by laser to the earth, and how oxides in the soil on Mars would guarantee oxygen for breathing for tens of millions of years, and so on.[103]

Concerns over the safety of cosmonauts persisted throughout the early days of the space program. Frequently the press discussed the question of whether an automaton could accomplish what man could, with less risk. But most authors assured their readers that safety was not a real problem. Granted, the tremendous speeds and vast distances of cosmic travel limited man's usefulness in the role of navigator of a spaceship. Man's sensory organs were not agile or quick enough and his reactions were too slow to deal with cosmic speeds. Hence automatic equipment was needed to control and operate the ship. But did this mean that man had been replaced? No. The problem of "man or automaton" did not exist. Man was the creator of even the most complex machines, which could fulfill only a limited number of problems that man had given them.[104]

The qualities of the typical cosmonaut and the advantages of his intellect over automation notwithstanding, journals regularly addressed issues of safety in space. They reported the preliminary experiments on flora and fauna and the development of space medicine, including new training regimens to accustom man to weightlessness and protect him from radiation and meteorites. Only after the safety of space flight was assured would man be sent into orbit.

The first flights with laboratory animals conducted in the early fifties pointed out the major problems facing future "stratonauts": g-force, weightlessness, radiation, technical problems in construction of a hermetically sealed cabin and a good space suit, protection during reentry, and the need for some escape mechanism. Nonetheless, the results of experiments on animals in hermetically sealed capsules launched to an altitude of 200 kilometers demonstrated that manned flight was possible.[105] The sealed cabins provided a solution to the problems of weightlessness, acceleration, vibration, and noise, and even such potential psychological difficulties as loneliness and isolation, especially during long flights to Mars and Venus. There might be some danger of psychophysiological effect in long-term space flight.[106] But while the first flights would be emotional and traumatic, "fear in the face of danger will give way to the emotional lift in connection with the recognition of the duty before the motherland and the great scientific significance of flight in the cosmos for all humanity."[107] By the fall of 1959 Soviet science writers were convinced that the fantasies of Jules Verne and H. G. Wells and the scientific theories of K. E. Tsiolkovskii had become a reality. The fears of some biologists were unfounded; experiments had shown that there were no "insurmountable obstacles for the penetration of man into the cosmos."[108]

In addition to issues of medicine and psychology, Soviet writings often addressed the problems of radiation, meteorites, soft landings, and the danger of accidents. The Soviets undertook detailed studies of the effect of alpha and cosmic rays, protons, charged particles trapped in the magnetic field surrounding the earth, and other radiation on cosmonauts in preparation for a lunar landing. They had measured the radiation which the early Vostok spaceships encountered, calculated the average potential exposure in space for a two-week trip, pointed out the possibility of termination of space shots if increased solar activity warranted it, and developed medicines to limit the effects of radiation.[109] In addition, studies had shown that there was only a very remote danger of meteorites striking a spaceship.[110]

Soviet citizens expressed other concerns as well. Hadn't the heavens become overcrowded with satellites? Wasn't a collision inevitable? And what about the danger of a satellite crashing to the earth, killing or injuring people? Science writers explained that this would rarely, if ever, occur. Orbits were calculated to take into consideration the effect of gravity and friction of the atmosphere on the rocket. As V. Bazykin explained to the readers of *Agitator*, there were over six hundred satellites in space as of 1967 whose orbits were known, although not precisely charted, in addition to stages of rockets and other man-made debris. Bazykin admitted that some might eventually reenter the atmosphere, especially those closest to it, and that others were already inactive. But he discounted the danger, since burn-up on reentry would destroy these vehicles and, as fantasts predicted, soon the rescue and repair of failing systems would be possible.[111]

Finally, Soviet writers focused on the technical problems of flight, including proper calculations for speeds and escape velocity, trajectory, geostationary orbit, and so on.[112] These calculations showed a determination to visit the moon and even other planets as soon as technology permitted.[113] One of the most serious problems concerned the development of modern fuels. Since fuel added to the weight of the rocket, it limited the size of the payload. But most scientists agreed that the development of liquid fuels and multistage rockets, and eventually space stations, would overcome this problem.[114] Some of the more futuristic treatments of rocket power addressed the need for atomic- and even photon-powered engines.[115]

In sum, like atomic culture, space culture was based on utopian visions of the role of cosmonautics in the construction of communism. It confirmed the advantages of the Soviet system, both for scientific research and development and for society. The Soviet citizen was grateful to learn that the USSR was ahead of the West—finally—in a leading area of science. After years of sacrifice, a payoff seemed to be at hand. Despite concerns, space research seemed scientifically sound. What is more, since a human being and not a dangerous technology like nuclear power stood at the center of attention, there was a good deal of support and interest, the lack of immediate application notwithstanding.

But in fact the potential for applications in meteorology and radiocommunications, cartography, and geophysics was considerable. With satellites, arctic ice floes and forest fires could be detected and tracked.[116] To some the achievements of rocket technology promised a major step toward human control of nature.[117] Cosmic research would lead to interplanetary travel, unlock the secret of subatomic forces, and even help to verify relativity theory.[118] One author predicted that gigantic passenger jets, victorious Antarctic expeditions, and the construction of the largest and most powerful atomic power stations and particle accelerators would follow on the heels of the sputnik successes by 1970.[119] Others spoke hopefully of the discovery of other life forms in the universe. Space was the key to the Soviet future and the construction of communism.

Remarkably, Soviet scientists and engineers overcame significant obstacles after 1945 to help rebuild a war-ravaged economy and reestablish normal progress in scientific research and development. Taking advantage of postwar political, ideological, and sociocultural trends, they embarked on almost fantastic construction projects of large-scale technologies, drawing on Party support, their own initiative, and the atomic and space culture. Within five years they had built an atomic bomb, within ten years nuclear reactors and a hydrogen bomb, and within fifteen years had put a man into space. These achievements led to the rediscovery of utopian and constructivist visions which had been so prevalent in the 1920s for the construction of communism in post-Stalin Russia. In particular, communism would be built on such advances in science and technology as rockets and reactors.

The nuclear reactor and space research and development programs share a number of features. First, for the scientific leadership and the Party, successes in these areas confirmed the advantages of the socialist system. After years of lagging behind the West in spite, or perhaps because of, Party involvement, the achievements signaled that greater scientific autonomy within the confines of central planning would produce results. The USSR could compete with the West, and soon would overtake it in many areas: in peaceful competition the communist system would win.

For the Soviet system as a whole, successes in large-scale technologies had a more fundamental purpose. After years of sacrifice, oppression, and coercion, the average Soviet citizen was finally seeing some return on his efforts. He had labored throughout the Stalin years to turn a backward peasant economy into an industrial power; his country had survived the Nazi invasion. But what did he have to show for his work? There was little tangible result in terms of consumer goods, although huge factories spewed smoke, produced iron, steel, and concrete, and generated electricity. A harsh Stalinist reality had supplanted the utopian dreams of the 1920s.

Now, however, constructivist visions were resurrected. The citizen had only to pick up a journal or newspaper, see a feature film, or listen to the radio to find out about Soviet scientific achievements in reactor and rocket technologies, electrification, or Metro construction. Unlimited supplies of energy from fusion and fission would permit the full construction of communism by 1980. Permafrost would be conquered; deserts irrigated; dams built; and nuclear-powered transport would free Soviet industry from the shackles of prehistoric Ford tractors, clunky ZALs, and creaky trains. Soon nuclear-powered rockets would climb into the cosmos to other planets and solar systems.

But it was *Sputnik, Lunik,* and Gagarin that convinced Soviet leaders and the public alike that these dreams could be achieved. For the first time, and in a new and prestigious area of science and technology, the USSR had beaten the West. The Soviets were the first into space, to the moon, to the solar system! America, and its exploitive capitalist monopolies, had failed on the launching pad. Filled with pride, few questioned the value of massive expenditures on space research, in spite of the low standard of living.

Space research was more readily accepted than nuclear power. While many expressed concern over the safety and cost of both technologies, it was clear that the destructive potential from space shots paled in comparison with radiation danger from the energy of the atom and isotopes. What was more, space research seemed more personal. Nuclear power stations, by contrast, were ominous, and may have reminded many of Stalinist gigantomania. And as yet there was no need for massive reactors. Because of concerns over the well-known dangers of nuclear energy, and the understanding that the United States was the leader in this field, it was more difficult to gain public support for atomic energy. In fact, the creation of broad national support had to precede the placing of nuclear power on the political agenda. For space, the public buildup was gradual to a point, but an explosion of popular support followed Laika, Belka, and Strelka, *Sputnik, Luna I* and *Luna II* to space. Over the next eight years the Soviet press was filled with evidence of self-adulation and fascination, as scientists, Party officials, and the common Soviet citizen alike became convinced that their sacrifices of the past forty years had not been in vain. Soon communism would be constructed.

Epilogue

This chapter has focused primarily on the period 1945–1975. In the last fifteen years, however, in books, plays, and short stories, [120] as well as in official pronouncements, enthusiasm for many of the technicist attitudes I have discussed in this chapter has waned. Public criticism of the Soviet nuclear power program predates the Chernobyl disaster by some eight years, and many scientists and engineers now openly bemoan such problems in the industry as the absence of a formal licensing procedure, the lack of Western-style containment on many of the Soviet reactors, the siting of reactors in massive blocks close to population centers, and fuel-cycle issues. The Soviet space program is virtually free of this kind of criticism, although many still question the efficacy of huge expenditures on research—so much so that Soviet scientists now actively pursue co-operative and cost-sharing projects with the United States and are trying to sell payload space to private customers. And, in spite of the fact that for nearly ten years the Soviet space program has followed a cautious policy of relying on proven technologies and missions of limited novelty, the public has become more aware of accidents, fatalities, and other failures of the program.

The change of political climate in the Soviet Union will most likely contribute to a growing circumspection about large-scale technologies. To date the policy of glasnost has helped to develop among the Soviet public a more realistic attitude toward the potential of science and technology and also a greater awareness of their danger. For the first time in Soviet history the news reports that the citizen hears and reads chronicle the sinking of ferries, crashes of airplanes and buses, and past cover-ups of man-made technological disasters. Even the events surrounding Chernobyl were reported quite openly and with relatively little delay or deceit. Up to this time, the absence of a vocal, popular antinuclear movement permitted the Soviet nuclear power industry to move ahead in construction of dozens of 1,000-megawatt reactors within the constraints of material and equipment supply, geography, and meteorological conditions. If Gorbachev's glasnost creates citizens with greater political sophistication, who oppose such large-scale technological projects, then the industry will face a more fickle and intractable obstacle to its plans.

To a great extent, nevertheless, technological determinism remains a major force in the contemporary Soviet Union. One of the primary reasons for this is the fact that the official ideology of the USSR, Soviet Marxism, is inherently technicist. Soviet leaders have revered science and technology since the first years of the Soviet state. Indeed, Gorbachev has surrounded himself with a handful of prominent science advisers. Faith in technology as the highest form of culture, the chief mode of modernization of the economy, the answer to standardization and mass production—the panacea for all social and economic problems—continues to hold sway. Atomic and space culture, which were built on forty years of achievements, are so deeply rooted in popular and scientific perceptions of Soviet leadership in space exploration and nuclear physics that it will take more than Chernobyl to shake this faith.

· PART FOUR ·

Philosophy of Science and Technology

· 7 ·

Biomedical Ethics

Richard T. De George

The hypothesis that the discipline of biomedical ethics is similar across borders is an attractive and plausible one. Strachan Donnelley begins the Hastings Center report "Biomedical Ethics: A Multinational View" thus: "Not surprisingly, biomedical ethics has become truly international. Medical technologies and scientific developments know no national boundaries and have spread rapidly through the world, especially in advanced industrial countries. Peculiarly modern moral dilemmas are never far behind."[1] Thus, once heart transplants become possible, each nation must ask whether to perform them or not; and if it does decide to do so, then the country must face the question of how to allocate fairly the limited resources for such transplants. The questions are the same, even if the answers may be different.

The hypothesis can be compared to the convergence thesis, which claims that the development of technology carries with it certain inherent values, commitments, and demands that will force the United States and the Soviet Union to become more and more alike, despite their different ideologies and despite the wishes of their leaders. The rise of Gorbachev, and the introduction of glasnost and perestroika, have added more credence to the convergence thesis than it had before. And this suggests that a similar type of convergence might be found in the area of biomedical ethics. While a few early signs have emerged to support this hypothesis, the differences in the existing literature indicate that the hypothesis is still mostly false when applied to the Soviet Union. The variations in both approach and practice with respect to bioethics in the two countries are more striking than the similarities. Nonetheless, this does not mean that Americans have nothing to learn from Soviet discussions

195

and practices; and some Soviet observers have closely followed, reported on, and learned from our discussions and practices.

Biomedical ethics comprises three areas: philosophy, biomedicine, and politics. Bioethical issues always arise in a certain social setting, and what constitutes an issue or problem is in part a reflection of the society in which it arises. Hence what may be an issue in one society may not be, or be recognized as, an issue in another society. How the problem is handled, discussed, and resolved may also vary considerably from society to society.

In the Soviet Union theoretical bioethical discussions take place in the professional philosophical, biological, medical, and legal journals. On this level we find statements of official Marxist-Leninist ethics and the derivation and justification of official socialist morality. Although one might expect that debate on biomedical ethical issues would be flourishing in the new period of openness, this is not yet the case. As I. N. Smirnov notes, Soviet philosophers are still sorting out what kind of enterprise bioethics is,[2] and the literature is still little developed. Discussions taking place on the nature of values and ethics will probably, if the present trend continues, lead to more work in this area.

In official statements and legislation we find the resolution of those issues that are debated. These are in turn justified on the theoretical level and implemented on the popular level. Since 1986 some medical officials have been surprisingly open in their portrayal and criticism of the existing medical system.

In the popular literature of newspaper and journal reports, articles, and letters to the editor, we find the interplay of real conditions and practical bioethical problems, which are often very different from the discussions in the professional theoretical literature. Since the introduction of glasnost in 1986, the popular press has carried a variety of articles, exposés, and letters dealing with the current state of medicine in the Soviet Union and with a variety of ethical complaints concerned mostly with abuses of the system by medical practitioners. The moral beliefs implicitly or explicitly expressed do not always coincide with the values preached on the official level.

Let me set the stage for discussing the differences between bioethics in the Soviet Union and the United States by turning to two areas in which those differences are dramatic. The first concerns the ethics of psychiatric hospitalization for what critics claim to be political crimes or dissent; the second is the area of abortion.

If we begin with a search of the American literature on Soviet bioethics, we find that 68 of the 131 English-language articles dealing with Soviet bioethics from 1973 through 1986 deal with the institutionalization in mental hospitals of so-called Soviet dissidents.[3] This is obviously at once a political, an ethical, and a medical issue. A large number of American and British as well as many other psychiatrists have charged that the Soviet government places political dissidents in psychiatric wards of hospitals instead of jails, and have ethically criticized Soviet psychiatrists for taking part in what they perceive to be an abuse of psychiatric medicine. The charges include diagnosing "non-existent illness in healthy people"[4] (such as "sluggish schizophrenia," an illness found only in the Soviet Union), and the administration of drugs with no therapeutic value without the patient's consent and with punitive intent.[5] The discovery of sluggish schizophrenia, which manifests itself in an obsessive desire for social justice in individuals who evidence otherwise normal behavior, is attributed to the Soviet psychiatrist Andrei V. Snezhnevskii, who died in the summer of 1987.[6] People so diagnosed were typically sent to one of the eighteen psychiatric hospitals that, until January 1988, the Ministry of Internal Affairs reputedly controlled.[7] A law passed on 5 January 1988 gave the Health Ministry final authority over those hospitals, made it a criminal offense to commit to a psychiatric hospital a person who is known to be mentally healthy, and detailed a set of rights and appeals procedures for mental patients.[8] While recognizing this as a step in the right direction, skeptics are waiting to see whether these changes in the law will have any effect on what, up until 1988, had been the practice of using psychiatric hospitals as a repository for dissidents, and on the maltreatment they received in such hospitals.[9]

Although there are reports of Soviet psychiatrists who have protested the criticized practices,[10] none have seen print. Since 1987 a number of articles and letters have appeared in the Soviet press concerning abuses by the police and psychiatrists in committing eccentrics and people who complain about the system or the corruption they see around them. One, which appeared in *Izvestiia,* criticized the use of psychiatry with respect to two women who had "pestered" authorities.[11] *Komsomol'skaia pravda* reported the case of Marina Pristavka, who was "always seeking the truth" and "isn't afraid to come into conflict with the bosses."[12] The official medical position is that there are no systematic abuses, even though there may be individuals within the system who perpetrate abuses.

In 1983 the All-Union Scientific Society of Neuropathologists and Psychiatrists officially withdrew from the World Psychiatric Association (WPA), charging that leaders of the American Psychiatric Association and the British Royal College of Psychiatrists had, through fabrication and falsification, waged a campaign of slander against psychiatry in the USSR. A British report indicates the Soviets withdrew because they were about to be voted out of the WPA on a motion from the Royal College of Psychiatrists.[13]

In their 1983 defense the Soviet psychiatrists claimed that all patients sent to psychiatric hospitals were mentally ill; that those who were prisoners had been tried and convicted of a crime against the state or had violated the criminal code; that many had a history of psychiatric treatment; and that, although frequently they and their families considered them healthy, this was a delusion not shared by those medically competent to judge.[14] One might even be tempted to argue that anyone who continues to challenge the Soviet system at great personal cost is not normal, and some might claim that putting such people in psychiatric institutions is more humane than putting them in jails or condemning them to hard labor for disrupting society. The official position remains that the diagnosis of mental illness can be made only by qualified psychiatrists, and that their diagnoses cannot be second-guessed by the general public or foreign psychiatrists unable to examine the patient.[15] Nonetheless, abuses of psychiatry are perceived by many in the Soviet Union as well as abroad. The government has taken some action to address the abuses. But as yet, no ethical concerns about psychiatric practices have been raised in print in the psychological, medical, legal, or philosophical—as opposed to the popular—literature in the Soviet Union. Even more significantly, there is no published discussion of the use of psychiatric hospitalization for political dissidents.

The use of psychiatry in the United States as a tool of social inculturation has also been criticized, especially by Thomas Szasz, who wrote against the "myth of mental illness" and the "mental health ethic" that state mental institutions take as the norm.[16] Michel Foucault has demonstrated how notions of insanity and the treatment of the insane are normative judgments that vary from society to society and age to age.[17] But the volume of literature claiming unethical use of psychiatry in the West is comparatively small as compared to the volume concerning the issue in the Soviet Union, and the ethical charges are much less severe. In the United States the issue of whether to hospitalize forcibly people who live on the streets—many of whom have been released from overcrowded state hospitals and mental institutions and who are unable to care for themselves in a conventional way—has led to a debate about the rights of such people and the responsibilities of state and local governments. From the Soviet point of view the existence of street people is evidence of the injustice of the American system, the inability and unwillingness of the system to care for people, the breakdown of the social order—all negative moral judgments. From the American point of view, personal freedom from involuntary institutionalization is pitted against the obligation of the state to provide care and the question of the proper use of psychiatric diagnoses.

Obviously there is more at issue here than simply a debate about the ethics of psychiatry and its use as a means of either reform or punishment. For our purposes it is interesting to note that although there has been much criticism in the United States and Great Britain of the ethics of Soviet psychiatry, there is virtually nothing written about it in the Soviet Union. This is an indication that the subject matter of bioethics is not simply a function of biology or medicine or ethics. What is perceived as a problem, as well as the way problems are treated, is fundamentally different in the West and in the Soviet Union. It is worth noting in particular that these issues are debated in the American bioethical literature while they are not—at least not yet—in the Soviet professional literature. Soviet government policy is by definition moral. Although abuses of such policy may be challenged—and we are hearing more such challenges since glasnost was introduced—the morality of the practices is not challenged. Since many aspects of biomedicine are determined by government policy, there is simply no ethical discussion of those issues in the theoretical Soviet literature.

The second area in which we see a striking difference is that of abortion. In the United States this is the single most written-about issue in biomedical ethics in the philosophical and popular literature. The arguments pro and con fill many volumes, and the political dimensions have split the country and the Supreme Court. The morality of abortion, the morality of state-financed abortions, and the morality of state interference in abortion are hotly debated issues. The topic arouses fierce passions and unites otherwise very different religious groups. Hardly a single introductory anthology in ethics fails to include a section on this topic, for it touches on the right to privacy, the right to life, the right of a woman to control her body, the nature of the person, and the moment of origin of human life.

By contrast, the issue is rarely mentioned in the Soviet literature. I could find no philosophical article in which it was discussed. The popular literature was almost as quiet. Abortion is legal in the USSR as long as the fetus has not reached viability, "which is defined as a weight of 1,000 grams," [18] even though fetuses of lesser weight (500 grams at twenty-two weeks)[19] are considered viable in other countries. Early in 1988 the Ministry of Public Health authorized abortions up to twenty-eight weeks on a variety of nonmedical grounds, including the death of one's husband.[20] As Dr. Boris Urlanis notes, "The right to have an abortion is an important right. Every woman may exercise it. But I want to emphasize: exercise it, not abuse it." The arguments Urlanis gives against abortion are made not on moral grounds but on medical grounds of adversely affecting the woman's health, and on pragmatic grounds of unduly restricting "the economy and future labor resources." He advocates that unwed mothers bring their fetuses to term because "they're raising a citizen." His argument is linked to the social need for more children to increase the population rather than to any ethical compunctions about abortion.[21]

Candidate of medicine Klara Bronislavovna Segieniece writes, "In my view, refusing to bring one's first pregnancy to term is a crime against morality," and she complains that 60 percent of first-pregnancy abortions occur among women aged nineteen to twenty-six, the prime childbearing years. But her moral concern is also not the right of the fetus but the social good of increasing the population. She claims, "It's purely an affectation to accuse women of murder when they have an abortion."[22] The latter remark is an indication that some people may hold the view that abortion is murder; but she does not take that view seriously, as needing refutation rather than simple rejection.

The history of abortion in the Soviet Union is a checkered one. Following the October Revolution abortion on demand was recognized as a woman's right. The significant population drop, however, prompted officials in 1936 to pass an Anti-Abortion Act, making abortion, except for therapeutic reasons, illegal. The concern was not that abortion itself was immoral but that its practice threatened the needs of the state for a larger population and workforce.[23] Owing to the large number of illegal abortions and the resulting incidence of infection and death, that act was repealed in 1955,[24] and the 1969 Public Health Act left the decision on pregnancy up to the mother, until the time when abortion might threaten her health.[25]

Abortion as a legally protected right is quite heavily exercised in the

Soviet Union. Some estimates, based on figures for Leningrad, put the ratio as high as 3.2 abortions for every live birth in 1962, and 4.1 for every live birth in 1967. If one includes the countryside, where the rate of abortion is lower, the generally accepted ratio is slightly more than one abortion per live birth.[26] Larisa Remennik, a demographer writing in *Nedelia* in 1987, remarked: "It is a sad fact, but our country has one of the highest abortion rates in the world. In 1978–79 the number of abortions per 1,000 women of reproductive age was: the FRG—5.9, England—11.4, the U.S.—27.5, Czechoslovakia—28.9, Hungary—36.9, Bulgaria—68.3, and the Soviet Union—102.4. In the Russian Republic in 1985 there were 123.2 abortions per 1,000 women."[27] Interestingly, she adds, "It seems to me that our society's widely held view of the psychological acceptability of abortion, of the 'normality' of the artificial termination of pregnancy, indicates the crumbling of certain moral foundations that have a common meaning for people in any society."[28] The arguments she gives against abortion are not based on any rights of the fetus but on the damage to a woman's health and, again, on the decrease in the birthrate.

Thus the complaints about abortion that see print concern the small number of children that couples are willing to have and the possible negative effects on the future of Soviet society. Since abortion is more prevalent in the large cities, where housing is in short supply, there is also concern on the part of some that the Russian population is decreasing while the population in other, especially Muslim, republics is increasing. Women complain about the primitive methods used to perform abortions and the need to pay bribes to receive adequate treatment.[29]

Despite the widespread use of abortion, Soviet statistics recorded more than 500,000 illegitimate births (9.8 percent) in 1987. Of first pregnancies, 411 of every 1,000 babies were born out of wedlock or in the first few months of marriage. Although illegitimate children receive the same benefits as legitimate children, and officially suffer no stigma, the weekly journal *Semia* calls the number of illegitimate children " a huge social problem."[30] The incidence of abortion and of illegitimate births among teenagers and young unmarried adults indicates that premarital sex is far from unknown. The preferred solution to abortion is the use of contraceptives. But Soviet women do not trust the pill, and condoms are of poor quality and in extremely short supply. A letter to the editor of *Meditsinskaia gazeta* claims, "It has been a year since the pharmacies in Lvov have stocked condoms."[31]

At the same time, Soviet society has been, by American standards, prudish. An article in *Meditsinskaia gazeta* illustrates the point when it asks, "Now how does one write about THAT [referring to condoms]? It's not done in our country."[32] An article in *Komsomol'skaia pravda* states: "For some incomprehensible reason, there is a taboo in our country on certain words and on a whole series of 'shameful subjects' about which there is essentially nothing shameful. For example, French television can for some reason say unabashedly: 'Mesdames, messieurs: Use condoms!' By contrast, the word causes people in our country to take a fit."[33]

Early Soviet experiments with free love, easy divorce, and deemphasis on the family have given way to a view of the family as the building block of society. Sexual modesty is the officially preached norm, and official attitudes toward sex are close to puritanical. Texts on morality, on the family, and on marriage stress the virtue of chastity. The Moral Code of the Builders of Communism, promulgated by the Party in its 1961 Party Programme, calls for "moral purity, modesty, and unpretentiousness in social and private life," as well as "mutual respect in the family, and concern for the upbringing of children."[34] In the words of V. I. Pokrovskii, president of the USSR Academy of Medicine and director of the Central Epidemiology Research Institute, in an interview published in *Izvestiia,* sex education is timorous and euphemistic, "complete with everything but storks and cabbage leaves," and "in need of serious revision."[35] As in most puritanical societies, such suppression does not necessarily guide individual practice—as the prevalence of abortion among unmarried women and the high incidence of illegitimate births indicate.

According to Marxist-Leninist theory, personality is a function of integration into society. Hence the fetus has no moral status. And since rights are grounded in society, the fetus has no rights. Religious arguments are given no official credence. There is little basis, therefore, for the kinds of moral debates about abortion that we find in the United States. A Soviet philosopher could, in principle, within the Marxist-Leninist framework, develop a theory of potentiality on the basis of which a fetus could have moral standing. But none has done so thus far. Abortion is an officially accepted practice but obviously not an officially mandated one. Thus the absence of discussion of abortion, unlike that of psychiatry, does not stem from a reluctance to criticize the government's activities.

The dissimilarities between the Soviet Union and the United States with respect to the use of psychiatry and abortion illustrate the differences that we can expect to find, and do, in the two countries in the realm of biomedical ethics.

The Character of Soviet Ethical Theory

According to Soviet Marxist-Leninist theory, ethics is the study of morality. It provides the philosophical underpinning for communist morality, as well as spelling out the techniques for deciding what is right and wrong, moral and immoral. Ethical theory, therefore, plays a major theoretical role in helping decide the content of communist morality, as well as in justifying that content.

Marxist-Leninist ethical theory, however, does not have a long history. The first attempts to construct anything like a theory did not appear until after 1945,[36] and it was not until after 1956 that ethics became a separate discipline instead of being subsumed under historical materialism. In the 1960s the field blossomed under the impetus of the Twenty-Second Party Congress and the Third Party Programme, which promulgated the Moral Code for the Builders of Communism and put an emphasis on morality as a motive force in the development of communism. But obviously Soviet ethical theory cannot be as fully developed as theories that have been formulated over centuries. This may in part explain the fact that the extension of the theory to relatively new issues, such as some aspects of bioethics, has not been as swift or as extensive as in the United States. The revised Party Programme of 1986 dropped the Moral Code as such but kept some of its principles while softening others.

In the United States the Vietnam War and a host of protests and scandals in the late 1960s, as well as the development of new techniques in biology and medicine, led to the growth of a significant body of literature in applied ethics. It was the critical aspect of ethics that came to the fore, rather than concerns for the justification of conventional morality. Critical ethics seeks to evaluate conventional morality and to replace conventional norms when these can be shown to be deficient or lacking in adequate theoretical support.

This critical approach to ethics was absent for the most part in the Soviet Union until 1986. With the introduction of glasnost the beginnings of critical ethics have appeared. But critical philosophical moral theory is still controversial and is approached cautiously in the Soviet Union because it is politically sensitive. According to Marxist theory, morality is always used as a means of social control. In bourgeois countries the ruling class uses it to protect its vested interests and to keep the masses relatively docile; but this use is veiled and denied. In the Soviet Union the Party leadership openly uses morality as a means of social control. Marxist doctrine justifies this because the Party's stated aim is to promote the welfare of the people and to guide them toward achieving what they, and implicitly all mankind, desire.[37] In criticizing past Soviet practices and theory from an ethical point of view one always faces the danger of going too far. For this reason philosophers, doctors, sociologists, and other academics tend to take their cue as to what is permissible from the top political leadership.

In the Soviet Union socialist morality is to be learned and followed, not questioned, for to question the moral norms of the society is to question the leadership of the Party and the means it specifies for attaining communism. To do so openly or in print is to undermine the confidence of the people in the Party's leadership and in the validity of the moral norms. This does not mean that the norms do not develop and change as the society develops and changes. But it does mean that the changes come from above—as the changes in the abortion laws came from above. Until 1986, discussion of the morality of new practices was not a proper matter for public debate but rather for debate among those qualified to answer the questions that would arise.

The role of philosophers in the Soviet Union and the United States is thus significantly different. In the United States philosophers are usually professors at universities. They are free to publish as they choose on any topic they wish to examine. Neither the government nor the scientific community can prevent a philosopher from writing about the ethics of medical or biological practices or techniques, and philosophers frequently examine such procedures as soon as they are suggested or developed. The bioethical literature then influences the thinking of scientists, government officials, and ordinary citizens.

In the Soviet Union most philosophers must work according to a state-approved plan, even if they have had a hand in developing it. The instructions given by Party congresses are transmitted down through the philosophical journals and institutes. This certainly inhibits freedom of research. Similarly, since all the publication outlets are government funded, implicit censorship and self-censorship operate. Whereas American philosophers conduct their debates in journals, newspapers, and in public, Soviet philosophical debates frequently take place in relative privacy within institutes or professional societies. Books are usually circulated, discussed, and criticized before publication, and controversial ones that do see print are published in small numbers. The extent to which these practices will change under the new policy of openness is yet to be seen.

Lenin defended the position that freedom of the press does not mean freedom to publish falsehoods or to confuse the general populace. Stalin set the pattern of protecting the public and of allowing publication only of those doctrines that were agreed on by those competent to decide. The debates of the 1920s gave way to the dogmatism of the 1930s and 1940s. Until 1986 post-Stalinist practice never achieved the pre-Stalinist degree of freedom, and Lenin's doctrine was enforced more than in Lenin's own time. Under Gorbachev the popular press has carried more protests about unethical practices, violations of official norms, and complaints about many aspects of life, and more and more issues are surfacing. Nonetheless, the theoretical bioethical literature still remains sparse and lags behind popular protests.

As we have seen, the foundations of morality in the Soviet Union are completely social, and the norms derived are also social. Marxist-Leninist doctrine holds that what ultimately makes an action right is whether it leads to the development of communism. If it does, the action is morally justifiable; if it hinders communism, it is immoral. Since the Party guides the society on the road to communism, what it decides or commands is ipso facto moral. There is little emphasis on individual conscience in opposition to accepted social norms. According to the Marxist-Leninist doctrine of the moral-political unity of the Communist Party, the rules and actions of the Party carry with them moral as well as political authority.

Within this framework, of course, a good deal of debate is possible. What actions in fact lead to communism, what actions hinder its development, and what actions are neutral with respect to it are all issues to be decided. The Party does not pronounce on everything, and even on those practices on which it does pronounce, discussion and debate are necessary and appropriate. The emphasis on communism as the moral ideal of mankind provides a basis for a humanistic as well as an allegedly scientific approach to morality. It also provides the basis for championing collectivism over individualism in dealing with ethical issues.

Since 1986 a subtle but potentially significant shift in emphasis has occurred, following the lead taken by Gorbachev in some of his speeches. Instead of emphasizing the class nature of socialist morality, discussions now include more talk about the universal quality of socialist morality, and hence about the universality of moral norms.[38]

Marxist-Leninist ethics, which includes the teaching of communist morality, is a required course in all Soviet institutions of higher learning, in addition to the mandatory classes in Marxism-Leninism. Hence Soviet engineers, biologists, doctors, and so on have been exposed to instruction that has as its intent not only the general teaching of Marxism-Leninism but also its application to their specialty. At least in theory, they have sufficient training to claim that there is no need for philosophical specialists to interfere in their discipline.

The desire of scientists to be free from philosophical interference has been reinforced by the history of the relation of Marxist-Leninist philosophy to science. In the 1930s and 1940s Marxism-Leninism was interpreted in a way that impeded the development of several areas of science. Many Soviet philosophers opposed both relativity theory and quantum mechanics based on Western or "idealistic" interpretations of those theories. Similarly they considered cybernetics idealistic. It was only after 1948 that the importance of computers was recognized and cybernetics was permitted as a field of study. Not until 1954 were relativity theory and quantum mechanics freed from interference by philosophers, when the idealistic interpretations of these theories were separated from the scientific data and the theories themselves. Since then a division of labor has been acknowledged: scientists do science and philosophers provide dialectical materialistic interpretations of their results.

The situation with respect to the development of biology and Lysenkoism is related to the power of the Party more than to Marxist-Leninist ideology or philosophical interpretations. But the relatively recent freeing of biology and especially genetics from Party control forms the background for the present, somewhat uneasy relations between biology and philosophy. All of these factors have conditioned the development of biomedical ethics in the Soviet Union.

If we compare ethical debates and discussions in the United States with those in the Soviet Union, a number of specific differences become apparent. The United States prides itself on being a pluralistic society. This is true in the area of ethics as well. We not only have a diversity of opinion on a number of issues—abortion being a clear case. We also have no agreement on ethical theory itself. Three approaches to ethical theory dominate: a utilitarian approach, which judges actions right or wrong by virtue of the consequences they produce; a deontological approach, which emphasizes duties, rights, and justice; and a variety of religious approaches. This leads to a good deal of disagreement and controversy, much of which appears in print.

In contrast, the Soviet Union is, or at least claims to be, an ideologically homogeneous society. Even if religious or other views are present as remnants of capitalism, Marxism-Leninism is the only officially approved approach to ethical theory and is thus the dominant one represented in the professional and official literature. Since differences on bioethical issues in the Soviet Union are discussed within a single framework, the result, predictably, is much less diversity and discussion than we see in the United States.

Another difference between the two countries is that, since Soviet physicians practice primarily in government-run institutions—hospitals, clinics, polyclinics—there is comparatively little private practice and thus little room for individuality in carrying out professional tasks. Private practice means less state control in the United States, and fewer imposed ethical norms.

Free medical care for all citizens is a right in the Soviet Union. Hence until recently there was no need for discussion about the many issues that private health care raises in the United States. Soviet articles typically criticized the inequality of care in the United States and the absence of treatment for those who cannot afford it. But with the advent of glasnost, the inequalities of care under the Soviet system have also surfaced, along with proposals for restructuring the health care system and charging for services.[39]

The idea of ethicists making hospital rounds with physicians has not yet reached the Soviet Union, although it is a practice that is now fairly widespread in major hospitals in the United States. When medical ethics is taught in a U.S. medical school, the emphasis is not so much on handing down a set of rules to be followed as on teaching students how to approach ethical issues in medicine and how to decide what to do on their own. Soviet medical students are taught medical deontology, the rules that physicians are to obey, which become so inculcated that the students do not question them in practice.

Some issues in bioethics have arisen from advances in biology: the use of organ transplants, genetic engineering, life-sustaining machines, and so on. Although these innovations may be debated before they are actually used, discussion of their morality takes on urgency when such techniques become capable of being widely put into practice. Hence we can expect more debate and discussion in the more technologically advanced societies. In less developed countries these issues are submerged by more pressing problems. In general, biomedical techniques are more advanced in the United States than in the Soviet Union, making many issues more pressing in the former than in the latter. But time lags may prove to be relatively short. Even more important, the Soviet government exercises closer control over scientific investigation and application than does the U.S. government. In the United States privately financed laboratories and research institutes and teams are able to pursue their research without government support or approval. Thus the government tends to react to, rather than promote, medical and biological developments. For example, the nest of ethical issues introduced in the United States by the practice of surrogate motherhood, as seen in the case of Baby M, has not arisen in the Soviet Union, and may never arise if surrogacy is not permitted by state officials.

Finally, in the United States we tend to make policy by consensus or compromise after public debate. Bioethical issues are increasingly being referred to special committees or commissions composed of experts and lay representatives holding diverse views. Recommendations from such commissions are debated by legislatures and in many public forums and publications. Even after legislative or judicial decisions are reached, criticism and debate frequently continue, as the Supreme Court decision of *Roe v. Wade* illustrates. In the Soviet Union, debate is more circumscribed; those appointed to commissions tend to be specialists and government officials; and Party or government decrees tend not to be questioned once issued.

Medical Deontology

After the October Revolution the Hippocratic oath was repudiated as bourgeois and individualistic. Its statements about teaching, fees, abortion, and secrecy were all out of place for a Soviet doctor. Although some of the values and norms are the same for both Soviet and Western physicians, the Soviet medical code, like the Soviet moral code, is socially oriented. The good of society takes primacy over individual good, if the two conflict, and the Party, as the official guardian of society's development and of its morals, is the ultimate determiner of right and wrong.

Although Soviet ethical theory is teleological in that it determines the morality of an action by its social impact and by its effect on the development of communism, in its practical approach to moral issues it is deontological in that it emphasizes duty: the duty of the citizen, the duty of the professional, the duties attached to each of the roles people play in society. Thus, the Soviet version of medical ethics consists not in debates about the morality of medical practices but in the delineation of the duties and obligations of doctors, medical workers, and patients. Doctors are not self-regulating members of a profession but employees of the state.

As employees of the state, doctors, at least in theory, are to treat all equally, and are to perform their duties not as they see fit in private conscience but as they are told or as they decide collectively. Doctors practice medicine primarily in clinics and hospitals. They are civil servants whose job is to protect the health of the workers.[40] This role, and the organization of Soviet medicine, make the ethical issues somewhat different from those relating to medicine in the United States, except, perhaps, as practiced in the U.S. military or in state hospitals.

Since there is little private practice, doctors are relieved of the difficulties and ethical issues relating to malpractice insurance. Malpractice, however, is still an issue. For instance, three doctors at Maternity Clinic no. 12 claimed that "outright inattention or carelessness" on the part of medical employees was responsible for up to 20 percent of infant mortality.[41] But malpractice is handled by medical or state authorities and not usually by insurance companies or individual lawsuits. Since the state pays for medical care and subsidizes the cost of medicine, the doctor is similarly relieved of some of the ethical conflict-of-interest issues that arise in the United States.

Nonetheless, other issues result from the nature of the system. Soviet medicine tends to be bureaucratic, and some patients complain about the "pro forma approach, inattention, and indifference of medical personnel." As a result, in Dnepropetovsk Province, a court authorized a medical ethics board to enforce the moral principles of the medical profession, which include "an attentive, sensitive attitude toward patients."[42]

Although doctors do not take the Hippocratic oath, in 1971 the Supreme Soviet Presidium established an oath for Soviet doctors. It includes the promise "to work in good conscience wherever it is required by society; . . . to relate to the patient attentively and carefully; . . . to preserve and develop the noble traditions of Soviet medicine, to be guided in all my actions by the principles of communist morality, and to always bear in mind the high calling of a Soviet physician and my responsibility to the people and to the Soviet state."[43] In addition, as of 1975 a mandatory course in medical deontology has been added to the Soviet medical curriculum. The medical professor E. Gabrielian, commenting on the need for such a course, indicated that such questions had received little attention,[44] although the attention they have received subsequently is prescriptive rather than critical. One Western commentator has observed: "Rather than practice medicine according to his own judgment, the average clinician is expected to conform to the received wisdom which—predictably enough—reflects the consensus view of influential senior specialists based on teaching hospitals and research institutes."[45]

The Soviet approach to medical ethics is in some ways very conservative, despite the fact that it is built on the new communist morality. Thus, although professional ethics is subordinated to general ethics and the good of society, the old paternalistic tradition of lying to a patient for the patient's own good, now vigorously challenged in the United States, is still stoutly defended and enforced in the Soviet Union. Dying patients are not told they are dying, since to do so, it is claimed, would serve only to depress them and cause them to lose hope.[46] And cancer patients are not told they have cancer, even when operated on; in such cases they are told the operation is for some other reason.[47] Other aspects of medical ethics, such as subordinating the patient's right to confidentiality to the interests of the state, are spelled out in the 1969 legislation on health care.[48] These are not issues for professional discussion but for implementation of state policy by doctors. Since morality is socially enforced and promulgated by the Party, there is no room for individual dissent. The emphasis in medical ethics, as in general ethics, is on conforming to socially promulgated moral norms rather than adhering to the dictates of one's conscience. For conscience is to be developed, corrected, and guided, and if the dictates of one's conscience contradict the social norms, individual conscience must give way.

Common complaints concerning violations of medical deontology include the falsification of records to cover up mistakes or malpractice;[49] the frequent necessity of bribes to doctors or medical practitioners to obtain service, operations, or medicines;[50] the paying of bribes for doctors' certificates excusing one from work;[51] and insensitivity of medical workers to patients' needs.[52] None of these complaints raises ethical issues requiring discussion or debate.

The allocation of scarce medical resources, such as dialysis machines, poses a problem for any nation that has the capacity to produce or buy such machines but cannot make them available to all who need them. The allocation problem is one of social ethics and practice, not an ethical medical problem for an individual doctor. In the Soviet Union the issue is not bound up with ability to pay, since medical care is provided for all. Nonetheless, there are complaints that the medical system is a two-tiered one, with special, well-equipped clinics that serve Party members, the military, members of the security police, and other members of the elite.[53] No ethical defense, much less critical discussion, of the two-tiered system has appeared in the Soviet professional literature, presumably because the system does not officially exist. Moreover, the problems of allocating relatively sophisticated medical treatment is not a pressing problem in a society in which, as Y. I. Chazov, minister of public health, in an interview in *Sovetskaia Rossiia,* acknowledges, shortages of the basics, including tweezers and disposable syringes, prevail. Artificial hearts and dialysis machines are purchased from abroad; the artificial heart developed by the All-Union Medical Equipment Research and Testing Institute has been approved by the Ministry of Public Health only for use on animals. Chazov also calls for attention to quality rather than quantity, noting that although the USSR leads the world in the number of hospital beds, "only 35 percent of rural district hospitals have hot running water, 27 percent have no sewerage system, and 17 percent have no running water. What good," he concludes, "are such hospitals for modern medical care?"[54] We might expect ethical discussion of either the need for advanced medical technology or the criteria for allocating such technology if it is not universally available. But thus far conditions have not prompted such discussion.

The reaction of the people in the Soviet Union at all levels to AIDS has been noteworthy. The level of popular knowledge about AIDS as late as 1987 can be gauged from a letter Dr. V. I. Pokrovskii, president of the USSR Academy of Medicine and director of the Central Epidemiology Research Institute, received from sixteen graduates of a medical institute, who wrote: "We intend to do everything in our power to impede the search for ways to combat that noble epidemic. We are convinced that within a short time AIDS will destroy all drug addicts, homosexuals, and prostitutes."[55] Reactions printed in *Komsomol'skaia pravda* to an article on AIDS ranged from support for the position advocated by the graduates, to denial that there was a serious threat, to calls for mandatory testing and vice squads.[56] In September 1987, in an attempt to provide information, an All-Union Scientific Research Institute concerned with public health mailed out 10 million copies of a leaflet, "This Is What You Should Know about AIDS."[57]

Pokrovskii, in an interview published in *Izvestiia*, stated that as of June 1987 in the Soviet Union there were 54 people infected with AIDS; but authorities did not know how many people were in the high-risk categories of prostitutes, drug addicts, and homosexuals. He said, "In our country, all those activities are punishable crimes . . . Whatever was declared illegal was simultaneously declared not to exist in the real world—at least where broad public opinion and the press are concerned."[58] By January 1989 the number of known carriers had risen to 146, including 27 children in the city of Elista in the Kalmyk Autonomous Republic.[59] How the Soviet Union will deal with the problem if it grows, given official reticence on sex, the shortage and poor quality of condoms, the lack of hypodermic needles, and so on, will be interesting to see. We can expect stronger condemnation, and perhaps control, of homosexuality and prostitution, both of which have been discussed in the press.[60]

In the United States AIDS is widely discussed, and the rights of those suffering from AIDS, as well as the rights of others who may be exposed to it, are debated in the professional, philosophical, and popular literature. In the Soviet Union there had been no discussion of such rights, and very little discussion of related ethical issues, until 23 February 1989, when David Kugultinov, the Kalmyk People's Poet, commented on the treatment by the public of the 27 children with AIDS. (They had not been allowed to enroll at preschool.)[61] He quotes their mothers as saying: "The U.S. recently enacted a law establishing tougher punishments for discrimination against AIDS patients. Maybe we should also enact a law of that sort to protect the right to life of these unfortunate people, especially the right of children and mothers to at least the life that fate has in store for them."[62] That suggestion has not been further discussed or implemented. Since public good takes precedence over individual rights, we can probably expect little defense of the rights of those with AIDS, at least not until the disease affects the heterosexual population more generally. The notion of privacy in the Soviet Union is also different from that defended in the United States. Whatever policy the government finally adopts, given past practice, it is likely to be accepted and approved by the general population.

On 26 August 1987, the Presidium of the USSR Supreme Soviet decreed that persons suspected of being infected with AIDS can be obliged to undergo testing, that anyone "knowingly putting another person in danger of infection with the AIDS disease is punishable by deprivation of freedom for a term of up to five years," and that "infection of another person with the AIDS disease by a person who knows that he or she is a carrier of the disease is punishable by deprivation of freedom for a term of up to eight years."[63] By January 1989 at least one prostitute had been sentenced under the law to four years in prison.[64] Although this law was considered unreasonable by many people in the West when the decree was first announced, by June 1988 the states of Washington, Indiana, Idaho, and Georgia had made it a felony for an AIDS carrier knowingly to transmit the disease, and in Washington persons who do so face sentences of up to ten years in prison. In Indiana the penalty for knowingly donating AIDS-contaminated blood is eight years in prison, or up to fifty years if someone is infected by the blood.[65]

Death and Dying

Problems related to death and dying are central to Western discussions of biomedical ethics. Marxism-Leninism, however, had long ignored questions about the meaning of life and death and criticized existentialism for its emphasis on themes of death and crisis. The need for some answers in this sphere was recognized by Adam Schaff in Poland in 1961, but attention to these issues is still rare in the Soviet philosophical literature. Discussions of these questions, as of many others in the area of biomedical ethics, appear only after the problems have been raised and debated in the West, and only after the pressure of practice has forced them to emerge. In the Soviet Union the question of the meaning of death and the proper attitude toward it did not enter the philosophical literature until after the practical problem of determining the moment of death for medical purposes had arisen.

The Soviet discussion has followed lines similar to those pursued in other countries. The debates have taken place in certain professional philosophical, medical, and legal journals. N. S. Malein, a doctor of jurisprudence, summarized an early discussion in his article "The Right to

Medical Experimentation." While N. Amosov argued for transplants from those who have suffered death of the cerebral cortex, M. D. Shargorodskii argued that life should be respected right up to the last moment.[66] In December 1982 a conference of lawyers, philosophers, and members of the Academy of Medical Sciences discussed this issue as well as others having to do with ethical and ethico-legal questions of medicine. Although most doctors and lawyers argued in favor of the brain-death definition, others maintained that irreversible brain death was a necessary but not sufficient condition of death.[67] In various debates some insisted on the ordinary criteria for determining death: cessation of heartbeat and breathing. While most others opted for brain death as the signal that human life had ceased, they were divided as to whether to adopt the definition of cortical brain death (the absence of consciousness) or that of complete brain death (the absence of heart function and breathing as well). The issue of reanimation and questions concerning the accuracy of tests used to determine brain death and guarantee that death is irreversible were also raised.[68]

From a Western point of view, what is curious about the debates is not the arguments presented, since they are the standard ones, but the manner of resolution. With no indication of how or why the decision was taken, Academician N. Blokhin, president of the Academy of Sciences, stated in November 1986: "We have adopted a decision on determining 'brain death' and have devised the criteria for defining it."[69] In December the presidium of the Academy of Medicine announced the criteria for brain death and approved a set of appropriate regulations.[70]

Two side issues emerged during the course of the debates and prior to the resolution of the brain-death issue. First, there was a limited discussion of death and immortality. In 1981 Academician V. A. Negovskii of the Academy of Science criticized an "idealistic conception of clinical death," which implied that some sort of life after death could be inferred from the testimony of those who have been reanimated after clinical death. He concluded that Marxist philosophy has resolved the question of immortality: "After death man remains alive in the results of his creation . . . Those who give themselves to humanity do not die. Science and materialist philosophy know no other immortality."[71] That position is consistent with Marxism-Leninism. Whether it is a satisfactory answer to the question of the meaning of life is debatable. But that debate has not emerged in the Soviet literature and is unlikely to do so.

Second, in 1983 I. T. Frolov introduced the question of the social aspect of the human being, thus implicitly opening up a whole range of issues that engage those in the West concerned with the issue of death. In particular, he raised topics such as the artificial prolongation of life and related problems of gerontology. In 1986 he argued,[72] citing V. A. Negovskii,[73] that human life is not just biological but social. Hence, when one is no longer capable of consciousness, truly human life is no longer possible. He did not press the question of whether those with advanced Alzheimer's disease, for example, are capable of social and therefore human life. Nor did he pursue similar issues about the implications of his position and how far the argument about the social nature of human life goes. He did note, however, that too few Soviet philosophers and ethicists take part in discussions about death. Among the unresolved problems, he said, is the issue of whether it is justifiable to protect the life of the aged who have lost their intellectual faculties. In a society of relative scarcity, he argued, it is more important to care for newborns than to prolong the life of the elderly.[74] That judgment is clearly in line with the notion of potential contributions to society. But left unanswered, and still to be debated, is the difficult ethical question of where to draw the line in prolonging life, in providing operations and medication for the aged and otherwise caring for them. The old cannot simply be written off, even when they can no longer contribute to society. In dealing with this question, how much weight should we place on past contributions to and engagement in society? How much respect does Marxist humanism demand for such people? And how many resources are to be expended in extending their lives? These issues are only slowly emerging in the Soviet literature, despite Frolov's writing in the area. An article in *Moskovskie novosti* examined the USSR's aging population and the supply of nursing homes, which had only 327,000 places in 1986.[75] Other topics concerning the elderly still lie submerged.

Similarly, Soviet philosophers have not entered the debate about the difference between killing people and letting them die. In 1982 A. D. Naletova reviewed the American bioethical literature on paternalism and touched on the literature of euthanasia, but without presenting any developed position herself.[76] But such reviews of debates elsewhere in the world are among the standard means of introducing discussions in Soviet philosophical circles. Marxist-Leninist moral philosophers might find palatable the position that no extreme measures are required to prolong life, even though thus far they have not articulated or defended that position, in that it is consistent with both Marxist-Leninist views of the relation of the individual to society and with the limited resources the Soviet Union can be expected to allocate to public health in the foreseeable future.

Soviet moral philosophers have not directly addressed, let alone spelled out, the principles that should apply in debating these difficult biomedical ethical questions in a socialist society, nor have they provided socialist society with much in the way of help in resolving them. The typical article ends with the statement that only socialism can solve these questions in a humanistic way and that the solutions will be amicably agreed on, a vision that contrasts with the fiercely opposing views with which the West struggles. At least in the West the problems are being raised and faced; the arguments presented by one side are challenged by others; and the process is open. This is not the way moral issues have, up until now, been approached or resolved in Soviet society.

Transplants and Experimentation

The problems of defining death and of performing transplants are connected, for if essential organs such as the heart are to be transplanted, they must be taken only from those who are legally dead, yet they must be fresh enough to survive being transplanted into another human body.

The Soviet Union performs many fewer transplants than other developed countries, but the inhibitions seem to be technical rather than ethical. According to Y. I. Chazov: "Today, kidney transplants could save the lives of thousands of patients in whom this very important organ has failed. But last year [1986] we performed fewer than four hundred such operations. In other developed countries, they do ten times that number. Our country has several research institutes looking for new antibiotics, but our public health system lacks the most effective drugs."[77] He complained that, as I mentioned earlier, the artificial heart developed by the All-Union Medical Equipment Research and Testing Institute has been approved only for use on animals. In March of 1985 O. Frantsen, writing in *Pravda,* claimed that using an artificial heart to maintain a patient's life until a suitable heart can be found for transplant is appropriate, and boasted that the Soviets had had success with using artificial hearts in calves, but that artificial heart implants as practiced in the United States were "not humane."[78] The ethical point he made was that experimentation with artificial hearts was not justified if it could prolong life only for months, not years—a plausible position. Chazov's more candid remarks since glasnost indicate that earlier moral declarations may have stemmed from a lack of technical ability, rationalized in ethical terms.

The first heart transplant in the Soviet Union took place in 1968, but the patient died after thirty-three hours. The next officially approved heart transplant was performed eighteen years later, on 27 October 1986. Unapproved heart transplants in the intervening years were not successful. On 11 March 1988 an article in *Pravda* announced that there were four people in the Soviet Union living after successful heart transplants. It also announced plans for an All-Union Transplant Research and Production Association to coordinate all the clinics involved in organ transplants in the Soviet Union.[79]

In the USSR the decision to transplant organs—for example, from accident victims—is considered a medical decision and may be made without the prior written consent of the victim and without informing, much less seeking permission from, the next of kin. Nor is permission required to perform an autopsy. The view that the medical profession has the authority to decide about the transplanting of organs and tissues holds sway. M. Kuzin, director of the Academy of Medicine's A. V. Vishnevskii Institute of Surgery, advises against discussion of transplant operations in the popular press, asserting the traditional view that "medicine is not a field that needs openness"—a premise more recently disputed by Leonid Zagalskii and Valerii Sharov, among others.[80]

Concerning not only transplants but experimentation in general, N.S. Malein acknowledges that the problem of medical experimentation has not received enough attention in the Soviet literature. He argues that in medical experimentation patients have the right to know of the risks, the potential side effects, the pain involved, and so on. He distinguishes between experimentation on the ill to improve their health and experimentation on the healthy for strictly research purposes. In the first type, he argues, patients have the right to complete knowledge, but physicians have no obligation to make full disclosure unless directly asked. In the second type, the burden is on the physician to make known all risks and other pertinent information.[81]

The right of patients to be told only if they ask is another instance of the kind of paternalism that is used to justify not informing a cancer patient that he or she has cancer. A patient is told only what the doctor thinks is appropriate and necessary. In the United States, by contrast, the patient's right to know is affirmed in the conditions of informed consent that must be met prior to any operation. Here the doctor is required, at least in theory, to make full disclosure of all risks and possible side effects.

Malein further argues for a prohibition on using prisoners as patients for experiments, even if they agree to serve. This is an acknowledgment that their consent, even if given, cannot be considered completely free. His is an interesting and uncharacteristic Soviet defense of individual freedom and individual rights, and it shows that there is room in Marxist-Leninist ethical theory for increased discussion of these subjects. There is no evidence, however, that his position has been adopted as official policy.

Frolov argues that the need to define the limits of experimentation coincides with the need to establish global criteria—that is, those contained in the Helsinki Declaration.[82] He wishes Soviet representatives to have a voice in establishing such criteria but seems to see international regulations as an acceptable alternative to unilaterally established political decisions.

The use of fetuses for experimental purposes has not been discussed in the Soviet literature, although there is some indication that fetuses are quietly being used in this way. The notion that fetuses are not persons because personhood, according to Marxist-Leninist theory, comes only with acceptance into the social community would seem to permit free experimentation on aborted fetuses. The fact that such experimentation is done quietly, however, seems to suggest that popular reaction might be negative. The attitude seems to be that the decision is one to be made by the competent experimenter and does not require popular acceptance.

The use of fetuses for organ transplants is a related topic that has also not been discussed in the literature. The issue of whether a woman should, through artificial insemination or other means, produce a fetus whose organs can be transplanted safely because of natural compatibility is unlikely to arise in fact—at least for some time—given the relatively small number of transplants performed in the Soviet Union and the need for permission from the appropriate ministry. It is easier to envisage this practice developing in a free-enterprise system than in a state-controlled system such as in the Soviet Union. Objections to the practice from the point of view of Soviet morality could not be based on the rights or personhood of the fetus; the practice would presumably be opposed on the grounds of adverse consequences for the woman who bore the fetus for this purpose and for society as a whole.

Soviet philosophers rarely speculate on topics of this type, which they consider abstract and divorced from the actual needs of society. If they follow past practice, as these questions become more and more discussed in the West, they will learn from the discussions; and if international standards are adopted, they will tend to accept them.

Nonetheless, new reproductive techniques are receiving some attention in the Soviet Union. On 13 May 1987 Public Directive no. 669 of the Ministry of Public Health expanded the permitted experimental use of artificial insemination in Moscow, Kharkov, and Leningrad. Any healthy man between ages twenty and forty can become a donor. The obligations of the husband and wife are spelled out, including signing a statement that reads: "We pledge not to file complaints against the physician in the event that the procedure does not take or that the child suffers from birth defects. We agree not to ascertain the identity of the donor. We pledge to keep secret how the child was conceived. We have been warned that we are responsible, before society and the child, for disclosure of that secret."[83] The article goes on to state that not all of the legal questions have been resolved, such as who raises the child in case of divorce (both parties promise to support the child), and other similar problems. The secrecy to which they are pledged does not preclude disclosure in later legal actions involving the couple and the child.

The use of artificial insemination and of "semen banks" for the eugenic purpose of increasing the number of offspring with certain positive characteristics is not presently defended or practiced. The possibility of future developments along these lines, however, cannot be ruled out.[84] Other issues, such as surrogate motherhood, have not yet arisen and are not yet being discussed in the legal or philosophical literature.

Genetic Engineering

One area on which a reasonably substantial body of literature has developed is genetic engineering. In the Soviet Union genetics suffered from the legacy of T. D. Lysenko and his supporters. According to Maksim Karpinskii, "only genetic engineering and biotechnology are being energetically developed in the country today," and even there the genes being worked on are imported from abroad.[85] At the 1986 congress on molecular plant genetics only three representatives of the USSR were present, as opposed to 1,800 from the United States; Karpinsky claims that ratio is accurate, with "barely fifty geneticists . . . who are doing research at a contemporary level."[86] A. A. Sozinov, director of the N. I. Vavilov Institute of General Genetics of the Academy of Sciences, writing in *Kommunist,* cites a shortage of skilled specialists and describes the situation as "alarming."[87]

An early related debate between V. Efroimson and B. Astaurov appeared in *Novy mir* in 1971.[88] Efroimson argued for the presence of altruism through natural selection. Astaurov, while not denying this, argued for the social factor in the development of ethics. The debate set the scene for the discussion of eugenics.

Loren Graham helpfully divides the Soviet discussion of genetics into two phases: a speculative phase, before the 1975 Asilomar Conference, and the practical period, following that conference.[89] In a series of meetings of and papers by philosophers and geneticists during the first period, two positions rose to the fore. A. A. Neifakh defended not only genetic engineering but cloning as well as a means of preserving the best of human genotypes. Efroimson argued in favor of genetic engineering but not of cloning. The major critic of genetic engineering for eugenic purposes was N. P. Dubinin, who argued that the moral consequences were totally unacceptable. In a series of articles he developed an extended argument based on the dual aspect of man—the biological, which makes possible the development of his ability to reason, and the social, which is transmitted not through genes but through education. Dubinin argued against eugenic utopias but in favor of work on medical genetics with the aim of promoting the health of people with adverse heredity.[90]

There now seems to be general agreement that eugenics, which aims at improving the general population through selection of certain genetic traits, is unethical as well as scientifically unsound at the present time.[91] The late philosopher A. F. Shishkin used Marx's claim against Hegel that it is the masses and not great men who make history to argue against any form of eugenics.[92] Although that argument is ingenious, other, more standard arguments have carried the day.

The Western fear, presaged in Aldous Huxley's *Brave New World,* that the Soviet Union might attempt either to create the new Soviet man through genetic manipulation or to breed a subservient subhuman population has no basis in present Soviet theory. Likewise, the Soviet fear that U.S. scientists, motivated by greed and lacking concern for the common good, might threaten human annihilation through unrestrained genetic experimentation should have been allayed by the self-imposed moratorium they have adopted. After the Asilomar Conference, at which were discussed the dangers of genetic engineering and the conditions under which it could be safely practiced and the self-imposed moratorium on such research could be lifted, the Soviet practitioners and theoreticians accepted the international guidelines for research that were developed.[93] In general, Soviet geneticists have been willing to adopt international guidelines; perhaps they hope thereby to keep their research free from political, philosophical, and popular intervention and control. Their fear, as we have seen, has historical roots.

Although the role of Soviet philosophers since the 1954 struggle with the physicists has been to interpret the results of science rather than to enter directly into what Marxist-Leninist philosophy allows or prohibits in the realm of science, with genetic engineering the lines are once more becoming somewhat blurred. For, as Frolov has argued, the relation between science and morality in this area is neither clear nor distinct. Yet even Frolov, who is the leading and the most prolific philosopher working in this area, is reluctant to make normative statements on genetic engineering, arguing that philosophers pursuing the ethics of science should not prescribe norms, but rather should clearly pose the ethical problems, throw light on the roots of the problem, and present norms that might be used to guide researchers.[94]

There is also fear on the part of some geneticists that emphasis on the social factors of development may go too far. Thus A. A. Sozinov argues against any taboo on studying the genetic basis of mental characteristics.[95] Perhaps partly to allay such fears, in 1986 Frolov reemphasized that the Marxist conception of the social essence of man is built on the dialectical interaction of the social and biological.[96]

Perhaps Frolov's most important contribution to the area is his emphasis on the global character of many bioethical issues. Such matters as the transport of human tissues across national borders, the control of communicable diseases, and the like can be adequately solved only through international cooperation. Frolov has reviewed and reported on both the Western literature and the comparatively sparse Soviet literature on biomedical ethics, and he has raised the right questions. But he and his colleagues provide few solutions of help to those in the West. Thus he says that under socialism "alternatives that are often disturbing to scientists and dangerous for mankind are avoided as a matter of principle, since a real and many-sided social control is provided over research activities, including activities in the field of genetic research."[97] He similarly quotes A. A. Baev: "We are convinced that common sense and good will will prevail in this area, at least in our own socialist country."[98]

Frolov convincingly argues that all must struggle against the possible utilization of biogenetic research for military purposes and for a prohibition against new biological weapons.[99] Yet he has no more control over Soviet policies than American philosophers have over American development of such weapons—and both the United States and the Soviet Union have indeed devised such weapons.

Frolov notes that the debates among Soviets are in many ways similar to those that take place in the West. The difference, he asserts, is that in the Soviet Union state control is real, the objectives and means used are profoundly humanistic, and the principles of socialism concern the full development of man. In capitalist society freedom of research and social responsibility are separate and conflict.[100] Even if his observations are correct as far as they go, they do not go far enough.

Despite areas of similarity and common problems, there are significant differences between biomedical ethics in the Soviet Union and the United States. One of these stems from the dissimilar philosophies. The rights and freedoms of the individual are considered paramount for many in the West; the dilemmas that ensue stem from opposing views and conflicting rights. The Soviet position claims to preclude this situation, and to a certain extent succeeds. Adopting the view that social norms should dominate, and that individuals are taken care of by taking care of society, keeps many disputes from even arising. The supposed moral-political unity of the leaders of the state and Party also makes many moral decisions into political ones, and vice versa. When one adds that disputes with state decisions have not been allowed free expression, and that many disagreements have been kept out of the press, the relative paucity of debate is understandable. This situation may change as glasnost gains more momentum. Nonetheless, many in the West would argue that the price one pays for the present degree of accord in the Soviet Union is the relinquishing of moral autonomy and of critical ethical thinking—a price that they consider too high.

If, in the field of bioethics, abortion and psychiatry are two areas of obvious difference between the Soviet Union and the United States, genetic engineering and transplants and the means of determining the point of death are superficially alike. Yet the problems they pose are seen differently in the two countries, even when the solutions arrived at are similar.

In general, Soviet discussions of such issues postdate discussions in the West. It is difficult to say whether these discussions would have arisen even if the topics had not already been raised elsewhere. In none of the areas of bioethics has Soviet debate as yet been initiated on topics not previously explored elsewhere. Soviet philosophers have also lagged behind other elements of Soviet society in articulating and addressing problems of biomedical ethics. I. T. Frolov is the outstanding exception. We can expect that with his example and encouragement, the needs of the society and the atmosphere of glasnost will bring about more philosophical activity in this area.

Despite the presence of a certain amount of discussion, it is not clear how political and legal decisions are made in the Soviet Union. We can assume that the debates and reports of them play some role, but there is no way to know exactly what that role is. The practices adopted are either spelled out in law or established by those in a position to carry them out. Justifications are not given, nor are decisions challenged in print. Frequently the ex post facto task of justification falls to philosophers. This is very different from the Western approach to similar issues.

The assertion that the problems that arise in Western countries will not affect the Soviet Union because whatever promotes the social good will prevail does not answer the question of how the social good is determined in specific cases, and it hides rather than clarifies a host of difficulties, some of which may be discussed in the future. This belief is based on the dual assumption that the social good is always apparent and always properly overrides individual good. There is as yet little in the way of a real grappling with issues such as gives rise to a great deal of the discussion in the United States. From an outside point of view, Soviet moral philosophers too docilely accept whatever is decided by the authorities, as if being so decided guarantees that the decision is automatically for the general good. One result is that the debates in the Soviet Union are not very informative for people elsewhere.

What those in the West can borrow from the Soviet approach is the concern for social consequences. Discussions there may serve as an antidote to what can sometimes be an overemphasis on individual rights in the West. But in the West no simple overriding of individual good by social good will be automatically accepted.

To the extent that Soviet solutions to similar problems differ from those we adopt, we might also learn from them, just as they might learn from us. Socialized medicine has its virtues, as does privatized medicine. As medical costs rise in the United States, we can ponder whether implementation of the right to medical care is compatible with totally private medicine.

The freedom that individuals enjoy in the West raises a number of issues that do not arise in the Soviet Union. Soviet ethical life is in some ways less complicated. Although the diversity and degree of freedom characteristic of the West are absent, they are not necessarily felt as an absence by those who are committed to the view that individual freedom or desires should give way to the social good. This is true in the realm of bioethics as well as in other areas of life. Soviet acceptance of pain and death might give Americans pause as they seek to eliminate the one and postpone the other at ever-increasing social and economic costs.

Although differences in approach make for different bioethical issues and, frequently, for different solutions, where we face common problems, we must learn to solve them together. Differences of approach are not incompatible with cooperation. But if mutually acceptable global practices can be developed only through mutual understanding, it is imperative that we learn to see clearly our differences as well as our similarities.

· 8 ·

Fact, Value, and Science

Bruce J. Allyn

There is an extensive literature in the West on philosophical questions concerning the relation of fact and value and issues of "value-free" science. These questions arise in debates over important social and political issues. In recent decades, those concerned about research and development in genetic engineering or nuclear physics, for example, have argued that science is not value free and should be subject to ethical review boards or public referenda. In addition to the individual researcher's personal and cultural values, there are external influences: funding for scientific activity is a reflection of governmental, societal, and corporate goals.[1] By contrast, those who wish to maintain islands of scientific activity outside social or political control have traditionally argued that they deal with "facts," that scientific knowledge and technology are neutral; it is the policy makers who determine whether technology is used for good or bad ends. While a strict positivistic fact-value dichotomy is intellectually untenable, the general belief that science and values are separate realms has had a strong influence on the historical development of Western science and society.[2]

Throughout Soviet history too there have been debates about the relationship of fact, value, and science among both social and natural scientists. The Soviet debate has been directly tied to central issues of Marxist-Leninist ideology, and has had significant political implications. In sharp contrast to the classic Weberian position of the value neutrality of both the natural and social sciences, Soviet ideology has championed the radical Marxian view that scientific cognition of facts *directly* produces value judgments. Lenin's principle of partisanship (*partiinost'*) formalized the view that no philosophical theory can be neutral in the class struggle. Along with the principle of the unity of theory and practice, this doctrine provided the philosophical basis in the early days of Soviet power for Party control of both social and natural science research.

These radical Marxist-Leninist doctrines were ideological tools particularly well suited to Lenin's revolutionary aims and Stalin's political goals of rapid mobilization and totalitarian control. The Marxist-Leninist view that the "correct" facts were those that facilitated the "progressive" movement of history opened the door to the ruinous influence of Lysenkoism. The political regime sought to determine the correctness of scientific research not only in genetics but also in physics, cosmology, chemistry, and to some extent even mathematics.

In the post-Stalin era those scientists who had learned the bitter lesson of Lysenkoism found it in their interest to assert that facts and values belong to different realms. This chapter will argue that in the post-Stalin era several new doctrinal formulations relating to morality and the role of science in socialist society have significantly undercut the theoretical basis of the Marxian linkage of fact with value. In the 1960s some Soviet philosophers began to make an essential distinction between fact and value, between what is (the subject of science) and what ought to be (the subject of ethics). In a few cases, Soviet scientists have tried to use a positivistic assertion of the logical incommensurability of facts with values to make a case for reduced political control over their activities.

Party efforts to determine the correctness of science in Marxist-Leninist terms diminished in the post-Stalin era. But the Party sought to maintain pervasive control of scientific research in an attempt to advance economic goals. Party leaders have had to rely increasingly on the advice of scientists to deal with the growing complexity of Soviet society. During the deradicalized, immobile period of Brezhnev's "developed" socialism, the leadership began to support professional ethics in hopes of increasing labor productivity and spurring intensive economic growth. The rise of professional ethics provided more politically neutral standards of behavior, evoking concern that the Marxian linkage of fact and value was being further pulled apart. In the 1970s Soviet commentators, expressing the fear that professional ethics in science could lead to socially irresponsible behavior, called for the development of a special discipline dealing with the ethics of science. The tremendous social impact of new technologies such as nuclear weapons and biogenetics has led—as in the West—to a new, not particularly Marxian emphasis on the linkage between science and values.

With the Gorbachev era the Soviet Union has entered a new period of mobilization and change. Scientific advances remain crucial to the future of Soviet power. But there has been an increased stress on morality and "universal" values as a way of achieving sociopolitical renewal and scientific progress. In Gorbachev's effort to restructure Soviet society, morality is increasingly defined in terms of the humanist elements of Marxism, as opposed to the determinist elements emphasized during Stalin's mobilization phase. Stalin used Marxist-Leninist doctrine to justify anything that advanced Soviet power in its progressive historical movement. Gorbachev's stress on Marxian humanism includes an increased emphasis on acknowledging personal dignity and guaranteeing the rights of the individual. At the June 1988 Party conference, Gorbachev made statements suggesting an increased commitment to protecting scientists from incompetent Party control and petty tutelage. He denounced the past "domination of command-administrative methods of management, when science often had imposed on it areas of research which did not stem from the logic of its own development." He further noted that expert opinion, "if it diverged from the interests of departments, was ignored and sometimes even persecuted."[3] Thus Gorbachev explicitly acknowledged that science has its own logic of development, which may diverge from the imposed directions of Party managers. The approach of the Gorbachev leadership seems implicitly to assert that facts and values may inhabit very different realms. This appears to be in line with the new effort by the Soviet leadership to stimulate further scientific advances by giving more leeway to Soviet scientists. At the same time, the emphasis on humanism and science, the linkage of science and values, has come increasingly to the fore.

Revolution, Mobilization, and the Marxian Moral Imperative

In the 1918 debate over whether to accept the extremely onerous terms of the Brest-Litovsk treaty with Germany, Lenin met with serious opposition. The left-wing communist Alexandra Kollantai asserted that it would be "traitorous" for the victorious proletarian state to sign such a treaty with the imperialists (giving up the Ukrainian, Polish, and Baltic regions of Soviet Russia). Nikolai Bukharin then argued that a Marxist does not make arguments for or against a given policy on the basis of ethical or moral grounds. He agreed that it was a bad treaty and the

Soviets should not sign it, but for different reasons: he argued that the treaty was a mistake because, as the evidence of objective scientific analysis indicated, a revolution was coming in Germany.[4]

As this story illustrates, Bukharin understood Marxism to accord no independent force or significance to moral argument. Scientific analysis of the facts would tell one how to act. In *State and Revolution* Lenin argues that "Marx treated the question of communism in the same way as a naturalist would treat the development of, say, a new biological variety, once he knew that it had originated in such and such a way and was changing in such and such a definite direction."[5]

By the standards of Western positivist philosophy Marx, Lenin, and Bukharin had all committed the naturalist fallacy of deriving value judgments directly from facts. But in Marxism the answer to the question of what should be done is given in terms of a direct "vision" of what is happening in the world. As Marx wrote in *The Poverty of Philosophy:* "In the measure that history moves forward, and with it the struggle of the proletariat assumes clearer outlines, they no longer need to seek science in their minds; they have only to take note of what is happening before their eyes and become its mouthpiece."[6] Marx argued that factual conditions would necessarily lead to the morally desirable end of communism.[7] Thus, participation in the proletarian class struggle needs no special, independent ethical justification. Moreover, Marx and Lenin disparaged morality, law, and religion as "bourgeois preferences" concealing class interests.

Numerous Western studies have endeavored to explain what it is about Marxism that leads it to reduce moral inquiry to scientific discovery. Charles Taylor argues that facts and values are intertwined in Marxism because certain undeniable moral conclusions are built into Marx's account of man's needs.[8] In Robert Tucker's view, Marx encountered no problem of conduct in the sense in which moral philosophers understand it because his mind was possessed by a mythic vision of what was "really" happening in the world and the imperative to participate.[9] Susan M. Easton argues that Marx transcends the fact-value distinction to embrace neither a scientistic approach nor a moral theory; rather, Marx gives a sociological account of morality illustrating that the descriptive and evaluative cannot be separated.[10]

In philosophical terms, Marxism is a kind of naturalism. As W. K. Frankena notes, naturalist approaches are based on a peculiar kind of vision (which critics call a blindness).[11] Marxism is also teleological, emphasizing the historical collective self-realization of the human essence. These general traits have been carried forward into Soviet ethical theory.[12]

These characteristics of Marxism make it an ideology particularly useful to any political party seeking to transform a society in a fundamental way. The vision of communism as collective human self-realization is portrayed not as a utopia but as an immanent objective, one that is historically necessary. For the Bolsheviks it was useful to argue that science had proven the historical necessity of communism; it increased the chance that the vision would become a self-fulfilling prophecy. To the degree that Lenin, Trotsky, and others inspired people with such a belief, and those people acted accordingly, it would help make the "scientific" projections come true. Moreover, the vague nature of the future goal allowed them to take liberties that could not be justified otherwise. The nature of Marxism-Leninism permitted the leadership to call for immediate sacrifices and to tolerate injustices in the present and near future.

Marxism-Leninism as an ideology was also particularly well suited to a regime that sought to harness science to effect sociopolitical change. The reduction of moral inquiry to scientific discourse meant there was no distinct boundary between fact and value. Empirical facts were highly subject to evaluation on the basis of political correctness. The goal of scientific activity, as commonly understood in the West, is to produce explanations of the natural world. A "good" explanation is one that combines empirical accuracy on the one hand with simplicity and breadth on the other. But, to paraphrase Marx's eleventh thesis on Feuerbach, the goal of Marxist scientific activity is not merely to explain the world but to change it. Hence the principle of the unity of theory and practice. In Soviet usage in the 1920s and 1930s this principle was reduced primarily to the idea that all theory should have practical applications.

The unity of theory and practice was in Soviet ideology directly linked to the principle of practice as the criterion of the truth of a theory. In Lenin's words, the practical "success" of a theory "proves" its truth (its "correspondence" to objective reality).[13] This Marxist-Leninist tenet provided a theoretical rationale for Party supervision of both the social and natural sciences. In Leszek Kolakowski's words: "The practical criterion can be applied equally in natural science and social science, where our analysis of reality is confirmed if the political actions based on it are effective."[14]

Thus these features of Marxism-Leninism all supplied a rationale for the Party to demand of its social and natural scientists "evaluative judgments" based on "scientific cognition." Facts and values were inseparably linked. To remove any possibility of political neutrality, Lenin had articulated the principle of *partiinost'*. As he argued in his polemical *Materialism and Empiriocriticism,* all philosophical theories were either idealist or materialist; all theories either facilitated or hindered the class struggle.[15] Given the Marxian view that present reality already has in its womb the moral ideal of communism, a scientific observer of the "real movement" of history can only become a partisan theoretician of the proletarian class. By the time of the First Five-Year Plan, the Academy of Sciences had been reorganized so that science could assist in the rapid industrialization of Soviet Russia.

Moral Gaps

As Western analysts have noted, there exists in Marx's writings a tension between the view of morality described in the foregoing, based on amoral historical necessity, and Marx's humanism.[16] Lenin's adaptation of Marxism emphasized the former. According to Lenin's oft-quoted explanation to the 1920 Komsomol Congress: "We say our morality is entirely subordinated to the interests of the proletariat's class struggle . . . We do not believe in an eternal morality, and we expose the falseness of all the fables about morality."[17] Lenin gave moral content to Marxism almost entirely in terms of historical progress "forward"—what Karl Popper has referred to as "moral futurism."[18] John McMurtry has similarly defined Marx's basic moral imperative: act so as to hasten historical development to the next highest stage.[19]

Although Lenin emphasized the subordination of all individuals to the progressive historical march forward, the humanist element of Marx was not completely lost—only submerged. As Eugene Kamenka has noted, the "truly human ethic seen by Marx did become part of the utopian component even of official Marxist-Leninist ideology."[20]

Steven Lukes has argued that Marxian thought about morality, given its long-term perfectionist character, and its judgment of actions by their consequences only, is highly insensitive to moral requirements of respecting the interests of persons in the present and immediate future.[21] This lack of restraints (these have been termed "side-constraint" principles in ethics) on the pursuit of historical progress explains some of the inhumane practices under twentieth-century Marxism, for Marxism-

Leninism accorded no intrinsic worth to such values as human freedom or justice during the transition to socialism.

The Stalin Period

Issues concerning the relation of fact and value, theory and practice, science and philosophy were argued during the 1920s in the "Deborin versus mechanists" debate. A. M. Deborin and his school were victorious over the "mechanists," who argued for a more independent stance for science vis-à-vis philosophy.[22] The Deborinists were later denounced; the official line stressed the view that science cannot be independent of philosophy. Both science and philosophy, along with law and morality, were said to be part of the ideological superstructure, which reflected the economic base. Thus science was "proletarian" in the Soviet Union and "bourgeois" in the West.

Party control of the social sciences began in the early 1920s. But, as David Joravsky documents, until 1929 there was no Party control of research institutes, scholarly societies, and natural science departments of universities.[23] During the New Economic Policy (NEP) there was even a tendency to regard Marxist philosophers of science as somewhat autonomous specialists.

Why, then, to use Joravsky's words, did the "dictatorship of Marxism" emerge from "the vanishing ambiguities of the 1920s"? The answer has little to do with changes in Marxist-Leninist theory and everything to do with the change in the mobilization strategy of the political leadership. In 1929 Stalin began the shift from the stability-equilibrium phase of NEP to a brutally rapid mobilization phase, which was characterized by two elements: (1) the denial of present empirical reality and its apparent constraints, and (2) the forced imposition of the Marxian moral imperative to assist the historical process.

The transition to the mobilization phase was symbolized by the victory of the "teleologists" over the "geneticists" in the industrialization debate of the 1920s. The genetic approach stressed the existing situation and its constraints: market forces, relative scarcities of factors, rates of return, profitability. The teleological approach embodied a desire to change the proportions and the size of the economy, to maximize growth, to emphasize a strategy of development rather than adaptation to existing circumstances. The teleologists denied the apparent constraints of existing empirical reality, proposing successively more outrageously ambitious variants of the First Five-Year Plan, claiming that for the Soviet Union to accept the direction of the market meant acceptance of the "genetical inheritance" of three hundred years of tsarism.[24]

No phenomenon was more a product of this rapid mobilization phase than Lysenkoism. Given the mobilization imperative, the rate of change of the genotype seemed too slow (even counterrevolutionary), and those who championed the "immortal hereditary substance" must be reactionary idealists. T. D. Lysenko's claim to transform nature over a short period of time quickly met the socioeconomic requirements of the day. Lysenko noted, "The Party and the government have set our plant-breeding science the task of creating new varieties of plants at the shortest date."[25] To focus on the empirical constraining factor of heredity, as did the formal geneticists, or the constraints of existing empirical market forces, as did the economic geneticists, constituted an unwillingness to see and further the nonempirical—yet true—developing "tendencies" of history. Although the economists of the "counterrevolutionary Bukharin right" might have been substantively correct in their statistical analyses, they did not interpret their facts from the point of view of the working class. Similarly, the "methodologically bourgeois science" of the formal geneticists did not select the "correct" facts—that is, the "facts" that served the moral imperative to assist historical development.

Under Stalin, the depreciation of empirical facts in favor of ideological goals led to further erasure of the distinction between fact and value. As Herbert Marcuse argues, there was a blurring between the cognitive and pragmatic meaning of language. Marcuse notes that the key propositions of Soviet Marxism "claim no truth value of their own but proclaim a pre-established truth which is to be realized through a certain attitude and behavior." Writing about the Stalin period, Marcuse elaborates:

> For example, Soviet Marxism is built around a small number of constantly recurring and rigidly canonized statements to the effect that Soviet society is a socialist society without exploitation, a full democracy in which the constitutional rights of all citizens are guaranteed and enforced; or, on the other side, that present day capitalism exists in a state of sharpening class struggle, depressed living standards, unemployment and so forth. Thus formulated and taken by themselves, these statements are obviously false . . . But within the context in which they appear, their falsity does not invalidate them, for, to Soviet Marxism, their verification is not in the given facts, but in tendencies, in a historical process in which the commanded political practice will bring about the desired facts.[26]

Thus the vanguard Party, the person of Stalin, became possessed of the magical ability to command false statements to become true. To put it another way, Stalin vested in himself the political authority to order people to bring about the desired facts.

State interference in the natural sciences reached its peak during the Zhdanov period of 1948–1953. Political pronouncements on the correctness of the sciences touched theoretical physics, cosmology, chemistry, genetics, psychology, medicine, and cybernetics.

De-Stalinization

With the death of Stalin, several new doctrinal formulations relating to morality and the role of science in society significantly undercut the theoretical basis for the fact-value linkage. As we shall see, this trend was predictable in terms of the logic of Marxism itself. The ideological debate in the 1950s and mid-1960s was shaped by the de-Stalinization campaign of Khrushchev, with its aim of protecting individuals in general, and scientists in particular, from the recurrence of such abuses of state power.

Two key doctrinal reformulations of the post-Stalin era were the increasing independence of morality and the new status of science as a direct productive force. Both make the linkage of fact with value increasingly tenuous in Soviet theory.

The "Relative Independence" of Morality

It is inherent in the logic of Marxism that as society approaches the utopian goal of communism, as the vestiges of capitalism are overcome, individuals become increasingly free, enjoying, to use Engels' phrase, a "truly human morality." According to Marx's own premises, as society reached the postalienation communist stage, Marxian "scientific" critical philosophy would wither away along with the state. [27] The true content of human interaction must be hidden for social science to assume a role; under socialism, social relations become "transparent" and directly intelligible, thereby eliminating the need for social science. Thus the forward development of socialism suppresses Marxian social science. Indeed, with each step on the path to communism Marx's eleventh thesis on

Feuerbach becomes anachronistic: the point is no longer to change society, or even to interpret it, but to know it as it is.

Marx noted that having reached communism, free human beings would still require natural science. Astronomy, physics, and mathematics would be needed to penetrate beneath such illusions as the appearance that the sun rises on the horizon instead of the earth revolving around the sun. But, as man overcame the social illusions of capitalist society, he would no longer need social science, and he would exit the prehistorical stage to enter a realm of freedom.

As I observed earlier, the humanistic elements in Marx were relegated to the utopian aspect of Marxism-Leninism. As Soviet theory has evolved through its various stages—dictatorship of the proletariat, socialism, developed socialism, and a brief courtship of integral (*tsel'nyi*) socialism—there has been increasingly more emphasis on normative statements and less on historical analysis of the world situation. Stalin's "Marxism and Linguistics" (1951) gave the green light to stress the "relative independence" of the ideological superstructure from the economic base. Following publication of Stalin's essay, and a Soviet-Czech conference on morality that same year,[28] an increasing number of publications appeared on the role of morality in socialist society.

In the years immediately after Stalin's death, the reaction to the inhumane, repressive character of his regime affected both Soviet law and ethics. Between 1953 and 1958 the Soviet government widened personal freedoms and increased guarantees against extrajudiciary repression.[29] According to the post-Stalin *History of Philosophy*, Stalinism had "retarded the scientific elaboration of ethical problems."[30] In the 1950s and 1960s new explorations were made of the theory of ethics and morality. According to Soviet writings, the key factor distinguishing Soviet ethics from all others is the classless nature of the socialist economic base, which allegedly gives new content to old ideals. In a 1955 article a Soviet philosopher, lauding John Locke as "a great thinker of the past," wrote that under socialism the terms "duty, honor, and conscience" acquire new content "because the will and conscience of man is subordinated to the great matter of service to communism, to the socialist motherland (*rodina*)."[31] A year later the noted philosopher A. F. Shishkin argued that Marxist theory sees the source of duty "in the objectively ripened demands of the progressive development of society."[32] These discussions marked the beginning in the Soviet Union of a study of ethical categories (duty, honor, conscience) that are independent in form if not in content from the historical process.

The 1961 Party Programme presented the new Moral Code of the Builders of Communism.[33] The first principle is "devotion to the communist cause." But many of the moral norms cited in the code, such as patriotism, love for work, truthfulness, and honor, are not so different from those of societies that espouse no ultimate goal of communism. The Party Programme included acceptance of the "fundamental all-human moral norms" developed by the masses in their struggle over centuries against exploitation.

During the 1960s a range of voices was heard in the Soviet debate on ethics. Iakov Mil'ner-Irinin, who worked as an editor for an Academy publishing house, proposed the grounding of ethics in a universal human nature where individuals are autonomous moral agents; but the Academy refused to publish his monograph.[34] Some Soviet philosophers were published who elaborated a conception of objective values grounded in human nature itself.[35] As Eugene Kamenka notes, there was among Soviet philosophers a growing interest "in the philosophy of values, in a desire to see ethics, like aesthetics, grounded in a set of categories and distinctions comparatively independent of politics."[36]

Beginning in the 1960s some Soviet philosophers began to adopt in their discussions of the fact-value question analytical distinctions that had previously been employed only in the bourgeois West.[37] A 1962 article by D. P. Gorskii argued that normative expressions "cannot be evaluated as objectively true or objectively false."[38] In a 1966 article "Truth and Value" O. M. Bakuradze writes, "A true judgment is a description of fact and consequently is distinguishable in essence from a judgment of value." A value judgment, Bakuradze argues, "expresses an attitude of a subject to an object"; furthermore, "a value is not what is, but what should be." Although Bakuradze concludes by asserting the standard line that "an absolute juxtaposition of a judgment of value and a cognitive judgment would be a crude mistake," he distinguishes between the descriptive and evaluative functions of propositions, as in the tradition of Western emotivist philosophers such as Oxford's R. M. Hare.[39] Bakuradze articulates the essential philosophical categories to describe the volition of the subject, his ability to prescribe. Prescription is a notion generally alien to the Soviet world view (historical necessity being the only legitimate source of initiative). Yet a book by Shishkin cites Bakuradze with approval.[40]

In addition to Soviet philosophers, there were also natural scientists who gave voice to the view that morality and science were basically different realms. As noted by Alexander Vucinich, the physicist E. L. Feinberg wrote in 1965 that "there is no room for the ethical element in a scientific system." A. N. Nesmeianov, former president of the Academy of Sciences, similarly argued that there was no connection between science and morality.[41]

The most politically interesting element represented by this new position that morality is not immanent in science was expressed in the Bakuradze article: "The existence of opposing moral principles is not the result of an inadequacy of cognition."[42] Bakuradze draws a line between fact and moral evaluation, treating the concept of value as a subjective representation. According to this view an infinite number of subjective evaluations are possible, and none are the result of unscientific cognition; they are due to the fact that different people evaluate the "facts" differently. This trend clearly provides the philosophical basis for resuming the "partisan struggle in philosophy" that Lenin had sought to end conclusively. Precisely for this reason the influential legal scholar M. C. Strogovich later criticized Bakuradze, reasserting that "a judgment which expresses an evaluation of something as worthy of approval or blame from a moral point of view may also be true or false.'"[43]

By articulating the essential philosophical terms to describe the volition of the subject, these Soviet philosophers raise the epistemological question for Soviet theorists of how man can be willful and cognitive at the same time. This had never been a problem for Soviet theorists as long as they accepted the resolution of the dilemma that the Marxist linkage of fact and value provided: visionary perception of—and consequent participation in—the "real" historical process makes man both willful and cognitive. Although the conservative Shishkin gave a favorable review to Bakuradze, he restated the politically correct Marxian linkage of fact with value:

> Marxist social science . . . points to the path of the transformation of society, relying upon the laws of social development and on the interests of the foremost class, directing the will of that class to the struggle of the masses to the preparation of the victory over capital. In that way the gulf is annihilated between mind and will, cognition and action, the present and the future, between what is and what ought to be, between the world of facts and the world of values.[44]

In a 1969 discussion O. G. Drobnitskii addressed criticism of Western philosophers regarding this linkage. "What is obvious to us," Drobnitskii writes, "seems to these people highly debatable." He claims not to understand why the basic principles of Soviet ethics were seen as "hypotheses in need of verification."[45] Philip Grier argues that the need to validate the ideals allegedly immanent in history is the main challenge of Marxist-Leninist ethics. But, as we have seen, naturalist theories are based on a peculiar kind of vision (evidenced in Shishkin's rhetorical flourish), which critics call a blindness. Drobnitskii adheres to the standard Marxian teleological, self-realization theory: communism is justified because it is the realization of man's collective essence.[46] Thus communism equals humanism.

In 1969 an article in the Party journal *Kommunist* put the lid on the discussion of the relation between fact and value. It criticized Soviet philosophers like Bakuradze and reasserted the principle of *partiinost'*, which "demands objective truth in evaluations of social phenomena. Not simply information about the facts of social life but their profound analysis and evaluation from the point of view of the interests of the working class." The article specifically criticized the Soviet authors who were opening the door to "subjectivism," "voluntarism," and the "free will of the individual."[47] Thus the Soviet debate has some potentially interesting political implications. Alan Montefiore has argued that the doctrine of the logical incommensurability of value and fact can serve ideological purposes, being associated with "a thorough-going individualism and anti-authoritarianism."[48] If a recognition of the "facts" were ever to carry an entailed commitment to any sort of approval or evaluation, he argues, the individual would lose his unconditional freedom to determine his values on the basis of personal decision alone.

It is difficult to say to what degree the trend toward a fact-value dichotomy in Soviet debates on ethics in the 1960s influenced the practice of deriving moral imperatives from factual analysis. The ethos at the time, with the beginning of the Brezhnev focus on "scientific decision making," was, in the words of one analyst, to take "a more open-ended empirical approach that did not excessively prejudge conclusions."[49] But by 1969, after the Kosygin reforms had been turned back and Brezhnev had consolidated his personal power, there was a return to the traditional normative view of scientific decision making based on the fact-value linkage. That same year the *Kommunist* article was published, denouncing the trend in philosophy toward separate descriptive and normative functions of propositions.

Science as a Direct Productive Force

Another major doctrinal reformulation, also initially spurred by the publication of "Marxism and Linguistics," related to the position of science in the ideological superstructure.[50] A lesson that should have been learned from the Lysenko period, Zhores Medvedev asserts, is that there cannot be two biologies, for "there is only one nature on earth"; but there can be two social sciences, because there "exists on earth two contrasting social systems."[51] As long as science was officially part of the superstructure, there could in theory be bourgeois biology and proletarian biology. When Soviet theorists began to speak of science as a "direct productive force," they took it out of the realm where its truth depended on the point of view of the subject, the class position of the scientist. Science now became fully "objective," outside the grip of the socioeconomic determination of any particular historical epoch.[52] This reclassification of the status of science, like the trend in philosophy toward distinguishing between the descriptive and the prescriptive, had the effect of making the Marxian linkage of fact with value even more tenuous.

The result of this ideological innovation was potentially twofold: (1) it could significantly undercut the theoretical position from which the Party could legitimately determine the correctness of scientific theories and research directions; and (2) it could bolster the theoretical position of those who wished to isolate pure science research from applied research.

Regarding the first possible effect, the question arises: if natural science is part of the economic base, what is its relation to Marxism-Leninism, the "scientific ideology" of the working class? How can the fact-value linkage retain its philosophical integrity if natural science is part of the base and Marxism-Leninism is part of the ideological superstructure? They become inhabitants of different realms.

As evidenced in the 1962 debate between I. V. Kuznetsov, B. M. Kedrov, and the British positivist philosopher A. J. Ayer, Soviet spokesmen had no inclination to agree with Ayer that philosophy was not a science.[53] This debate continued into the 1970s.[54] In a book published in 1968 I. T. Frolov reasserted the position that both natural science and Marxist-Leninist social science are "scientific"; yet they deal with different "levels of being"—the physical world and human society, respectively.[55] In effect, different principles govern the physical, biological, and social levels. Human societies are considered emergent phenomena that cannot be explained in terms of biology. Man has a social essence. (As we shall see, Frolov would later become extremely influential.)

Thus Marxian social science and natural science have different objects of study; but there seems to be an epistemological difference between the two sciences as well. The scientific cognition of Marxism-Leninism is a function of the working-class viewpoint with its own interests (albeit allegedly embodying historical necessity), while scientific cognition in the natural sciences is a function only of the scientific method, independent of class interests and historical epochs. This does assert that the Party is definitely not competent to comment on the validity of conclusions reached by scientific method. Soviet commentators still maintained that the social orientation of science cannot be separated from its contents: science provides principles that serve one set of purposes for some classes and different purposes for others in different historical epochs. But it is one science that can be used variously by different classes in different times.[56] By the mid-1960s the study of science (*naukovedenie*) had become a new discipline dealing with the logic and evolution of science itself.

During the post-Stalin thaw the Party ceased to attempt to determine the *correctness* of scientific theories according to Marxist-Leninist ideology, and Party interference in science declined for a period beginning in the early 1960s. In general, the role of primary Party organizations (PPOs) in research institutes in the early sixties was reduced owing to the disorganization of the regional Party apparatus resulting from Khrushchev's reform.[57] In 1964 Lysenko was dismissed and the reign of his pseudoscience subsequently denounced in several high-level Party statements. Writing about the period before 1970, Vucinich argues that "step by step, the Stalinist union of science and ideology was dismantled, giving way to new developments that freed science from the real and potential damages of ideological interference."[58]

As I have noted, the reclassification of science as a direct productive force could bolster the theoretical position of those who wished to isolate pure science research from applied research. This issue arose in the 1961–1965 reforms of science administration in the Soviet Union. Loren Graham has documented the debate, noting the views of N. N. Semenov, who sharply criticized the past interpretation of the Marxian principle of the unity of theory and practice, working to the advantage of applied science. Semenov argued that research in the two different areas of pure and applied research should be assigned to separate research organizations—to the Academy of Sciences and industrial ministries respectively. Semenov and his colleagues who advocated a role for the Academy of Sciences restricted to pure research did seem to influence the actual outcome of the reforms, which strengthened the position of the Academy in coordinating theoretical research throughout the Soviet Union.[59]

The Brezhnev Immobilism

As Stalin's mobilization phase showed, the interpretation of Marxian doctrines relating to fact, value, and science was influenced by the leadership's strategy and commitment to mobilization. The linkage of fact with value and the unity of theory and practice were doctrines particularly well suited to a society on the move, participating in the "real movement" of history. Khrushchev still proclaimed the revolutionary goal of attaining communism by 1980. But the Brezhnev regime was not possessed of great vision about the future. By 1969 the Soviets had practically ceased to write about the future communist utopia.[60] During the Brezhnev period the five-year plans no longer carried the outrageously ambitious qualities of the Stalinist mobilization, evidence of a reduced gap between aspirations and capacities. The political climate in general tended toward stability (Brezhnev's successors would say stagnation), relying more on material incentives than political enthusiasm and coercion as in Stalin's day.

The Brezhnev regime elaborated the concept of "developed" (*razvitoi*) or "mature" (*zrelyi*) socialism. Many would say this was to cover the failure to attain real communism. Under Brezhnev the point was no longer to change society so much as to make the existing structure perform better. The elite impulse to reshape society was supplanted by an emphasis on businesslike behavior and the desire to preserve the existing social model while steadily improving material living standards. Progress began to be seen less in historical terms—the march toward communism—and more in terms of rationalizing the existing economy, which required an increased reliance on the expert advice of natural and social scientists in order to understand the growing complexity of the Soviet system and to manipulate it more effectively. As Soviet society moved out of the Stalinist totalitarian mobilization phase, it developed different socioeconomic requirements for intensive as opposed to extensive growth.

Thus, after sixty years of the Party's claiming forward historical movement toward communism, the Brezhnev regime was ready to make the self-satisfied claim that the USSR had arrived—somewhere. The official theory of developed socialism suggested that Brezhnev's Soviet Union had come about as close to communism as possible without actually being there. If Marxist-Leninist morality is teleological, based on a collective self-realization theory, then what happens to the fact-value linkage as the collective self is realized? During the Brezhnev period a Soviet philosopher described what would occur:

> The differentiation between the good and the useful, between moral criteria and criteria of social progress, between happiness and virtue, will be removed when humankind enters its period of true history. Spiritual values—truth, goodness, beauty—will have self-sufficient existence; their separation from man . . . will disappear with the disappearance of the material roots of alienation. Service to beauty alone (the artist as a type, the "priest of the beautiful"), to truth alone (the scientist, the "wise man"), or to the good alone ("the saint," the "righteous soul") will cease to be different forms of existence of the spirit. "Moral" life, "the life of truth," and "the life of beauty" will come to coincide as notions. It stands to reason, the psychological structure of the individual and the nature of behavioral motivations will change. Humans will become genuinely human and will attain existence in the real sense. This goal and result of historical development must be taken into consideration when developing a contemporary ethic. [61]

Thus the rationale is present in Marxism for Soviet theoreticians to support a dehistoricization of Marxian ethics. The linkage of fact with value comes apart to the degree that the political regime chooses to claim that it has arrived at the historical goal. So there is justification for absolute values (the good without reference to its utility in promoting communism) and absolute truth (not immanent in history).

Given its conservative and change-resistant mentality, the Brezhnev leadership did begin to dehistoricize, ritualize, and absolutize its vision of the socialist way of life. In 1964 there began a series of conferences on socialist rituals. The effective content of the Moral Code of the Builders of Communism was reduced to conservative support for Soviet state and moral imperatives to work harder and more efficiently.

The official theory of developed socialism claimed all the fruits of communism, if not the reality. Developed socialism was officially said to be characterized by "the fundamental abolition of class barriers."[62] At the stage of developed socialism, V. Kudriavstev asserted, there is favorable soil for "the development and most complete realization of all those democratic norms and principles of just human interaction proclaimed by Marxism on a scientific basis, taking into account the finest ideals of human culture." Given that class barriers are fundamentally abolished, the state and law at the stage of developed socialism "reflect the will and the interests of the whole people," permitting the "widening and defense of the rights and freedoms of the individual person, immanently inhering in the socialist way of life."[63] This "immanent inherence" allowed M. Strogovich to conclude happily that "any decision reached by the organs of the developed socialist state should be legal and just."[64] Soviet writings stressed the new freedom brought to man by the "scientific-technological revolution" (NTR), emphasizing that man becomes the regulator (*naladchik*) of the technological process, not the element directly included in it (thus increasing the goal-suggesting function of man).[65] Moreover, the NTR was said to abolish the distinction between mental and manual labor (as Marx writes in "The Critique of the Gotha Program"), creating an increased need for specialists.

Officially most Soviet philosophers still adhered to teleological ethics. The criterion for moral progress was still the subordination of individual interests to the "historically necessary" interests of society, which were still to be determined on the basis of purely "factual" data. In his 1978 study Philip Grier noted no inquiry into the intrinsic worth of goals, which is the object of interest in standard Soviet accounts of Marxist ethics.[66] For reasons of ideological continuity and legitimacy, it is unlikely that Soviet philosophers would officially embrace anything but a teleological ethic. A 1982 book by Iu. Davydov praising the "immutability of moral absolutes" in Tolstoi and Dostoevskii was met with a blistering attack in the Party journal *Kommunist.*[67]

Theory aside, the actual situation in Soviet society was something quite different. Under Brezhnev there was a perverse conflation of fact and value in the sense that the moral norms of socialist society became simply the established social facts. Moreover, official ideology served to mask the facts. Far from the Marxian "transparency" of communist society, in Brezhnev's developed socialist society there was a profound distance between reality and appearance. Given the officially progressive nature of developed socialist society, there could be no prostitution, drug use, or other such features, which were by definition characteristic of decadent capitalism. Such (actually existing) phenomena could not be "facts." The purge of the Institute of Concrete Sociological Research in the early 1970s was perhaps a means of ensuring that the right facts—the historically progressive as opposed to the empirically true ones—were published.

Debate over Professional Ethics

By 1969 Brezhnev's scientific decision making had begun to move away from its new empirical meaning toward its more traditional meaning based on the linkage of fact and value. The Kosygin reforms had been rejected as fostering spontaneity; the PPOs in research institutes received the formal right of control in 1971; and the number of ideological seminars that scientists were required to attend increased dramatically, lest narrow pragmatic inclinations get further out of hand.

The development of Soviet ethical theory began to be heavily influenced by the Party's need for intensive economic growth. While ethics had been developing in Soviet theory as an increasingly independent factor, the Party had been set on gaining control of it. By the 1970s the creative development of Soviet ethics had been superseded by the Party's emphasis on the functional role of communist morality in Soviet society in increasing labor productivity. This emphasis was in line with the post-Stalin mobilization strategy involving a campaign of moral education (supported by the theory of the heightened role of the moral factor) rather than physical coercion. The Party undertook a series of initiatives to use labor collectives to instill moral consciousness and strengthen discipline. The Institute of Philosophy of the Academy of Sciences established an independent section dealing with ethics. In 1969 a department of ethics was established for the first time in the faculty of philosophy at Moscow State University. Through the 1970s the department had only five professors and five staff members, but they have played a significant role in the debate, publishing several textbooks on Marxist ethics.[68]

In the 1970s and early 1980s numerous Party pronouncements enjoined ethical philosophers to connect their work with practical economic and social tasks. In 1981 the former ideology chief Mikhail Suslov asserted: "Science is called to more active educational work, to arm the ideological cadres with clear, well-founded recommendations."[69] Suslov presupposed the linkage of fact and value: science yields "recommendations." During this time there occurred a debate among Soviet ethical philosophers over the relation between philosophical ethics and applied ethics, especially professional ethics.[70] V. T. Efimov, chairman of the department of philosophy and scientific communism at the Moscow Automotive Institute, responded on cue. What is "most important," Efimov wrote, is "the connection between theory and practice of communist socialization."[71] Here we see the unity of theory and practice (in the limited sense that science must have practical applications) being reasserted to harness social scientists to the Party goal of the intensification of labor. Efimov proposed the establishment of "morality studies" (*moralevedenie*) as an independent discipline, focusing on applied research—in particular professional ethics. This was in line with a trend under Brezhnev toward a greater demarcation between applied social sciences and theoretical ones.[72]

L. M. Arkhangel'skii, a department head at the Institute of Philosophy of the Academy of Sciences, disagreed. He put forth the Marxian view of the fact-value linkage, maintaining that Marxist philosophy and normative ethics cannot be separated. Moreover, Arkhangel'skii recognized that the advancement of "morality studies" and "professional ethics" independent from Marxist philosophy raises the danger of fostering political neutrality and independence. "Certain applied studies taking place outside the general context of the system of ethical knowledge," he wrote, "are losing their ultimate orientation points."[73] The apparent neutrality of professionalism makes it to some degree a competing ideology to Marxism-Leninism. In the West, professionalism and specialization have historically been important factors in the effort to eliminate normative statements from scientific discussions.[74] L. M. Kosareva, a senior research fellow at the Institute for Scientific Research of the Social Sciences (INION), argues that the conception of value neutrality in the development of natural sciences was a result of the historical epoch of the "early bourgeois revolutions." The age was characterized by the personal search for meaning in the framework of professional activity.[75]

Other Soviet ethical philosophers similarly warned of the risks of basing applied, professional ethics on empirical data without the orienting perspective of Marxism-Leninism. A. A. Guseinov, a professor in the Moscow State University department of ethics, wrote that if ethical analysis is to remain at all "scientific," it must not build its conclusions on "isolated facts."[76] Another philosopher argued that the "practical ethics of a businessman" are indeed real, but they are arbitrary and have "no relation to the notion of moral good," which is known in historical context.[77]

Thus tensions appeared in the Soviet debate over the social implications of professional ethics. A. I. Titarenko, the head of the department of ethics at Moscow State University, concluded that morality studies cannot be separated from philosophical ethics. Titarenko stressed the urgent need for empirical studies of the actual morals prevailing in Soviet society—to respond to the Party's call for philosophers to be less abstract and have a practical impact on increasing labor productivity. But methodology in ethics must have "historical vision" (*videnie*).[78] These points were reviewed in subsequent debates among Soviet philosophers.[79]

The Brezhnev era saw considerable development of professional ethics. In March 1971 a Soviet doctor's oath was legislated; the physician's primary obligation, however, was to "communist morality" and secondarily to "the perfecting of professional knowledge."[80] For the first time a conference on professional morality was held, in Vladimir in September 1980; another followed in Tiumen in April 1981. The role of professionalism in insulating a group from political and social control reached its height under Brezhnev in the Soviet military. The professional military exercised an almost exclusive prerogative over decisions involving sophisticated, technical matters of weapons systems and nuclear strategy. (The Gorbachev leadership would later remark on the dangers of becoming "captive to technology and military-technocratic logic" and would greatly increase civilian control over the military.)

Although there are risks to promoting professionalism, the Party found reasons to do so. Western analysts have suggested that one important factor in achieving the objectives of "intensive" economic growth is to promote increased professional autonomy and lateral cooperation among specialists, areas against which Stalin's mobilizational system was biased. Professional ties among natural scientists do provide an alternate source of identification and standards to Marxism-Leninism. Stephen Fortescue notes that secretaries of research institute PPOs are almost always rank-and-file or middle-level professional researchers, with a primary commitment to their professional colleagues or institute management rather than to the Party.[81] Moreover, he argues that PPOs are incapable of acting as a force for Party-conceived change, particularly if the change is opposed by non-Party hierarchies and bodies. A study of the views of émigré Soviet scientists concluded that the Party is regarded as having a potent influence on personal careers and the internal life of research and development institutions, but substantially less influence over technical matters and the formulation and conduct of concrete research and development programs.[82]

The Ethics of Science

During the Brezhnev era Soviet theory also began to emphasize the importance of the "scientific-technological revolution." But the initially optimistic tone about limitless potential for scientific progress began to be affected by certain troubling phenomena, such as advances in nuclear weapons and genetic engineering. I. T. Frolov, who emerged as a leading theoretician, spoke of the ethical responsibility of scientists and the linkage of science and values. Like many in the West, Frolov argued that the classical view of the value neutrality of science can disorient scientists and relieve them of the burden of moral responsibility for complex situations involving "antihumane" technological applications of science.

Frolov and B. G. Iudin explicitly addressed the problem of professional scientists seeking further autonomy by espousing an ideology that divides normative, value-laden questions from factual, value-free ones. They write: "In current times one not infrequently encounters socially irresponsible behavior, seeking justification in the ethics of professional science."[83] In certain areas of scientific activity in the 1970s, specifically in recombinant DNA research, Soviet scientists would affirm with particular passion the logical incommensurability of fact and value in order to avoid reimposition of political controls over biology such as were known in Lysenko's day. Unlike in the United States, Soviet biomedical research managed through the early 1980s to remain relatively free of regulation.[84]

Thus, in the 1970s Soviet society began to encounter the linkage of science and values, which no modern society can avoid. Soviet commentators called for the development of the ethics of science as a "relatively independent" sociophilosophical discipline.[85] A new journal was founded in 1980 that dealt frequently with issues of science and values.[86]

Gorbachev: Science and Marxian Humanism

As Mikhail Gorbachev wrote in his 1987 book *Perestroika:*

> One may argue that philosophers and theologists throughout history have dealt with the ideas of "eternal" human values. True, this is so, but then

these were "scholastic speculations" doomed to be a utopian dream. In the 1980s, as we approach the end of this dramatic century, mankind should acknowledge the vital necessity of human values, and their priority.

It may seem strange to some people that the communists should place such a strong emphasis on human interests and values . . . Marxist philosophy was dominated—as regards the main questions of social life—by a class-motivated approach. Humanitarian notions were viewed as a function and the end result of the struggle of the working class—the last class which, ridding itself, rids the entire society of class antagonisms.[87]

A year earlier Frolov had remarked: "The sharpest paradoxes arise when we turn to these 'eternal' problems. They also include those that are today described as the socioethical and humanistic problems of man and of scientific cognition."[88]

It does seem strange to hear the Soviet General Secretary talking about the priority of values common to all mankind. He refers not only to the international arena, where nuclear weapons produced by human science place an "objective limit" on class confrontation. He refers also to Soviet domestic life. Since the April 1985 Central Committee Plenum, and the Twenty-seventh CPSU Congress, there has been a new emphasis on the humanistic aspect of Marxist-Leninist ideology. The 1986 Party Programme advocated the rule of "a genuinely humanistic Marxist-Leninist ideology."[89] For perestroika to succeed, Soviet leaders argue, there is a need to activate the "human factor," the creative initiative and intensified productive activity of people as the main engine of social progress.

Does Gorbachev's new emphasis on "humanitarian notions" mean that Soviet society has rid itself of class antagonisms, that Marxian social science can finally wither away in favor of the humanism promised at the end of historical struggle? Not quite. In an attempt to explain the new emphasis on humanism, P. N. Fedoseev, former vice president of the Academy of Sciences, admits that, owing to the demands of the revolutionary and industrialization periods, the most fully developed aspects of Marxism-Leninism were the theory of class struggle and economic principles. The question of humanism had remained "in the shadows." This fed the "enemies of Marxism" and led to claims that communism was incompatible with humanism. Moreover, Fedoseev notes, there occurred "deviations from humanistic ideals" during the collectivization of agriculture, Stalin's cult of personality (with its "false" thesis about the heightening of the class struggle as socialism develops), and the Brezhnev period of stagnation.[90]

Although Marxism-Leninism and the role of the Party have not withered away, Gorbachev's new mobilization strategy may be based on principles of greater independence for science within the framework of a more open society. In the Gorbachev period a more distinct line has been drawn between fact and value. On the one hand, there is an emphasis on empirical facts (through polls, social science data banks). On the other hand, there is a new emphasis on normative statements (discussions of universal values and ideals) with less reference to scientific analysis of the historical situation. At the same time, there has been an increased stress on the linkage of science and values, emphasizing the ethical responsibility of scientists. Frolov, long an advocate of a full theory of Marxian humanism in science, has played a major role in this latter trend as a personal aide to Gorbachev and chairman of the national committee on the history and philosophy of science and technology. The committee includes a range of natural and social scientists.[91]

The Rehabilitation of Empirical Facts

It had been a basic doctrine of Soviet ideology that practice was the only criterion of truth.[92] Empirical facts were merely one argument for the truth—to be proven historically. But the Gorbachev era began with the announcement that practice is not always the criterion of truth—at least not the practice of the Brezhnev era (denounced as the "period of stagnation," or *zastoi*). Gorbachev repudiated the concept of developed socialism, arguing that the forms of organization of society that had developed were in practice translated into absolutes and considered immutable, leaving "no room for objective scientific analysis."[93] A 1988 article by D. P. Gorskii argues that the obsolete (*ustarevshaia*) practice of the Brezhnev era could not provide a criterion of truth for the ideas needed for perestroika. Gorskii puts forth a second criterion of truth—a new "theoretical" criterion of truth. (Gorskii had been among those in the 1960s to make a distinction between fact and value.) He notes a series of cases in which practice cannot serve as the touchstone of truth: future social development, mathematics, analytical propositions, logical concepts. Theoretical knowledge, he argues, is a product of practical activity which acquires a "relatively independent" life and begins to anticipate (*predvariat'*) social practice.[94] The publication of Gorskii's article can be seen to represent a significant doctrinal modification: it partially disconnects the unity of theory and practice.

In the Gorbachev era the mass media, activated by glasnost, have been assigned the task of watchdog of the truth. Academician G. A. Arbatov suggested that past campaigns against genetics and cybernetics were possible because, "among other things, we lived under conditions of lack of glasnost and an obedient press."[95] During the Brezhnev era a huge gap existed between the official facts and the truth. With glasnost even unpleasant, "unprogressive" facts have come to light. Roald Sagdeev has asserted, for instance, that "for too long Soviet science has hidden its inadequacies behind official panegyrics to its success."[96]

There is a new call for accurate facts, social statistics, and public opinion polls (as during the 1960s, which led to the founding of the Institute for Concrete Sociological Research in 1968). Academician V. I. Goldanskii has stated:

> Glasnost plays a major role in securing an effective feedback mechanism, a truly scientific sociological analysis of the responses and reaction of our people to various party and governmental initiatives, to international events and scientific discoveries. Objective and impartial analysis is needed—analysis which does not accept wishful thinking for reality. To this day, alas, this is still a rare phenomenon.[97]

This is a far cry from the previous official view that statistical data that is merely "formally" correct is not true unless it serves progressive historical goals.[98] Soviet commentators have condemned the methodology of science in the Stalin and Brezhnev periods as "dogmatic" and "scholastic," as giving the "theoretical" world priority over the "real"; there "was a departure into a removed past or future."[99] Goldanskii calls for scientists to have full access to the empirical facts, decrying censorship of articles from *Science* magazine distributed to Soviet scientists, and noting that "unknown bureaucrats painstakingly keep us from important information which we urgently need and which, of course, is completely accessible to our foreign colleagues."[100]

Some of the most striking attacks on past abuses of the fact-value linkage have been made by writers and artists, who have played a major role as advocates of Gorbachev's reform. Anatolii Rybakov's novel *The Children of Arbat* makes a direct attack on the abuse of the doctrine of practice as the criterion of truth (when used to mean that whatever serves the Revolution is true). Rybakov exposes Stalin's claim that he was expelled from the theological seminary for propagandizing Marxism when the real reason reportedly was his failure to pay his fees and his lack of desire to finish. "But the version about expulsion owing to propaganda of Marxism is the correct version," Rybakov writes. "It works for the image of the chief [*vozhd'*] and, consequently, serves the matter of the Revolution." Later in the novel Stalin says to Kirov: "Historians may err. They are captive of the naked historical facts. Historians, as a rule, are poor dialecticians and go nowhere as politicians." [101]

Discussion of "Eternal" Problems, Ideals, and Universal Values

The 1986 Party Programme contains no Moral Code of the Builders of Communism. The new section on communist morality notes first of all that it is collectivist and secondly that it is "humanistic." [102] For reasons of ideological continuity and legitimacy, it is unlikely that Soviet ethical theory would ever officially cease to claim that it is based primarily on the advancement of the collective social good.

The 1980s saw a resurgence of interest in the philosophical literature on the question of values and the value aspects of scientific cognition—as in the early 1960s. An article by Iakov Mil'ner-Irinin was finally published that treats ethics as a purely normative science grounded in a universal "human nature." (As I noted earlier, his 1963 monograph dealing with the theme was not published.) Editors of *Voprosy filosofii* included a note that they did not share several of the author's views. Mil'ner-Irinin argues that his approach is consistent with Marx (he quotes Marx in *Capital* on the development of "human nature as an end-in-itself"). He concludes by noting that there is "no need to demonstrate that this view of man, i.e., as a real spiritual being, has become possible only from the heights of the practical construction of socialism." Thus his approach is in line with the interpretation of Marx and Engels that a "truly human morality" can emerge only as society has advanced considerably toward communism. [103]

Numerous Soviet philosophers were published in the 1980s who made a sharp distinction between fact and value, between cognition and evaluation. [104] In 1982 Bakuradze published a book, *The Nature of Moral Judgment*, [105] which develops the same themes as his 1966 article. The debate goes on: authors such as Bakuradze, who put forth emotivist theories of ethics, have again been criticized by other philosophers as "psychologizing" the concept of evaluation. [106]

There has also been a new discussion of the category of the "ideal." [107] A. A. Novikov notes that there are many "blank spots" (*belye piatna*) on this theme. Now the category of the material is paralleled by a category of the ideal in Soviet philosophy. There has been no official acceptance, however, of an ideal entity existing independently of matter, which would signal a total separation of fact and value. Such a position was put forth by the prominent literary critic M. A. Lifshitz and criticized by I. S. Narskii of the Academy of Social Sciences of the Central Committee. [108]

Beyond the debates among Soviet philosophers, there is among Soviet political and social leaders an increasing amount of discussion of values not as class based but as universal and, in some cases, absolute. Whereas Marx had explained alienation as a result of the capitalist mode of production, Gorbachev has admitted that socialism is not immune to this universal affliction. He has stressed the need to overcome the "alienation of the worker from social property and government," which in the Soviet Union had been produced by the deformation of socialism to bureaucratic centralism based on "command and administer" methods. [109] Politburo member Alexander Iakovlev referred to the goal of creating a political system "aimed at overcoming the age-old alienation of man from power." [110] In Gorbachev's book *Perestroika,* there is a section entitled simply "Alienation Is Evil." [111]

In Soviet debates on many social issues the terms of discourse are the terms of moral evaluation, *good* and *evil,* as opposed to the terms of historical evaluation, like *progressive* and *reactionary.* Following the trend toward "morality studies," Tatiana Zaslavskaia advocates scientific study of the notions that the principal social groups and strata have about social justice or injustice. She argues that "frequent encounters with injustice, with a discrepancy between words and deeds, and with the defenselessness of good and the impunity of evil engender disillusionment in social values, indifference, and withdrawal into personal interests." [112]

Again, some of the most powerful voices are those of the writers and filmmakers. In the film *Repentance* (Pokaianie), for which Tengiz Abuladze received a Lenin Prize, the dictatorial abuse of power is portrayed as a universal evil. While the primary reference is to Soviet history, the dictator in the film is a composite of Hitler, Mussolini, and Beria. The dictator expresses the political program of the party: "We'll turn our city into a paradise on earth." But the apolitical artist hero replies: "Only the spiritual leader, the moral hero, can enlighten the people."

Repentance also dramatizes the linkage of science and values. Scientific experiments are being conducted in an ancient church (*khram*), a symbol of truth, love, and goodness. The church frescoes portray the expulsion of Adam and Eve from the Garden of Eden, after eating of the tree of the knowledge of good and evil. If the scientific experiments continue in the church, we are told, the city will blow up. In the background we hear an address by Einstein on the social responsibility of the scientists who gave us nuclear weapons. With regard to the message of this scene, Abuladze speaks of the dangers of the unrestricted pursuit of science, quoting Ecclesiastes I:18, "He that increaseth knowledge increases sorrow."[113]

The increased discussion of universal values and ideals in Soviet society appears to represent a further step since the 1961 Party Programme first acknowledged the "fundamental all-human moral norms" realized through centuries of struggle against exploitation. The realm of values is still considered to be the historical outcome of man's own creative activity. But now, after much suffering, humankind has arrived at communist ideals. Moral content is to be found not only through Marxist-Leninist social science based on historical movement "forward." Ivan Frolov argues that human culture needs both scientific cognition and art for man to develop as an active being: "Science cannot fill man's soul to the full . . . art opens to him a world of mystery that does not lend itself to rational scientific cognition . . . Because of art's orientation to man's moral and emotional world, it is precisely in art that the moral-philosophical and humanistic problems of life and death, good and evil, human freedom and honor are raised in their most acute forms."[114] This "new synthesis" of science and humanism represents a step away from the long-standing official Soviet cult of science. It also enables the Soviet people to reclaim the greats of Russian literature (although this effort had been initiated during the 1960s thaw).[115] Frolov lauds the wisdom of Dostoevskii and Tolstoi on "eternal" problems. In one of a series of articles on the "meaning of life" in *Voprosy filosofii*, O. S. Soina calls Lev Tolstoi an "ally in the struggle for deeper comprehension of man," the meaning of life, death, and immortality. The author asserts that Tolstoi's doctrine on the meaning of life obviously reflects the complex crisis atmosphere that enveloped Russia in the last quarter of the nineteenth century. Soina fails, however, to suggest that the renewed attention to the "meaning of life" in the Soviet Union in the early 1980s may have reflected the acknowledged "precrisis" state of Soviet society.[116]

To his discussion of the new synthesis of science and humanism, however, Frolov adds a caveat: We have not yet come to the end of history. "In recognizing historical progress, and consequently making a value judgment with regard to human history, we by no means interpret social development in providential terms or attach an absolute value to current value orientations."[117] Thus Frolov urges caution about dehistoricizing ethics, or adhering to ideas about divine guidance or destiny. Gorbachev notes that religion is still "unscientific."[118] Some Soviet philosophers criticize writers such as Chingiz Aitmatov, Vasil' Bykov, and Viktor Astaf'ev for deviation from Marxism-Leninism: "They seek means of moral purification, of perfection of man, not in contemporary society, but beyond its limits—in religion."[119]

Gorbachev's strategy of mobilization requires the activation of science and culture. Yet, unlike in the Stalinist days, Gorbachev's leadership has evidenced a recognition that genuine scientific development is impossible without a high level of scientific independence. Politburo member Aleksandr Iakovlev has made some strong statements on this score:

> The Party today has once again confirmed the course of restoring Lenin's norms to its policy on science and culture. In brief, this can be expressed this way: the creation of the most favorable opportunities for the progress of the human spirit, its creative basis, support for the whole diversity of scientific and artistic creativity, the development of an atmosphere of healthy competition, and the exclusion of incompetent interference. The only thing that is unacceptable here is that which is aimed against man, against the moral health of society, and against the individual's dignity.[120]

At the June 1988 Party conference G. I. Marchuk, president of the Academy of Sciences, stated that scientific activity "must be free of petty tutelage and any kind of control by the management apparatus, and it must be entirely able to act independently on all questions relating to its own process and to the entire socioeconomic and political process. And if today we are making it our aim to turn working people into the genuine masters of production, it is just as important to make scientists their own masters in science."[121]

Soviet leaders have in the past made statements in support of the independence of individual scientists. But to protect his dignity and independence, an individual must have rights. Marxism has, however, lacked a conception of rights—rights of the moral personality—in which subjects have definite claims against one another and against government. Under Gorbachev there has been a trend toward the emergence of "civil society"—that is, the emergence of social, economic, and political activity independent of the state and guaranteed by the rights of individuals vis-à-vis the state. In 1987 a Soviet commission on humanitarian affairs and human rights was created, with Fyodor Burlatskii as its chairman. Some steps have been taken to protect the mass media's role as watchdog of the truth. In February 1988 a new All-Union Council on Professional Ethics and Law was created under the USSR Union of Journalists. Among other things, the council was called on to "assist in providing legal aid to journalists in their professional activities and to concern itself with the protection of their professional honor and dignity."[122] The council was to draw up a code of ethics to be presented for public discussion. In January 1988 a new law was promulgated on the rights of individuals to defend against unjustified commitment to psychiatric hospitals.[123]

At the June 1988 Party conference, some interesting concrete proposals were made to grant greater autonomy to Soviet scientists, including one to give laboratories the right to control their own financial and material resources. S. N. Fedorov, the famous eye surgeon, stated that the Party is "drowning in matters of economic management," and arranged, with the assistance of the chairman of the Council of Ministers, Nikolai Ryzhkov, to lease his whole institute.[124] In addition, proposals have been advanced to increase funds for basic research (in 1987 only 6.5 percent of the funds allocated for scientific research went to the academic institutions that carry out most of the country's basic research), and to initiate a weekly newspaper devoted especially to science. Trends suggest increased opportunities for Soviet scientists to have professional contact with international science.

These developments have, of course, been balanced by Party calls for scientists to be receptive to society's problems and needs. The 1986 Party Programme warns that "both scholastic speculations and passive registration of facts . . . are contradictory to science."[125] Marchuk stated that "in conditions of perestroika it is more important than ever for scientists to have high civic responsibility and to be acutely receptive to society's problems and needs. And this cannot be achieved without consolidating our science's healthy forces on the basis of perestroika and preventing lack of coordination based on disciplinary, regional, or other narrow departmental interests."[126] The Party is still concerned that professional and disciplinary allegiances will push toward value neutrality and hinder Party efforts to achieve scientific progress in critical areas such as computers and high technology.

In sum, Marxist-Leninist theory on fact, value, and science has undergone significant change. This has been due to the logic of Marxism itself, the lessons of Lysenkoism, and the steady deradicalization of Soviet society with the passage of time. In the broad sweep of Soviet history, the trend has definitely been to pull apart the celebrated Marxian linkage of fact with value. The interpretation of this doctrine has varied depending on the mobilization strategy of the political leadership. In the 1970s Brezhnev's strategy led to a backlash toward a more normative view of scientific decision making, reasserting the fact-value linkage. Gorbachev's strategy for mobilizing science and culture appears to be based on a pragmatic recognition of the utility of separating fact from value, with a new emphasis on the independent role of the arts and non-Marxian classics in realizing the human spirit. At the same time, a new and not particularly Marxian linkage of science and values has come increasingly to the fore in Soviet debates on such issues as genetic engineering and nuclear power—just as it has in postindustrial Western states.

· PART FIVE ·

Literature and Art

· 9 ·

The Changing Image of Science
and Technology in Soviet Literature

Katerina Clark

In the Soviet Union there is a power in the topic of science and technology that makes it a privileged theme in literature. A major reason for this power is Marxism's claim to be a science in its own right, and the proclivity of the Bolshevik founding fathers for Prometheanism and technological utopianism. In addition, in the nineteenth century, even before Marxism became a factor in the Russian radical movement, "science" had been the banner under which socialists had rallied to confront an "obscurantist" tsarist state and the official Russian Orthodox church.

At the time of the Revolution, the Bolsheviks frequently talked of the total transformation of man and his environment. The means suggested for bringing about such a transformation included reason and political consciousness, to be sure, but also modernization and technological progress. Indeed, over time the emphasis they placed on technological progress was so intensified that it became *the* measure of advance toward communism. In effect, science and technology became inextricably bound up in Soviet culture with the recipe for political progress.

The cornerstone of Bolshevik plans for modernizing Russia was widespread electrification, which was closely allied to the second major policy for modernization, the transformation of the backward countryside by making it conform more to the urban model.[1] Electrification was given cardinal importance, however, as is dramatized in Lenin's famous slogan of 1920, "Communism Equals Socialism Plus the Electrification of the Entire Countryside." Electrification had long been a dream of Lenin's. In a 1913 article, for instance, he describes how it will transform factory workshops from "dirty" and "disgusting" places to "clean and light-filled laboratories worthy of man," and will "liberate" householders from "cold

259

and drudgery."[2] But in Bolshevik culture electrification became an icon not only of the way forward through modernization and education (made more accessible to the ignorant peasantry as the electric light enabled them to read in the evenings), but also of an entire, and somewhat Manichaean, epistemology (electricity brings light and dispels darkness).

The years immediately after the Revolution were the most utopian period in Soviet history. Much as the Bolshevik leadership scorned utopianism (Lenin once remarked that Marxist theory "was the first to convert socialism from being utopian to being a science",[3] it is clear from their heady statements that they had succumbed to the millennial pathos of a revolutionary age. Trotsky, for instance, sets out in the closing pages of his *Literature and Revolution* (1924) dizzying panoramas for the world of the future, when "through the machine [and science] man . . . will command nature in its entirety."[4] Then "mankind will become accustomed to looking at the world as submissive clay for sculpting the most perfect forms of life," for transforming nature, and "rebuild[ing] the earth," right down to changing the shape and location of mountains, rivers, steppes, and even oceans, whose present form "cannot be considered final."[5] As with all Bolsheviks, however, Trotsky sees such a radical transformation of the environment as but a means to the ultimate end of transforming man himself. Man, he maintains, will work on himself, and try to subordinate even his bodily functions, such as breathing and digestion, to the control of reason and will. The technological environment in which the new man operates will impress its qualities of efficiency, spareness, control, and regularity on him as well; in the end, "the average human type will rise to the heights of an Aristotle, a Goethe, or a Marx."[6]

It is a sign of this time of great millenarian fervor that, soon after the Revolution, state publishing houses put out Russian translations of several of the classic utopias, which advocate founding kingdoms of science. These include, in 1918, More's *Utopia* and Campanella's *City of the Sun* (both rushed into print by the publishing house of the Petrograd Soviet) and Bacon's *New Atlantis* in 1923 (which took longer to appear because it had not been published in translation in Russia before).[7] An analogous position informs many other utopias and science fiction works published in these early years, including A. Bogdanov's *Red Star*, first published in 1908, K. Tsiolkovskii's *Beyond the Earth*, and the many novels by H. G. Wells that appeared in Russian translation in the first few years after the Revolution.

Not one of these works was ever given any sort of official recognition. Lenin, however, is known to have been enthusiastic about Campanella; he saw the frescoes about science and its heroes that adorned Campanella's City as a model for the agitational poster displays of the Revolution. A. V. Lunacharskii wrote a trilogy of plays about Campanella in the twenties (it has been speculated that Lenin suggested the topic to him),[8] and Marx once hailed Bacon as "the true founder of English materialism and of all contemporary experimental science."[9]

In these early years a broad spectrum of intellectuals, Party and non-Party, cherished a vision of a new society not unlike those depicted in Bacon and Campanella. In their utopias science was essentially a new religion that brought man to higher levels of development than had been known hitherto, enabling him to prolong life and triumph over mental and physical disorders. Such a vision was bound to appeal, for instance, to intellectuals influenced by the "Godbuilding" heresy (as Gorkii and Lunacharskii had been in their time) or by the concern of the thinker N. F. Fedorov, the distinguished geologist V. I. Vernadskii, who formulated theories about the biosphere, the pioneer of Russian rocketry and science fiction author Tsiolkovskii, and others to overcome human mortality and even resurrect the dead via science.

I do not want to suggest that these classic utopias had any direct influence on Bolshevik policies. They do provide, however, useful heuristic models for highlighting the contradictory role envisaged for science in the new Bolshevik society.

A red thread common to all these utopias is the privileged role in the hierarchy enjoyed by the scholar-scientist. Campanella, for instance, has a spokesman for his City of the Sun criticize the widespread practice in the world outside the utopia of having a man of affairs rather than a scholar-scientist rule a country; his criticism is made on the grounds that it is very harmful to have the ignorant rule.[10] Each of these utopias sets as a criterion of eligibility for leadership the extent to which an individual has mastered a fixed body of knowledge, which is called science, but which has a quasi-religious aura. Thus each utopia provides for rule by what might be called an illuminatocracy, or knowers of "the light."

In a sense, the Bolsheviks had an analogous vision of their future society. They believed that as society moved toward communism, those who held power would be experts not only in traditional science and learning but also in Marxism-Leninism, the new science that had superseded religion. Lenin, in his address to the Eighth All-Russian Congress of Soviets on 22 December 1920, talked of how science and technology would assume an increasing role in society: "In the future at All-Russian Congresses of Soviets you will see not just politicians," he said, "but engineers and agronomists. And this is the beginning of that most happy of epochs when you will find fewer and fewer politicians, and there will be fewer and shorter speeches about politics, while the engineers and agronomists talk more and more."[11] Of course, the engineers and agronomists could talk "more and more," but only as authoritative representatives of Marxism-Leninism could they truly lead the way to communism. When society attained communism, no one would really lead, in the sense that all would be "illuminati," but the topics of debate would concern major construction projects and the like rather than political issues.[12]

Although there were definite elements of utopian illuminatocracy in the Bolshevik vision of the way forward, these were potentially in conflict with other aspects of their program. The Bolsheviks' emphasis on proletarian hegemony and the class struggle set them apart from most utopian writers and made the ideal of an illuminatocracy problematical for them. Most utopias provide for some class of "drones" who will labor to provide the utopia with goods and services. Marx, of course, deplored all such rigid divisions of labor. Campanella in *The City of the Sun* perhaps comes closest to the Marxian view in that he provides for citizens who toil in the fields, but for only four hours a day. The remaining time is spent learning "joyously," debating, reading, reciting, walking, and exercising the mind and body with play.[13] Nevertheless, the idea that the drones might become sufficiently illuminated to *rule* the educated, or that the educated classes are manifestly inadequate (both of which notions are commonly found in the theoretical writing and fiction of the Soviet 1920s), represents a radical departure from the illuminatocratic utopian tradition.

There has been a tension in Soviet history between the impulse to privilege the scientists and technocrats and a mistrust of them, coupled with a tendency to favor either the proletariat or the warrior class (in the Soviet context, the military and security forces). All the classic utopias stress the role of the warrior class; but at the same time they—perhaps unrealistically—expect that the warrior class will naturally subordinate itself to the illuminatocracy. In the Soviet Union this tension has never been resolved, and has been a major factor in Soviet intellectual history. Perhaps as a consequence the intelligentsia has been both one of the most privileged and one of the most persecuted groups in Soviet society. At times the contradiction has resulted in gross distortions, such as the Stakhanovite movement of the mid-1930s, when official rhetoric insisted that the Stakhanovites were epistemological titans (as it were, the Aristotles, Goethes, and Marxes whom Trotsky had foretold). Although the Stakhanovites were humble workers with only rudimentary education, thanks to their proletarian nature, their strong will, and their closeness to the illuminati in the Kremlin, they could defy the prognostications of experts (engineers and such) and act on their own, higher knowledge to dazzle the world with their superhuman achievements. [14]

This major qualification notwithstanding, the Marxist-Leninist blueprint for the way forward has many points in common with those utopian works published by the government in the first years after the Revolution, and with those of Bacon and Campanella in particular. These two share with Marxism-Leninism a prejudice in favor of applied rather than theoretical science (which in the Soviet Union was to become particularly marked under Stalin). They also share a belief that it is the role of science to keep making new discoveries that will improve the lives of their citizens rather than to uncover the mystery of the universe. In other words, for them science is not a matter of open-ended inquiry but of unfolding existing truths and establishing further ramifications or new applications for them.

The restrictive view of science to be found in Bacon and Campanella can be interpreted in terms of their rejection of commerce. The free-market system offends these authors because it leads to exploitation, and thus to educational deprivation for the drones, who merely service the market. As a solution, they advocate something similar to Karl Popper's "closed society." Each of their utopias is relatively cut off from the outside world (in Bacon this occurs through an elaborate system of restrictions and prohibitions, in Campanella by immense fortifications); citizens do not encounter outsiders except in war, and dissidents are punished by expulsion from society. The utopias are not to be cut off from *all* contact with the outside world, however. Envoys are sent to other countries both to trade and to learn of their latest scientific and political developments. But for other citizens, commerce with the outside is neither direct nor free.

When the Bolsheviks made their revolution, they did not anticipate its leading to a closed society. Rather, they expected that their proletarian revolution would be followed by analogous revolutions in many other countries, until the entire world had Marxist governments. In the meantime, however, they wanted to guard their citizens against the bad influence of bourgeois, commercial civilization while they worked to realize a radical transformation in society through education, science, technology, and a "scientifically" based political and social order. Lenin's disparagement of utopian schemes notwithstanding, it can be said that they wanted to create a utopia *in place*—that is, in Russia rather than on some remote island or in some remote future time. This dream has haunted Soviet policy makers ever since, so that spectacular progress in science and technology has remained an issue and a yardstick for measuring the progress toward communism.

In consequence one can, by using the particular focus of the changing image of science and technology in Soviet literature, chart the major shifts that literature has made over the course of Soviet history. The classic utopias provide useful heuristic models for systematizing these changes.

The entire history of Soviet literature can be analyzed in terms of two alternating patterns, one comparable to the patterns found in the classic utopias, the other favoring the opposite pole in the great contradiction of Russian intellectual history—that is, privileging the worker or the warrior. The first pattern is of a literature that is forward-looking, that stresses transformation, values science, technological progress, and efficiency, and is obsessed with the West. The second is of a literature that has turned inward, focuses on the present (or the past), and is not particularly concerned with science and technology but emphasizes rather the class struggle or some other struggle with enemies of the state.

What I am proposing here is not another of those "pendulum swings" so often used to describe the vagaries of Soviet history. The alternating patterns function as a sort of dialectic rather than an either-or. Thus, in any given period both patterns will be present in fiction, but one will predominate. Each time a given pattern predominates, however, it appears in a configuration in some ways unlike its previous configurations. The changes in configuration both reflect changing times and result from the fact that the pattern comes into prominence largely in reaction against the specific configuration that the pattern representing the other side of the dialectic took during the preceding literary period. Of course, to reduce all the complexity of Soviet literary history to these two alternating patterns is to generalize egregiously. This model is nonetheless a useful one in that it is not only largely true empirically, but also captures the way Marxism-Leninism's scientism interacts dialogically with some of the other currents in the Russian intellectual tradition.

This chapter traces the alternation of these patterns through seven periods in Soviet literary history: War Communism, the NEP, the First Five-Year Plan, the thirties and forties, the late Stalin and Khrushchev years, the Brezhnev era, and the period that began in the late seventies. Given the immense scope of the undertaking, I have had to limit the range of topics covered and have largely ignored science fiction (discussed by Richard Stites in Chapter 10), samizdat and tamizdat literature, and the important topic of the influence of nineteenth-century intellectual thought on the treatment of science in Soviet literature. From the Stalin period on, my coverage focuses on the most significant and relevant novels of mainstream Soviet literature—that is, on works that typify a period, have become classics of Soviet literature, or have occasioned a great deal of controversy and critical attention.

War Communism

The first phase in this series of alternating patterns corresponds roughly to the period of War Communism (1918–1921) or a little beyond.[15] These years saw an intensification of the prerevolutionary debate about the way forward for Russia as writers reacted to the Bolshevik commitment to modernization and urbanization. Since both sides of the debate were well represented in the fiction of these years, this was the one time in Soviet history when both patterns were strongly present. But the first pattern predominated. In *theory,* this was a time of millennial fervor and futuristic zeal. In *practice,* much literature of this period was backward looking, reacting against the times. But it was the futuristic utopias that set the stage for this phase. Even the most retrograde works were haunted by the vision of a future society in which science and technology would have triumphed, and man would be at one not with nature but with the machine.

Many diverse literary groups in this period were infatuated with technology, which they saw not just as a means toward amelioration but as an end in itself. Most of these groups had been active (or had antecedents) before the Revolution, such as the sophisticated constructivists, who had links with the prerevolutionary futurists and with the pan-European avant-garde. A less sophisticated group, but one also with prerevolutionary and pan-European ties, was the self-styled proletarian cultural organization the Proletcult. For most of its members the machine was an ideal. The machine did not merely symbolize, as it did for the Bolsheviks, such cardinal values as efficiency, perseverance, rhythmic work, an even pace, rationality, and, as a corollary, lack of chaos, emotionalism, and impulsiveness. Many Proletcult writers believed that man should subordinate himself to the machine until it impressed its inexorable rhythms on him and he himself became machinelike. In other words, the assembly-line automaton that Charlie Chaplin parodied in *Modern Times* was for them an ideal. In part these sentiments were influenced by the current vogue for Frederick Winslow Taylor's theories about the scientific organization of labor, which were championed by the Proletcult poet Aleksei Gastev, who conducted research on efficient labor methods. The Russian poets, however, carried the identification of man with machine farther than Taylor had envisaged, as can be seen in the Proletcult poet Vladimir Kirillov's famous poem "We," which celebrates that "We have become like metal, our souls are at one with the machine." [16]

Before long, those who sang the praises of the smokestack and the blast furnace were being confronted by opposing voices. [17] A nexus of antiurbanist movements, which were also anti-Western and antimodern in their orientation, emerged in these years, such as *muzhik* socialism, Kliuevism, and Scythianism (associated especially with the writers A. Blok, R. Ivanov-Razumnik, and B. Pilniak). These movements had a broad following. For example, in 1922 Pilniak was the most popular writer in the Soviet Union. [18] The antiurbanists rejoiced in the fact that in all the upheaval and violence of the Revolution, the peasant had virtually dismantled fledgling urban and industrial Russia and returned her, willy-nilly, to her "natural," anarchic state. They contended that the way forward lay with the *old* Russia; the great industrial cities should not be restored but, on the contrary, should be dismantled further.

Such sentiments inform the several peasant utopias written in this period, including S. Esenin's long poem "Inonia," written in January 1918, and A. V. Chaianov's *Journey of My Brother Aleksei to the Land of the Peasant Utopia* (1920). [19] Chaianov's utopia, which offers the most comprehensive account given at the time of an ideal society, effectively recommends dismantling the cities even further and largely restoring in Russia the old, pre-Petrine way of life (with, however, as befits a utopia, more emphasis on education and culture). In many respects Chaianov anticipates here the ideal for Russia posed by Solzhenitsyn in such works as his *Letter to the Soviet Leaders* (1974).[20] These peasant utopias were country cousins, as it were, of the several antiutopias written at that time, which also attacked the urbanist ideal, such as E. Zamiatin's *We* (written 1920–21 but published in the Soviet Union only under Gorbachev), and L. Lunts's *City of Truth* (published posthumously in 1924).

Despite the intensity of the debate between these opposed positions, it cannot be said, as it could when this debate resurfaced during the First Five-Year Plan, that those for and against the technological idyll were irreconcilable opponents. The debate was conducted with a degree of hyperbole and intensity characteristic of the first flush of revolution, but the two parties were essentially in dialogue. For one thing, most opponents of the industrial idyll did not reject the dream of the new technology completely. On the contrary, Chaianov, for instance, recognized that the only way to realize his dream of urban dispersion was to utilize futuristic means of transportation. Even the more militant Nikolai Kliuev sometimes published Scythian tracts in Proletcult journals.[21]

It is not surprising that writers like Chaianov advocated using advanced technology in Russia in the same works in which they argued against modernization. Under War Communism, a time when the novels of H. G. Wells and other varieties of science fiction were very popular, the notion of daring scientific discovery captivated the imagination of most intellectuals, whatever their field. Actors and artists began to call their studios laboratories, where they were to study the "science" of color and of movement, or the "deeper mathematical laws" of their art. Linguists and literary scholars joined the formalist movement because it promised to put their disciplines on a scientific basis for the first time; and the formalists' writings, in turn, influenced writers of fiction, who tried to apply the new "scientific" principles to their own work.[22] Much of the new terminology found in theoretical journals of the arts and humanities came from the world of science or of engineering (such as Eisenstein's theory of montage in the cinema). And several members of the avant-garde began to divide all of humanity into two categories, based on the metaphor of scientific and technological discovery: people were either inventors *(izobretateli)* or consumers *(priobretateli)*. The inventor is the rare genius who makes a radical discovery in some field, thus revolutionizing human knowledge, technological development, or the aesthetic sensibility, while the consumer is the ordinary citizen who merely benefits from the discovery.[23]

This distinction implies an individualistic conception of scientific and creative work that was at odds with the collectivist ethos prevailing after the Revolution. Those who made the distinction also championed free inquiry (a point of view that comes out more strongly in the alternative translation of "inquirers" and "acquirers"). Zamiatin, for instance, using a scientific metaphor once again, this time of the laws of dissipation and conservation of energy contained in the concept of entropy, advocated a sort of perpetual intellectual revolution whereby no truth or norm should be allowed to become fixed and unquestioned. He looked to the inventors to keep shattering all truths and norms in fiery explosions or revolutions.[24]

For every intellectual who wanted a free and nonconformist climate of inquiry, there were others who regarded culture and science with awe and dedicated themselves to their dissemination among the populace. Most such intellectuals were convinced that a market situation corrupted science and culture, and, like Campanella, wanted to preserve the people from its dangers. Many a non-Party intellectual favored an epistemological "dictatorship" in Soviet Russia to keep science and culture pure, which of course often meant maintaining the status quo and curbing new developments.[25]

In the first years after the Revolution, then, there emerged in Soviet literature an agenda of issues concerning scientific and technological progress, many of which had been inherited from tsarist times, but were made more urgent by the mandate to effect a "revolution." These issues are, first, the benefits and dangers of technological progress; second, whether scientific inquiry should be open-ended or whether science's role was merely to uncover further scientific truths and discover useful applications for them; and third, what attitude should be taken toward the scientific work of the West. For most of Stalin's time these issues received very one-sided coverage, and there was little public debate. The questions were still sufficiently alive by the time Stalin died, however, for the debate to resume. But the revival of these topics was not just a matter of a more liberal literary climate permitting freer discussion. It did not coincide with the end of Stalin but actually preceded it, and must therefore at least in some measure be ascribed to the diastole and systole of Soviet intellectual history. One mark of this periodic shift from dominance by one of the patterns I have identified to dominance by the other is the fact that the burning issues concerning free intellectual inquiry were not prominent during the next phase of Soviet literary history, even though the return then to a freer economic situation and some, albeit limited, private publishing gave an opportunity for a broader forum than had been available under War Communism.

The New Economic Policy

During the NEP Soviet literature switched from the first of our alternating general patterns, which had largely prevailed under War Communism, to the second, in which literature focuses on the present or past and is concerned not so much with the issue of technological development as with struggles with enemies. This development seems almost perverse: under War Communism the country had been preoccupied with fighting the Civil War, but its literature had been obsessed with the issue of technological progress; under the NEP, however, the country was concerned with the reconstruction and further development of industry, but its literature was obsessed with the Civil War.

Although the cult of science disappeared from most NEP fiction, the issue of technological advances did not, of course, disappear from literature.[26] It was, however, felt more in the minor fiction and poetry of this period, much of which was structured as parables about how modernization came to the village. One of the few major NEP works concerned with modernization is F. Gladkov's *Cement* (1925), indeed one of the most seminal novels of the socialist realist tradition. *Cement,* which concerns the postwar restoration of a cement factory in Novorossiisk, became the prototype for that backbone of Stalinist literature the production novel. Many a Western critic has dismissed the production novel as a simplistic tale of "boy meets tractor." In actual fact, statistically boy more often meets that marvel electricity in these novels than he meets tractor. In this novel, even though the main task of the hero is to restore the cement factory, Gladkov makes sure that he electrifies the town as well. The bringing of electricity became the main, although not the only, convention for all Soviet novels of technological advance, whether set in the rural or the industrial sector, with making or "meeting" the tractor coming second.

The production novel, then, is heavily clichéd, and its clichés serve as a means of celebrating the Soviet Union's general technological progress. (The novel illustrates this progress as it is effected in a small part of the Soviet Union, which stands for the progress of the whole country.) At the same time, since technological advance has been inextricably bound up in Soviet culture with the political, the hero's progress in a given novel toward modernizing his small pocket of Soviet reality (be it by building a power station, by producing more tractors, or by harvesting more grain) both parallels and symbolizes his own political maturation, which, in turn, stands for the reaching out of the country as a whole toward communism.

Although *Cement* was to become *the* model for the production novel, it was also not unrepresentative of NEP fiction in general. In particular, the hero of the novel, Gleb Chumalov, is also a Civil War hero, and this enables Gladkov to infuse the plot with the symbols and ethos of that war. In consequence, *Cement* is far from a production novel in the sense one might expect from the quip "boy meets tractor"—that is, a pedestrian account of the organization and processes of production, without any alleviating adventure, romance, or excitement (novels of the Five-Year Plan period better answer this description). In fact *Cement* errs in the other direction: it is positively melodramatic.

That *Cement* should have been adopted in the Stalinist 1930s and 1940s as a prototype for the production novel is not surprising, given that it exemplifies many of the attitudes toward science and technology that marked those years. The very fact that production novels became the main fare of Soviet literature under Stalin is itself indicative of a highly utilitarian attitude toward science. With few exceptions, it was not until the very end of the Stalin period that the scientist became a hero of literature (and those exceptions, such as L. Leonov's *Skutarevskii* of 1932, did not involve a particularly positive scientist as hero). *Cement* itself has a militantly pro-proletarian and anti-intellectual ethos. In particular, it denigrates theoretical science (caricatured as the futile endeavors of an unworldly but anti-Soviet engineer who shuts himself away from the factory in a musty world of books and keeps drawing the same geometric designs). Indeed, the novel plays down the necessity for expertise of any kind in working with technology, emphasizing instead worker enthusiasm and, above all, individual superhuman effort.[27] As is typical of the production novel, *Cement* foregrounds the hero's feats as he, military-style, mobilizes the masses and charges the enemy (technological backwardness, overcautious administrators and specialists, and so on), routing them all and winning by sheer zeal and will. Even Marxism-Leninism is represented less as a science than as a mystical form of knowledge.[28]

Cement exemplifies an attitude toward modernization that we have come to identify with Stalinism—that is, a faith in the absolute value of tempos, of pushing the pace of technological change, and of radically altering the landscape in doing so. The novel contains a topos that was to become a convention for all production novels, and that is the moment when the hero proposes to the local bureaucrats a plan for technological improvement, usually to be achieved within a certain time frame (in this case restoring the factory). The bureaucrats and experts invariably oppose his plans, arguing that it is not feasible to work for such rapid change; usually (as in *Cement*) they *explicitly* call his plans utopian.[29] The hero wins, of course, and his "utopian" scheme is realized. Usually toward the end of the novel his modest feat is placed in the broader context of the radical transformation that all of society is to undergo as the hero either experiences or describes a vision rather like the one in Trotsky's *Literature and Revolution*—a vision of a future in which the environment will be completely transformed. At the end of *Cement,* for instance, the hero's wife advises him to be prepared for the coming day when their whole world will have to be "burnt off" to make way for the new.[30]

The First Five-Year Plan

It was this emphasis on radical change through rapid industrialization that became a hallmark of fiction in the next phase, that of the First Five-Year Plan. A major topos of literature in that period was some mention of the millennial change that had occurred thanks to technological progress. In V. Kataev's novel *Time, Forward!* (1932), for instance, the hero finds that the terrain alters so radically every day that he has to keep changing his route to work.[31] What was to be reconstructed was not just the terrain, of course, but above all man himself. Industrialization, however, provided the controlling metaphors for changes both in social structure and in individuals. Society was conceived as a machine whose separate parts were harmoniously interrelated and regulated, and in which the Party itself might be seen as a "great conveyor belt" or a "driving axle" (to use the titles of two typical works of the period),[32] and the middlemen who implemented policy as "levers." Social institutions were seen as a sort of assembly line for retooling a human product. Machines were useful for this purpose, too; in several works industrial machines impress their own rhythms on the psyche of the workers who operate them and turn out the new Soviet man.[33] In other words, this was a time of utopianism when it was considered that, to use *the* catchphrase of the period, "technology is the answer to everything."

Five-Year Plan literature was, then, in terms of its preoccupations and cardinal values, somewhat like the urbanist and Proletcult fiction of the period of War Communism with one small difference: that other, antiurbanist side of the great debate of those years was no longer heard.[34] Also, in this later literature the dream of an industrialized utopia took a more definite shape than in earlier counterparts.

A new topos that entered fiction during this period was the building of some giant construction project, or even an entire city in a remote wasteland. The phantom rise of Magnitogorsk on the plains of northern Kazakhstan, the Soviet answer to the building of Petersburg, became a dominant symbol for the way brute nature can be transformed at a dizzying pace once the new technology is mobilized (see Kataev's *Time, Forward!*).

The fiction of the First Five-Year Plan also laid a good deal of emphasis on the wonders of electricity, the traditional icon for the transforming powers of science. Recurring themes were the building of a hydroelectric dam and the mechanization of industry or agriculture through electrification (with conveyor belts, milking machines, and so on). Symptomatically, Gladkov went from a 1925 novel about rebuilding a cement factory to a 1932 novel, *Energy*, in which his earlier hero becomes a minor character, and cement is merely the stuff for making a great hydroelectric dam. The one major novel about science dating from the plan years, Leonov's *Skutarevsky*, deals with research on electrical transformers.

In the fiction of this period science regained some of the status it had lost under the NEP, but it was a very reduced, utilitarian notion of science that was championed, one oriented toward practical tasks. During the plan, the ideals of rational organization, efficiency, and positivism reigned. Facts and statistics were granted a charisma unequaled before or since. Thus writers no longer minimized the know-how needed for technological achievements, as Gladkov had in *Cement*; indeed, they were now expected to be conversant with technology and would be attacked in reviews for any mistakes they made in describing machines or work processes.

Looking at this period in terms of our two alternating patterns, then, it could be said that fiction reverted to the first, forward-looking pattern. But, it will be recalled, this pattern entailed, in addition to an active commitment to technological advance, an obsession with the West. This was true with a vengeance of Five-Year Plan fiction. With all the emphasis on technology and modernization, it was inevitable that the West should loom larger in fiction than it had for some time. After all, Western technology, engineers, and even workers were being used on some scale in the industrialization effort. Most novels of this period include some foreigners among their characters, but these foreigners are not particularly active protagonists. Their function is to serve as symbolic demarcations of the boundary between "us" and "them." They are generally not villains, as they were so often to be in later fiction, but more foils to the Soviet viewpoint, devices for motivating long-winded philosophical dialogues that are essentially digressions from the main development of the plot.

As this description suggests, Five-Year Plan fiction tended to be mired in facts and figures. Novels were fairly slow-moving, in sorry contrast to the dizzying pace they sought to celebrate.[35] Such suspense and pace as was to be obtained in fiction was generally derived from the master theme of these years, that of the struggle with nature, the saga of man, science, and technology pitted together in deadly combat against the near-intractable forces of the natural world.

The 1930s and 1940s

By 1932, when socialist realism was mandated as *the* method for Soviet writers to follow, authoritative figures in Soviet literature were concerned by low reader interest in Five-Year Plan fiction. They largely bypassed it in drawing up a list of models of socialist realism for Soviet writers to follow in their work, preferring instead works from the twenties such as *Cement*. Although the novels of the plan years were rejected, their one convention that was successful with readers, the theme of struggle with nature, was not. It was developed further in the thirties to become the central topos of socialist realism. But as this theme was later developed, stress was placed on the aspect of struggle rather than on modernization, and hence drew more on the conventions of the committed fiction of the NEP period than on the literature of the plan years. In other words, most classic, or Stalinist, socialist realist fiction is, the commitment of the production novel notwithstanding, not really involved in a fundamental way with the issues of science and technological progress.

As the foregoing implies, during most of the rest of the Stalin years fiction reverted once again to the second of our two alternating patterns. Whereas before it had been forward-looking, obsessed with the West, with dramatic change, and with technological progress and efficiency, it now looked more to the struggle with internal enemies, and primarily to the great struggles of the past or to those being waged in Soviet institutions. Ostensibly fiction was still concerned, as it had been during the plan years, primarily with economic progress and technology's contribution to the struggle with nature. Technology soon faded from prominence, however, as superhuman heroes became the mainstay of fiction. When technology played any role at all, it was usually in romanticized accounts of Arctic exploration or aviation feats.[36]

More often than not, in fiction the struggle with nature was now present metaphorically rather than literally. Indeed, it became *the* controlling metaphor for political maturation.[37] Thus the great socialist realist classic of the thirties, N. Ostrovsky's *How the Steel Was Tempered* (1934), was not actually about steel but about how the hero achieved political maturation—became "steeled" or "tempered"—through a series of encounters with class enemies, primarily in the Civil War. Another common metaphorical use of the struggle with nature was the struggle to master man's willful, immature, or egocentric tendencies. In the forties the struggle was often transposed to the world of the bureaucracy and used to frame encounters between the hero and self-seeking, "careerist" bureaucrats.

The symbolic struggle with nature became a controlling metaphor that shaped the plot of socialist realist novels regardless of their subject matter. Thus, in some sense Stalinist novels invariably, by their very structure, celebrate that original Soviet dream of progressing toward a kingdom of science on earth. In any classic of socialist realism, all characters are divided into categories of positive and negative. A positive protagonist will earn his right to this status by proving himself a champion in the struggle with nature and a negative character will represent those dark elemental forces that must be vanquished in this struggle.

It was not true, of course, that the theme of industrialization and modernization faded out entirely in the Stalin period.[38] At the end of the thirties some important works on industrial themes were published that stressed efficiency and tempo (such as Iu. Krymov's novels *The Tanker Derbent* of 1938 and *The Engineers* of 1938–1940, which shows greater respect for engineering expertise than had been seen in literature for some time). The outbreak of war in 1941, however, soon put an end to this revival, and the theme of struggle with the nation's enemies came back into vogue again.

After the war, as the country began to recover from its devastation, literature returned to the production novel and the theme of technological progress. The production novels of this time (usually referred to as the Zhdanov era)[39] were strongly influenced by Gladkov's *Cement* (no doubt the fact that through most of the forties he was director of the prose section of the Literary Institute, which trained writers, had something to do with this). *Cement*'s influence extended even to writing on rural life (such as S. Babaevskii's infamous pastoral novels, in which all the problems and shortcomings of a *kolkhoz* region are solved by the building of an electrical power station).[40] As in *Cement* itself, emphasis was placed on the tempo of development, and on the extraordinary hero who defies realistic estimates of when a particular project might be completed.

V. Azhaev's *Far Away from Moscow* (1948) provides a classic example of this theme. A construction team that has been sent to build a gas pipeline in the Soviet Far East disregards the advice of its engineers on when and how it should be built. Despite a severe winter, they meet Moscow's wishes and complete the pipeline in one year instead of the projected three. This outcome can be predicted: just as the directors and Party officials assigned to the team are embarking on the project, they are written up in *Pravda* as "People Who Can Do the Impossible."[41]

It is not true, however, that the fiction of the forties adopted unchanged the attitudes toward science and technology that inform *Cement*. There were signs of change even in the novels of Azhaev and Babaevskii. Gladkov had played down the need for any expertise, but these authors of the forties did not. Although Azhaev in *Far Away from Moscow* predictably deprecates the project's chief engineer (the "bourgeois specialist"), he also insists that its administrators need to complete some technical training; and Babaevskii has the fiancée of his *kolkhoz* hero put off their marriage because she feels keenly her lack of education in agriculture and wants to go away and take some courses.[42] In V. Panova's *Kruzhilikha* (1947) the factory director even marries a bluestocking engineer whose "companionship" helps him achieve a better understanding of what he should do in the factory. By the time Khrushchev came to power, it was no longer just the case that the positive characters in a novel would realize they needed more training. Rather, a common trait of the *negative* character was that he did not have enough education and did not keep up with the latest developments in his field, while the positive character did.[43]

The Late Stalin Years and the Era of Khrushchev

This increasing emphasis on the role of expertise in achieving the goal of a modernized economy eventually ushered in scientists and engineers as accepted heroes of literature, a development that did not begin with Khrushchev but can be sensed in the growing stress on science typical of the late Stalin years. In Stalin's *Economic Problems of Socialism,* for instance, he speaks of the need for a higher general educational level as a necessary precondition for the transition from socialism to communism.[44] Of course, the late Stalin years also saw a renewed interest in science and scientists of a most insidious kind, reflected in the anticosmopolitan campaign and the doctrine of the two sciences, both directed against Western, "bourgeois" science. Repressive though these policies were, they also made it clear that science itself, including Western scientific achievements, had assumed importance again in the eyes of the leadership.

Official directives of the late Stalin era were by no means *only* repressive and retrograde in their implications for science. Stalin's famous essay on linguistics of 1950, for instance, condemned the stranglehold of dogmatism and hallowed authorities on Soviet linguistics, asserting: "It is generally recognized that no science can develop and flourish without a battle of opinions, without freedom of criticism."[45] He also attacked persons who try to establish themselves as absolute authorities in their discipline, stifling all theories that do not support their own, and described their sway as that of an "Arakcheev-like regime," "a self-contained group of infallible leaders . . . which has begun to . . . behave in an arbitrary manner."[46] Such remarks were seized on with alacrity by writers, who began to produce works in which Lysenko-like dictators in some field of science or engineering are exposed as retrograde tyrants.[47]

The Khrushchev era (1953–1964)[48] brought Soviet literature as close as it has ever come to realizing Lenin's dream of a time when, in discussing issues of national interest, it was the "engineers and agronomists" who talked rather than the politicians. During these years Soviet literature reverted to the first of the two patterns I have described. Under

Khrushchev most literature was preoccupied with how to modernize and make the economy more rational and efficient, and much of it was concerned specifically with the work of scientists, agricultural specialists, and engineers.

By 1953 (if not before) Soviet society had reached a point where, thanks to long-standing educational policies, the technological qualifications of the labor force were dramatically higher than before and the majority of the populace was conversant with science.[49] One might have expected a new kind of fiction for this new mass reader. But although the extent to which literature was able to go beyond the conventions of socialist realism was as yet limited, there was a distinct shift in the sociological profile of the writer who typified the age. Most of those who achieved prominence under Khrushchev had some form of scientific or technological higher education. Those who did not had entered literature from a background of reporting for the Party or Komsomol press about *kolkhozy*, construction sites, or factories. The majority were also Party members.[50] Thus, to a greater extent than before or since, the writers represented a kind of Leninist illuminatocracy; they combined in their backgrounds a knowledge of *both* science and technology, *and* dialectical materialism.

In these years a vision of science not unlike that of Bacon and Campanella inspired Soviet literature. Writers were idealists who believed that the kingdom of light could still be achieved in the Soviet Union; society had merely strayed from its path under Stalin. Writers saw the road leading through greater respect for the demands of science, and the creation of a more efficient and rationally organized economy in which citizens would be allowed more initiative to participate in the process of economic rationalization. Under Khrushchev this change of direction was perceived as part of the task of overcoming the Stalinist heritage. In actual fact, however, the movement for change had begun before Stalin's death, although it was more forcefully and explicitly expressed in fiction published later. Let us look at two main trends, the first representing, more or less, the "agronomists," and the second the "engineers."

The first Party plenum after Stalin's death (September 1953) focused on agriculture. It is not surprising, therefore, that the agronomists were the first to achieve prominence in post-Stalin fiction. This trend was dominated by the "Ovechkin school," led by Valentin Ovechkin, and including V. Tendriakov, S. Zalygin, P. Nilin, and G. Troepol'skii (a genuine agronomist). Ovechkin, a Party member and economic journalist, had been publishing sketches on rural themes since 1927, but his impact had never been remotely as great as it was to be between the years 1952 and 1956; after 1956 he faded out completely as a significant writer.[51] Thus he was a man of the hour. He could even be called prose's answer to poetry's Evtushenko as a poet laureate of post-Stalin change.

In 1952 Ovechkin published the sketch, "District Routine," the first in a series published between 1952 and 1956 under the same general title.[52] After March 1953 his sketches began to receive official attention. His subsequent ones, most of which were published in *Novyi mir*, were all published in *Pravda* as well in at least extract form, and their appearance in *Pravda* was often followed by an article expressing views similar to those in the sketch.[53] Also, many of the concrete proposals Ovechkin made in his sketches soon became official agricultural policy.[54]

The sketches of Ovechkin and others in his school spearheaded the movement for post-Stalin change generally, not just in agricultural policy. Yet the first sketch in the series, "District Routine," actually appeared *before* Stalin's death. Moreover, two other writers in this school, Troepol'skii and Tendriakov, published sketches expressing a similar point of view at the beginning of 1953; in other words, they were accepted for publication before Stalin died.[55] Indeed, A. Nove has argued that the policies announced at the 1953 Party plenum were planned well before Stalin's death.[56] In short, "post-Stalin change" began when Stalin was still alive.

The fiction of the Ovechkin school marks a transitional phase between works exemplifying the second of our two alternating patterns, which had dominated most of the Stalinist period, and works exemplifying the first pattern, which was soon to enjoy a resurgence in the Soviet fiction of the late fifties and early sixties. Writers of the Ovechkin school argued for a more "scientific" (more rational and efficient, more scientifically based) organization of agriculture. They also privileged modernization and technological improvements. But by Soviet standards they were not particularly forward-looking. As the very title of Ovechkin's series of sketches, *District Routine*, suggests, these writers were concerned with the prosaic, practical tasks of running *kolkhozy* and rural districts.

Pedestrian and routine as the writing of this school might appear to be today, it was considered daring in its time because it questioned the imperative for superhuman, radical transformation, which we associate with Stalinism but was in fact an ideal of many of the early Bolsheviks. Through the negative example of one of the series' characters, Borzoi, a rural administrator, Ovechkin attacked the practice of pressing for modernization and economic achievement at a dizzying pace, and with complete disregard for its human cost. In the pages of *District Routine* Gleb Chumalov and his many avatars in Soviet literature were laid to rest, never to be introduced again with the same confidence as before. Soviet literature has continued to this day to question forcing the pace of industrialization, although in ever more sophisticated formulations of the perennial problem. [57]

Ovechkin, then, had a very modest, reduced sense of the kingdom of science, which for him meant primarily introducing rationalization into the economy. Fiction of the Ovechkin school in general was characterized by narrow horizons of both space and time. These limitations were less characteristic, however, of the other kind of writing prominent in the fifties and sixties—fiction about scientists, inventors, and engineers. During the first five years after Stalin's death, the overwhelming majority of fictional works were on rural themes. [58] The Ovechkin school faded from prominence after 1956, however; and thereafter, for the remaining Khrushchev years, fiction about engineers defined Soviet literature. No doubt the fact that the 1955 plenum was on industry was a contributing factor in this development, although most of the issues raised in this fiction had been taken up in important works published either in the last years of Stalin's rule or during the 1954 and 1956 "thaws."

The reemergence of the scientist and inventor as a literary hero did not mean just a change of subject, but rather marked a reaffirmation of science and its values, and a return to the notion of professionalism in science, which had received such a battering in the thirties and forties. From the late forties to the early sixties, fiction took issue with three basic principles that defined the approach to science usually associated with Stalinism. The first of these was the requirement that Soviet science subordinate objectivity to the interests of dialectical materialism, and the assertion that no Soviet scientist could be politically neutral for everyone must be active politically. [59] The second was that long-standing prejudice against scientists and intellectuals as being "bourgeois," and not part of "We Soviet People." [60] The third and related principle was actually a distortion or extreme application of Marxism-Leninism's preference for applied rather than theoretical science: it was held that scientists should be occupied only with projects that feed directly into the current needs of industry, and that theoretical science involves purely abstract, egoistic, and sterile work that cannot contribute to the way forward.[61]

Overtly, the main preoccupation of fiction about scientists and engineers was not so much the rightness or wrongness of these Stalinist principles as with how certain abuses had harmed Soviet science, and with the country's great potential for achieving technological superiority. In this sense, fiction about engineers was more forward-looking, and more concerned with the West, than was fiction about agronomists. The broader horizons of fiction after Stalin were particularly marked after 1955, when large-scale reading of Western scientific and engineering journals was reintroduced. In other respects, however, many of the differences between late-Stalinist and post-Stalinist fiction about scientists and engineers could be ascribed to the different conventions in each case for legitimizing an argument rather than to any fundamental difference of values.

The major changes in attitude toward science and technology can be found not by comparing Stalinist with post-Stalinist fiction about scientists, but by comparing the relevant values in most of the fiction published in the forties with those found in certain books published in the very last Stalin years. In both the late-Stalinist harbingers of post-Stalin change and in post-Stalin fiction generally, abuses in the spheres of science and technology generally involve rigid enforcement on the part of some scientific bureaucrat of one or another of the three Stalinist scientific axioms. Moreover, the treatment of this topos was frequently structured along lines standard for all Stalinist fiction of the forties, regardless of subject matter: as an opposition between self-seeking "careerists" on the one hand and those truly committed to the cause on the other.[62] At the end of the Stalin period, careerists in science and engineering were represented as examples of the phenomenon Stalin had attacked in his essay on linguistics—that is, as empire-building scientific authorities who retard the advance of science in the selfish pursuit of power, establishing themselves as absolute authorities in their field and stifling all those whose discoveries or inventions challenge their own.[63] After Stalin died, all careerists were represented as Stalinists, and those who opposed them as champions of the post-Stalin movement for change. The difference is not absolute, however: in V. Dudintsev's notorious *Not by Bread Alone* (1956), for instance, when the hero, Lopatkin, tries to get his invention adopted, his opponent, the rival inventor Avdeev, is represented as having established an "Arakcheev-like regime" in Lopatkin's branch of engineering so that Avdeev's invention is still used throughout the Soviet Union long after it has been superseded. [64]

V. Kaverin's trilogy about immunologists, *The Open Book* (1949–1956) is an excellent source for looking at the way attitudes toward scientific work changed over the course of the forties and fifties. The first part of the trilogy, *The Open Book*, was published in 1949, before Stalin's essay on linguistics. In it the author defends the view that the interests of science must be subordinated to those of dialectical materialism and to the immediate needs of Soviet industry, and he shows the necessity for scientific workers to engage in political activity. It is only self-seeking careerists who oppose these principles. [65] The second book, *Dr. Vlasenkova* (1952), was published after Stalin's linguistics essay, and expresses the opposite views on these points, although in the opening (and presumably first-written) sections there is a residue of the contrary attitudes. [66] In the later chapters Kramov, an "Arakcheev-like" villain, musters his sycophants to reject the dissertation of a character who challenges his theories, and who insists that "one should conduct the investigation only on the basis of *universal* biological laws." [67] The heroine is torn between submitting to Kramov's will and following her conscience, but finally she decides to stand up for intellectual truth (which triumphs, of course). [68] In *Searches and Hopes* (1956) the new attitudes discernible in *Dr. Vlasenkova* have become bald statements as the heroine draws a distinction between the "pseudoscientists"—the parasites of the scientific world who enjoy wealth and power and will distort scientific truth any way necessary to achieve their ends—and the "serious scientists," who persist in true academic inquiry, using traditional academic discipline. [69]

The theme of Stalinist abuses in science, which was prominent in *Dr. Vlasenkova*, persisted in Soviet literature for some time to come. As the literary climate worsened, however, it was taken up more in samizdat and tamizdat. Literature reverted to the tamer theme of careerism and intrigue in the scientific institute and research team, a perennial topic since the late forties, and one that persisted through the eighties. [70]

As even my brief synopsis of *Dr. Vlasenkova* makes apparent, fiction of the late forties and fifties reopened the question of independent intellectual inquiry, which had been common in the literature of War Communism. Indeed, it has often been suggested that much of this fiction about scientists and engineers was not just about Soviet science and technology but about intellectual and creative freedom. It was easier to treat these delicate topics in terms of scientific inquiry because writers could then support their arguments with examples of direct and obvious utilitarian benefit to the state. It was rare for any writer to suggest that intellectual inquiry was valuable in itself. And so, in Dudintsev's *Not by Bread Alone,* the battle for the right to intellectual independence revolves around rival inventions of a machine for the centrifugal casting of sewerage pipes; the author keeps emphasizing how much metal is saved, and so on, by his hero's machine as compared with that of his rival.[71]

It would be wrong, however, to assume that the theme of the inventor was merely a sort of Aesopean language for talking about the situation of the writer or intellectual. The cult of science was strongly felt in Soviet society of the fifties—the age of *Sputnik.* Indeed, *Sputnik* brought with it a revival of science fiction, a genre that had been very poorly represented in the thirties and forties.[72] Even as *Sputnik* was being launched, I. Efremov's *Andromeda Nebula* appeared, a science fiction novel written in the tradition of Bogdanov's *Red Star* and other early Soviet science fiction. In a preface written for the 1958 edition, Efremov tells his readers that, now that *Sputnik* has been launched, he will have to advance the time predicted in the earlier, journal edition for realizing his scientific intergalactic utopia: "I miscalculated the pace [*tempov*] at which technological progress will gather speed and especially that gigantic potential, what is effectively the boundless power which communist society will give mankind." Efremov goes on to say that he offers no apologies for clogging his narrative with scientific information, concepts, and terminology, for he believes it is only by making no concessions to reader accessibility in this regard that he can convey what it will be like at that future time when "science will profoundly pervade all concepts, conceptions, and language."[73]

This cult of science reached its zenith in the early sixties, when it was reflected not just in literature but in official policies as well. In 1961 the term "scientific-industrial revolution" (originally proposed by Bulganin at the Party plenum of 1955) was included in the Party Programme. In that year, too, official ideologues changed the status of science in the base-superstructure model of classical Marxism, identifying its place as the base rather than the superstructure.

Yet at the very time when faith in science and technology was at its height, opposing trends had begun to appear in literature that would eventually express themselves in an explicit reaction against the cult of science and technology. Initially this reaction was directed largely at the simplicity of the prevailing account of science.

For all the apparent dissidence of most of the fiction published under Khrushchev that purported to be arguing for "truth," it was largely informed by the somewhat simplistic, scientistic conception of science that had dominated Soviet culture up to that point. Despite such suggestive titles as *Those Who Seek* and "One's Own Opinion," the heroes of this fiction were not really "questing" in a very meaningful way. Their authors, generally speaking, believed, like Lenin, that there was *a* science that was known, or knowable, *a* truth or set of truths that were objective, and which Stalin had violated. V. Dudintsev in *Not by Bread Alone,* for instance, makes heavy use of the same dark-light symbolism that Lenin was so fond of using in describing the wonders of science. And D. Granin's *Those Who Seek* is actually about inventing an electrical cable fault detector—in other words, working in the service of electricity, Lenin's great icon of science. Several other novels about engineers (such as G. Nikolaeva's classic *Battle en Route* of 1957) are about making tractors, another traditional Soviet symbol for modernization and technological advance.

After the launch of *Sputnik,* Soviet literature was able to break into a fresh theme—cosmic flight. It would be too glib to claim that this switch in subject matter enabled Soviet literature to reconsider its stock patterns and symbols. Indeed, as we have seen in the case of Efremov, *Sputnik* inspired many a writer to proclaim that the kingdom of science was nigh. Nevertheless, at about this time we find in Soviet literature a major reevaluation of its traditional symbols for the way forward via science and technology, such as electricity and the almighty tractor. On a superficial, thematic level we find the first questioning of electricity itself in A. Bek's novel *The New Appointment* (written in 1964 but not published in the Soviet Union until the advent of Gorbachev), which ridicules Stalin's insistence that Siberian blast furnaces be powered by electricity rather than coal. And Granin's *I Go into the Storm* (1962), in which a young scientist who flies into the eye of a storm to perform scientific experiments is tragically killed by *lightning,* presents a more complex account than *Those Who Seek* (1954), or even "One's Own Opinion" (1956), of the moral and philosophical issues posed by trying to push forward the frontiers of science. (He also questions the ideal of rationalism.)

As such axioms of Soviet literature were being challenged, reverberations were felt at a less superficial level. Most 1950s fiction that decried abuses of truth spoke of *two* truths—the false, Stalinist truth and the true, Leninist truth of the post-Stalin era.[74] But in the 1960s the notion that truth is complex, and perhaps even varies according to point of view, became a common theme of fiction. This more modern standpoint had an effect in turn on the very way novels were put together. Through most of the fifties the old formulas of Stalinist fiction were still being used, albeit adapted so as to convey an anti-Stalinist political message. But in the sixties the socialist realist conventions began to disintegrate as writers experimented with complex points of view, irony, and other forms of self-conscious narration.

The Brezhnev Era

For the decade or so stretching from the mid-sixties to the mid-seventies, Soviet literature, in reaction to the cult of science which had characterized the Khrushchev era, reverted again to the second of the two alternating patterns. The three main trends in fiction were *byt* prose (about the everyday lives of insignificant individuals), war fiction (about the Soviet role in the Second World War), and "village prose" (which generally dealt with life in some remote rural spot cut off from the mainstream of Soviet life). We have tended to look at these three trends as if they were different. In fact, however, each represents a piece of the same whole. In a sense, war fiction was the guarantor which ensured that something as relatively unconventional as village prose could exist. At the same time, village prose was in some respects a kind of *Heimatsliteratur* for the returning soldier (several of its authors had fought in the war, and all were of an age to have gone through that experience and remembered it). It is not surprising, then, that these seemingly distinctive trends share certain features that define their work as exemplifying the second of the patterns I have identified: their fiction is not forward-looking and shows no interest in technological progress (indeed, if anything, it questions its value), looks inward at the Soviet Union rather than at the West (this is true even of war fiction), and is often preoccupied with the struggle against the nation's enemies, sometimes perceived as the Stalinists, sometimes as the invading Germans.

The trend that undoubtedly best defines literature in this period is village prose. It was here that, for the first time since the beginning of the twenties, writers questioned on any scale the technological ideal and the view that material progress is all. They also bemoaned the threat to the environment and the traditional way of life posed by advances in technology. In a sense, then, the duel of utopias (urbanist and antiurbanist) that characterized the earliest years of Soviet literature, under War Communism, resurfaced at this time. There was not, however, even in official pronouncements, much left of the utopian enthusiasm for the transforming powers of technology that one had so often heard during that earlier period. Trotsky in *Literature and Revolution,* for instance, had insisted that "the machine is not in opposition to the earth," and claimed that "man will do so well" with the machine that the creatures in nature "won't even notice the machine, or feel the change, but will live as [they] lived in primeval times."[75] Society at large had begun to reexamine its blithe confidence in technological progress, which was signified by the emergence of village prose.

Although village prose blossomed in the sixties and seventies, one can identify earlier postwar antecedents. Even before Stalin's death, in works such as L. Leonov's *Russian Forest* (1953), writers had begun to show how economic policies could have a detrimental effect on a local ecology, and even to question whether that effect might not reverberate throughout the "organic whole" of the country.[76] After Stalin died, the trend, although initially far from dominant, gathered strength. In crucial respects the values of village prose countered those of the intervening Ovechkin school of rural writing (which supported modernization); yet village prose essentially grew out of the Ovechkin school. Many trace its origins to Vladimir Soloukhin's sketch "The Hamlets of Vladimir" (1955), which at the time seemed to represent just another contribution to the Ovechkin school, but retrospectively seems distinct from it because of a crucial moment when the narrator rounds the bend on his rural ramble to be confronted by the horrifying pollution emitted by a provincial factory.

The contrast between writers of the Ovechkin school and those of village prose is a matter not just of countervailing ideals but also of the sociological profiles of the writers themselves. If writers of the Ovechkin school tended to have a common background of engineering or agricultural training, economic journalism, or Party work—that is, they were men of "socialist practice"—those of village prose had a more humanistic background, usually involving study in the humanities or the arts.[77] This difference in background was no doubt a factor in shaping the fiction produced by the two groups: thus village prose is rather more sophisticated and better crafted than the journalistic fiction of the Ovechkin school, and it tackles broader and more fundamental issues. It is not just about "district routine"—about, for instance, an agronomist who recommends to the *kolkhoz* head a date for sowing corn that is different from the one mandated from above, but also more reasonable. Instead, village prose engages some of the most important philosophical issues raised by the Promethean ideal to which the Soviet Union has so long been committed. Indeed, many of its writers present in their accounts of Russian rural life what is effectively a counterideal: the village becomes in their works a humanist's idyll, a paradise (which they variously see as lost, to be regained, or to be preserved), which stands for wholeness in an age of alienation.

One of the major examples of village prose is V. Rasputin's *Farewell to Matyora* (1976), a work that represents some sort of apotheosis for the trend. It should be obvious even from my sketchy plot outline that this novel is a parable about the dangers of wreaking havoc with the old traditions and sense of community in the name of the Promethean tomorrow. *Farewell to Matyora* concerns the inhabitants of an island on the Angara River which is to be submerged in the building of a hydroelectric dam, that favorite symbol of Stalinist fiction for the new age. On this remote island the inhabitants have been able to preserve the old Russian customs and Orthodox rituals to a remarkable degree, but now they have been ordered to evacuate to the mainland, where they will be housed in ill-constructed, inhospitable apartment buildings, and where it will be difficult to maintain the old traditions.

At the center of the novel are three generations of the Pinigin family—the old matriarch Darya, her son Pavel, and her grandson Andrei. Andrei wants to work for the hydroelectric station, to keep pace with the dizzy new times. Rasputin has Darya enter into several philosophical dialogues with him over the course of the novel in which Andrei argues for the new way of life, Darya for the old. Andrei, echoing Gleb Chumalov and others who have gone before him, maintains that "man is the tsar of nature and there is nothing which humanity cannot do with its wondrous machines."[78] Darya, however, sees human beings as small, weak, and deserving of pity. Modern man, she believes, has grown proud, and has sought to clamber out of his "human skin," uprooting himself in those frenetic cities from the earth, from nature, and from God; he has forgotten that he has a soul and a conscience.[79]

Thus Rasputin, his periodic disclaimers that he is not against technology notwithstanding,[80] implicitly calls for a reordering of official priorities away from the goal of technological advance at any cost, urging instead that more emphasis be placed on moral and spiritual values, on paying attention to preserving the environment and its aesthetic qualities, and on maintaining continuity with Russian traditions. These views are, however, not as radical as they might seem, but rather represent an established position within one of the major debates of Soviet intellectual life, which emerged in part from under the coattails of village prose but continues today, when village prose has long since faded from prominence. Indeed, the seventies saw a reevaluation of hydroelectric power as the cornerstone of Soviet energy policy; thus *Farewell to Matyora* cannot be regarded as daring by virtue of the fact that it raises questions about that old Soviet sacred cow, the hydroelectric dam.[81]

The Late 1970s and 1980s

Village prose itself arguably began to wither away after *Farewell to Matyora* was published in 1976, although not without some help from on high. It was clear by the late seventies and early eighties that that aspect of village prose which Konstantin Chernenko was to characterize in his speech to the June 1983 Party plenum as "God-seeking motifs and the idealization of the patriarchal order" was troubling many in high places. Undoubtedly, sectors of the literary community had also begun to tire of its contrived quaintness and its provincialism. In any event, beginning in the late seventies there was a broad-based reaction, reflected in both criticism and official speeches, against what was now labeled a literature of nostalgia, refusing to accept the mandate for modernization.[82] Authoritative voices pointed out that recent fiction had tended to be about inconsequential people in situations that were equally inconsequential from a political or economic point of view, and called for a more impressive "scale" (*masshtabnost'*) in fiction. Official spokesmen insisted that, as the title of the 1982 Central Committee resolution on literature put it, literature "Increase Its Creative Ties with the Practice of Communist Construction"; that writers devote themselves more to public affairs and economic journalism, as they had done in the past; that they depict the scientific-technological revolution.[83]

In a sense, then, the authorities were calling for a return to the sort of writing that characterizes the first of our two alternating patterns. As Soviet literature reacted against the hermetic, parochial, and Russocentric world of village prose, it in some strange sense returned to the opposite pattern, although the literature of the eighties bears little resemblance to the literature of the twenties and other earlier periods that was inspired by the Promethean ideal. Literature typical of the late seventies and early eighties represents the first pattern in that it is more forward-looking than that of the sixties and seventies, is concerned with scientific and technological progress (albeit in large measure in order to question the old Soviet axioms about it), and is positively obsessed with the West. Science fiction, which had all but disappeared from Soviet literature during the heyday of village prose (some would say it was kept out), enjoyed a resurgence at this time.[84] Also, there was a sort of mini-revival of the highly journalistic fiction of the Khrushchev years, the last time that literature about science and technology had enjoyed any prominence. Publishing houses began reprinting the works of the Ovechkin school,[85] while in criticism and official pronouncements it was declared that their literary approach and general values are "even today still resonant."[86] Even Dudintsev's *Not by Bread Alone* was republished in 1979.

This revival was not, however, able to change the direction of Soviet literature to a very marked degree. More production novels and sketches on rural and industrial themes appeared in the early eighties than had been the case in recent years, but they did not have much resonance.[87] Indeed, while the official platform called repeatedly for writers to produce a sort of fictionalized economic journalism, critical attention, and even the literary prizes, were accorded to works written in a very different vein (it was almost as if the directives were being given out with one hand and the prizes with another).

If Soviet literature was not to return to being a sort of economic journalism, a palpable change did nonetheless occur in the late seventies—a change, moreover, that responded, at least in part, to pressures to give technological progress and the modern world their due. Fiction typical of this interstitial time conflates the conventions of village prose with those of an earlier variety of socialist realism, usually the production novel. A prototypical work exemplifying the shift from village prose in the last years under Brezhnev is Chingiz Aitmatov's novel *The Day Longer Than an Age Doth Last* (1980). When it first appeared, the book was hailed as a model for new literature (in a speech to the Seventh Writers' Congress in 1981, G. Markov, then head of the Writers' Union, enshrined it in the official canon as an exemplar of socialist realism).[88] And yet it was also extremely popular among intellectuals and even dissidents. Aitmatov's novel provided a model for combining both "the village" and the "greater scale" with its singular structure involving two plots that run parallel throughout and are alternately presented but come together at the end. The main plot concerns railway workers at a remote whistlestop in the desert of Kazakhstan (the small, hermetic setting, typically now no longer a village but a workers' settlement) but digresses into Kazakh lore. In treating this material Aitmatov, a former fellow traveler of village prose (being himself Kirghiz and not Russian), walks a fine line between using the conventions of that school, with its traditionalist ethos (he even uses as his positive heroes aging representatives of the folk, reminiscent of Rasputin's Darya from *Farewell to Matyora*), and showing how technology has ameliorated their lives. The second plot line concerns a joint U.S.-Soviet space mission which uncovers a superior civilization in another galaxy. When that civilization reaches out to earth, proposing scientific collaboration, the joint commission decides to prevent further contact by sending up two rings of rockets, one Soviet, the other American, to encircle the globe and destroy any vehicle approaching from space.

The civilization the astronauts discover in this other galaxy is particularly significant for us, because it has many aspects in common with Campanella's *City of the Sun,* including a cult of the sun called the *Derzhatel'* ("Holder"; solar energy is also the inhabitants' source of fuel). These inhabitants have attained a higher stage of development toward their ultimate aim of realizing a kingdom of reason, and, as in Campanella, have made considerable advances in controlling the climate, alleviating physical and mental suffering, and prolonging life for as much as two hundred years. They live in a high-density urban environment and yet are free from want or pollution, although their civilization is threatened by encroaching arid areas which they have been unable to check. Most characteristic of the civilization, however, are not its many achievements but the four things Aitmatov maintains they do *not* have: money, government, weapons, and war. The absence of money, of course, defined Bacon's and Campanella's utopias too; but these classical utopias, far from spurning government, weapons, and war, saw them as crucial to the maintenance of the utopia. Aitmatov, however, depicts the pacifism of his superior society as its crowning achievement, and earth's proclivity for wars as its greatest failing and the main impediment to the advance of civilization.[89] Indeed, earth's response in shutting out this superior civilization by military means is arguably a parable of the "closed society" of the Soviet Union.

The Day Longer Than an Age Doth Last was not the first book of the turn of the decade to be written according to this general recipe; but with its considerable official endorsement it became a banner for the new trends in fiction, sometimes labeled "fantastic" or "cosmic" realism. While Aitmatov's novel advocates founding a kingdom of Reason, many works representative of this trend champion various forms of nonreceived science, such as the occult and folk remedies. But they share with Aitmatov a fascination with notions that typify utopian literature and speculative science fiction, such as an obsession with prolonging life or even achieving life eternal—the Fyodorovian imperative.[90] Aitmatov is atypical, however, in his faith in achieving the kingdom of Reason. More typically, Anatolii Kim's novella *Gurin's Utopia* (1979) contrasts a utopian project—a large glass phalanstery in the middle of what appears to be an arid region of Central Asia—with the distinctly un-utopian reality of achieved Soviet life. Furthermore, the utopian dreamer's foil in the novella, an earnest and exemplary champion of the old ideal of a technological utopia, annotates the project with the observation that "any utopia or City of the Sun is just an attempt of one person to solve with his little brain the problem of happiness for all mankind . . . Do you have the moral right to do that?"[91] His observation may be seen as an echo of Lenin's scorn for utopianism, yet it also implicitly questions the attempt to found a City of the Sun in the land of the Soviets, a question raised with varying degrees of explicitness in much of the fiction of this period.

The end of the Brezhnev era and the successive changes of leadership did not alter significantly this new direction Soviet literature was taking at the end of the seventies or the agenda of issues it addressed. Initially there was some attempt to stem the tide of literature of overtly religious content. In 1983, under Iurii Andropov, for instance, directives went out to publishing houses stipulating that they must censor from texts any mention of various well-known Russian religious thinkers.[92]

If such restrictions were intended to redirect literature into something like the first of our patterns, they had little effect. Indeed, one would need a Procrustean bed to fit the kind of writing seen since the late seventies neatly into the first pattern. The serious questioning of the Promethean ideal which marked so much fiction of the previous twenty years could not dissipate overnight. The heady, innocent early days of the Revolution, and even the idealism of the Ovechkin era, were now well past. Most writers could no longer give completely unambiguous responses to the big questions about scientific and technological progress raised in the debates of that period, and made more urgent by the intensification of the arms race (whose shadow hangs over much recent fiction).

A de facto reordering of priorities had been taking place in Soviet prose over that twenty-year period. The critic A. Bocharov, in an article of 1982 on the state of Soviet literature, which effectively sums up a fait accompli, maintains that Soviet literature has outgrown the "Haileyism" of its past (the reference here is to Arthur Hailey's popular novels, such as *Airport,* some of which have been translated into Russian and are widely read, and which describe how a particular workplace is run).[93] It is now possible, Bocharov proposes, to dispense altogether with the old mandate to write about how the hero works with the new technology (or, as we might say, to use the old "boy meets tractor" plot). Rather, writers of today should concentrate on moral and spiritual problems.[94] Indeed, even Ovechkin's writings are seen in a different perspective from that generally found in reviews of the fifties. What reviewers find particularly instructive in his sketches, besides the general orientation toward greater efficiency and progress, is the negative example he gives of *Borzovshchina* (named after the protagonist Borzoi), illustrating the dangers of forcing the tempo of production and modernization. This theme has experienced a revival in recent fiction, but in many instances the writer attributes some factory disaster in the present to the disastrous policies of the Stalinist past, when the pace of industrial progress was forced and ill-advised, resulting in shoddy work that had to be paid for sooner or later.[95] Such parables are, of course, directed less at uncovering the past than they are at exposing policies in the present.

The debate over technology in the eighties was less likely to be conducted in the fiction of factories and *kolkhozy.* For a start, the intensification of East-West rivalries in the early part of the decade and the phantom of nuclear disaster proved more compelling topics for writers than, for instance, how to stimulate initiative on the *kolkhoz.* The typical novel of the eighties is centered on a contrast between the West, represented as riddled with moral degradation, a cult of violence, and militarism, and the Soviet Union, which is imperiled by the diseased West but nevertheless holds out the only hope for man's survival. As this trend intensified, Aitmatov's *Day Longer Than an Age Doth Last* faded from prominence and was replaced at center stage by the fiction of Iu. Bondarev, an extremely vocal champion of the environment, but from the conservative camp.[96]

Bondarev's novel *The Game* (1985) is of particular interest because of its bizarre account of modern science and technology. Actually, the intellectual who—quite typically for the eighties—is the novel's protagonist is not a man of science but a Soviet film director, Krymov, fresh from a Paris film festival, where his latest work has just carried off first prize. While there, he was drawn ineluctably (as have been so many of his counterparts in other recent fiction) into a philosophical dialogue with a worthy antagonist who represents the West, in this case John Grichmar, an American director who is more worthy than most because he is partly of Russian descent. Krymov has returned to Moscow early, fleeing the cocktail bar atmosphere of the West, where he is repelled by everything, right down to the very smells of the toiletries, but Grichmar follows him home, hungry for more dialogue.

Their conversations and Krymov's thoughts center on the idea that the technological revolution of the twentieth century has gone out of control and on forebodings of imminent apocalypse, the destruction of the earth. Grichmar asks Krymov to collaborate on a movie he wants to make, to be called *The Last Turtle*. In this film the world will be plunged into atomic war, which only a single turtle survives. The turtle then makes for the sea, but finds in its place a burnt-out crater. Overcome, it perishes, and that great red sun which is the explosion dies out too.[97] It is not clear whether Grichmar realizes that in Hindu mythology the turtle supports the earth, but it *is* clear that this somewhat crude and illogical parable is intended to warn of the death of the sun as symbol of knowledge, the future, and the life force.

Krymov rejects Grichmar's offer, for, as a Russian, he must have hope and faith. Indeed, his own apprehensions focus less on the danger of nuclear war than on the evils of the technological age in general, and of Americanization in particular. Taking a cue from Ronald Reagan, he describes America as the work of evil and the devil; he also calls American culture a "concentration camp of the everyday" (*bytovoi kontslager'*).[98] Krymov believes, as did so many of the idealists and zealots of the period of War Communism, that it is the commercialism and orientation toward entertainment that is so fatal in Western culture.[99] He seeks out places

that are as untouched as possible, but is dismayed to find even the re-
motest corners of the Soviet Union threatened by technological advance:
in his beloved northern area of Pechora the salmon are slaughtered by an
electrical device, and he predicts that in twenty years everything there
will be dead. His jaded American friend tells him the age of technology
cannot be stopped. Its express train has gathered speed. The passengers celebrate and drink cognac. The engine driver has gone crazy;
there are no brakes; but "you and I drink and know that ahead there
looms the abyss and destruction for all." Yet Krymov believes (and we
have heard all this before) that Russia is different—indeed unique—for
she has spirituality and has been "programmed as the conscience of the
whole world." "If anyone can save prodigal civilization it is Russia."[100] At
the end of the novel Krymov dies, but as he does so he sees approaching
him his spiritual mentor, Avvakum, the intransigent schismatic of the
Russian Orthodox church, founder of the Old Believers, who was burned
at the stake but defiant to the last.

In this novel, then, we see, besides the old myth of Russian particular-
ism, the extreme idealism and puritanism that, in various forms, has
dogged Soviet intellectual history since the very beginning. Like Aitmatov before him, Bondarev shows a fascination with quarantining and san-
itizing, but he would exorcise Western culture not in order to achieve the
old ideal of a closed society, but to establish an *enclosing* society where
everyone is protected from the modern world. For him, electricity and
the sun are no longer utopian symbols of a future where science and
reason reign and technology is their handmaiden. Rather, they are apocalyptic symbols portending the death of civilization.

Gratifying though *The Game's* rabid anti-Americanism might have been
to some, the novel must surely have given pause to those who still truly
believe in the Promethean ideal. When, in this sort of fiction, the writer
attacks "America" and its alleged rabid militarism, which threatens to
send a holocaust down upon the entire world, he—and even a writer as
rabidly anti-Western as Bondarev—could always make the point that
brakes should be put on the arms race in general. Moreover, this point
was often linked to a concern for the way unchecked technological expan-
sionism can lead to its own form of holocaust, as Mother Earth becomes
unfit for human life. A high percentage of the fiction on contemporary
topics that appeared in the eighties contains some doomsday scenario,
which normally conflates the imagery of nuclear holocaust with that of
ecological disaster. The "holocaust," however, is at base a metaphor for

the irrevocable destruction of the human moral and spiritual ecology (a "fire," in the words of the title of Rasputin's next novella, published in 1985). This position was strengthened in the wake of the Chernobyl disaster of 1986, which found both direct and indirect reflection in literature.[101] Indeed, expressions of concern for maintaining the delicate balance of the ecology, which in the heyday of village prose were sometimes reminiscent of the "precious bodily fluids" rhetoric in *Dr. Strangelove,* began over the course of the eighties to sound like that of the pan-European movement of the Greens.

The obvious question arises: What has been the impact on this debate of Gorbachev's accession to power, which after all antedated Chernobyl? Gorbachev became First Secretary more or less as Bondarev's novel was being published, while Rasputin's *Fire* appeared shortly thereafter (it was probably begun before Cherneko's death). There is, in other words, no magic dividing line—particularly not in terms of the Promethean issue—between much of what one finds in literature before and after his accession. Gorbachev presents himself as an enthusiast for modernization and efficiency, and under him, official voices have again called for fictionalized tracts written in the service of the "scientific-technological revolution."[102] Once again, however, the writers have been slow to produce it and seem to revel instead in creating doomsday parables.

In many respects, the kind of fiction that has been appearing under Gorbachev, striking though it is in its outspoken critique of Soviet society, represents in its treatment of the Promethean theme an intensification of trends that marked the eighties in general, rather than a dramatic change of direction. So many of the motifs and attitudes concerning science and technology that seem most striking in recent fiction can be found in *The Game.* These include the doomsday scenario of ecological disaster brought about by a humankind drunk with technology and military might;[103] a questioning of the long-standing program for the "conquest of nature";[104] a skepticism about institutionalized belief systems or mandatory accounts of "science," which in literature and film has been given full rein in pathos-ridden accounts of martyred intellectuals;[105] the recurrent theme that less attention should be paid to mere plan fulfillment and more to the ultimate aim of producing fine human "souls";[106] and even the cult of Avvakum, a figure ostensibly far from the mindset of a scientist, let alone a dialectical materialist.[107] Indeed, the main difference between Bondarev's presentation of these themes and their characteristic treatment under Gorbachev consists in the muting of his rabid anti-Americanism—but only a muting. Bondarev represents the conservative wing—some Soviets label it "fascist"—of a spectrum of intellectual positions.

As writers voted throughout the decade with their pens in favor of a more humanistic emphasis in literature, and of paying more attention to moral and spiritual problems, the implications of their vote reached beyond the thematic, beyond the mere matter of the topics generally chosen (tractors versus spiritual malaise). At some level their vote implies a reexamination of some fundamental Soviet assumptions and ideals.

It is in this area that the years of thaw under Gorbachev have played such an important role. A thaw essentially means a time of reevaluation accompanying a change in leadership, a time when the conventional symbols and clichés can be transvalued. This particular thaw has, to a more marked degree than the earlier ones under Khrushchev, been noted for its retrospectivism, for the republication on a huge scale of manuscripts and authors previously blacklisted, and particularly material from the 1910s and 1920s by authors who emigrated or were deemed "internal exiles." The country is also officially committed to filling in the notorious blank spaces in its past and has undertaken an extensive reexamination of history. In addition to the dramatic rehabilitations of figures like Bukharin, we have seen the mass media flooded with memoirs relating to all periods of Soviet history, material that seeks to correct false versions of events given in the now discredited official histories. The history of Soviet science, has, of course, been a beneficiary of this process, and some of the literary landmarks of recent years, such as Dudintsev's *White Raiments* (1987), on geneticists in the forties, and Granin's novel *The Aurochs* (1987), on N. V. Timofeev-Resovskii, have been concerned with the dark pages in its history.

Despite the obvious benefits for science and knowledge (*nauka*) of this public soul-searching, one must ask: Why is a country that is ostensibly acutely aware of the urgent need to modernize its economy so obsessed with the past? I would suggest that this is so at least in part because it seeks a more ambitious reevaluation or reorientation than it has entertained until now, even in the dramatic thaws under Khrushchev. In consequence, the mass media have been obsessed by the many fateful moments in the nation's past when alternate versions of the way forward were eliminated; much of its fiction set in the past is concerned with a point "on the eve" of a major shift in policy (to use the title of V. Belov's novel on the prelude to collectivization, the long-awaited second part of which was published in 1987). Hence, for instance, A. Rybakov's *Children of the Arbat* (1987), one of the most famous books of the thaw, is set around the time of the murder of Kirov, an event generally recognized as setting up a sequence leading to the Great Purge two years later.

Rybakov's novel is one of the many to appear during the Gorbachev thaw that show how, under Stalin, highly competent experts and scientists, many of them sophisticated and cosmopolitan, lost out politically to ill-educated parvenus and xenophobes who sought to impose policies and theories that were scientifically or technologically unsound; the losers were purged or silenced as the country marched on to implement disastrous policies. It can thus be linked to the return in literature in the eighties to the ideal of the intellectual as an axiomatic value. But this novel, like so many that make similar points (such as *The White Raiments*), was actually written in the Khrushchev years with the hope of publishing it then. There is, as these examples suggest, an aspect of belatedness in much of the thaw which makes its publications particularly difficult to categorize. Insofar as these particular works divide all characters into black and white, and show other signs that when they were written, the socialist realist tradition still had a considerable hold on literature, they now seem quite dated. They share, however, many themes in common with *The Aurochs*, which is quite self-consciously *not* a product of the Khrushchev period.

Granin's novel is an act of hagiography centered on Timofeev-Resovskii, a geneticist of world renown who was condemned to prison camp and exile because he would not give up his research on the effects of radiation on gene mutation.[108] His career becomes, in Granin's "documentary-novella," a parable of the true scientist and the potential savior from ecological disaster. Timofeev-Resovskii, called Aurochs in the text (although the reference is quite clear),[109] is identified with that ancient European bison, one of the last of a noble breed who are untamable and completely unconcerned with conforming to the regnant conventions. In presenting this somewhat romanticized image of his subject, Granin is at pains to differentiate between what his "aurochs" stands for and scientific thinking of the Khrushchev fifties, when Timofeev-Resovskii was rehabilitated and permitted to reenter the Soviet scientific world. *Then*, the narrator tells us, intellectuals were infatuated with the potential of physics (or, alternatively, of mathematics, its chief rival claimant to the status of queen of the sciences). They believed physics could provide keys to a rational organization of mankind in the future—something we have seen in Efremov's foreword to *The Andromeda Nebula*—and they adopted a condescending attitude toward the humanistic sciences, regarding them as somewhat outdated and of secondary importance. But when they confidently appealed to Aurochs for support, they were met with a bemused smile. [110]

Granin's novel provides an excellent test case for gauging the changing attitudes toward science and technology. He himself was trained as an electrical engineer, is a Party member, and has served for many years as head of the Leningrad branch of the Writers' Union. Thus he represents not only mainstream Soviet literature but also that category of engineers and agronomists, who were intended by Lenin to become more and more visible in the Soviet political arena; indeed, his field is even electricity. Several of Granin's earlier works, such as *Those Who Seek* (1954), seem committed to the Promethean ideal and to the crucial role the scientist-engineer will play in reaching it. Thus *The Aurochs* can be seen as a mark of the evolution not only of Granin's own thinking on the question of the role science and technology should play in the new society, but also of that of Soviet literature generally. Literary works now question the traditional Soviet opposition between the materialist world view and the religious or some other position considered antimaterialist, and argue that such an absolute opposition has been fateful for the Soviet Union; there must be some middle ground. [111]

The Aurochs is a sign of evolution not just in the literary treatment of this subject but also in Soviet thinking in general. There are suggestions that perestroika is increasingly conceived as entailing not just an economic or even a political rebuilding, but a fundamental overhaul of the Soviet cast of mind, leading to yet another "New Man." Even at the highest levels, concern has been expressed for the way the country is wedded to an uncritical faith in technology as a cornerstone of its world view. It is widely felt that, just as faith in science can devolve into scientism, the Soviet faith in technology has devolved into a kind of technologism. Articles have appeared recently in such authoritative places as *Kommunist* which argue that Soviet society should work toward a new mental outlook, one that is more humanistic and less crudely oriented toward sheer technologism.[112] There has even been an attempt to found a high-level think tank for this purpose, answerable only to Gorbachev himself.

Thus, in terms of the broad context of the diastole and systole of Russian attitudes toward science and technology, the Soviet Union seems to be experiencing a reaction against the idealization of science in its most militantly dehumanizing forms, and literature itself has been at the forefront of this reaction.

· 10 ·

World Outlook and Inner Fears in Soviet Science Fiction

Richard Stites

Much of the commentary on contemporary Soviet science fiction suffers from an insufficiency of historical understanding. Unlike similar genres in Western nations, Russian science fiction, known by the somewhat more lyrical name of *nauchnaia fantastika*—science fantasy—was born into a setting of backwardness in material life, small but flourishing islands of high culture and academic eminence, and a visionary and largely radical intelligentsia who dreamed of remaking Russia into a land of social justice and technical competence. In 1890, about the time that a continuous tradition of science fiction arose there, Russia was very far from either goal. With the onset of industrialization, the older dreamy and millenarian traditions of social fantasy gave way to a "harder" genre of scientific speculation. After a decade or so of technological speculation about such things as the wonders of electric power and modern weapons, Russian Marxism was linked with this technical genre in Aleksandr Bogdanov's *Red Star* (1908), a communist utopia set on the planet Mars. Out of it—as out of Gogol's "Overcoat"—flowed the main strands of revolutionary science fiction for the next two decades. Bogdanov's red and rosy vision was matched and challenged by a series of antisocialist works offering either the frightful specter of a totalitarian future or a counterutopia of antimodern, pastoral escape.[1]

This dialogical pattern was repeated in the revolutionary decade 1920–1930 under Soviet power in an environment of Party dictatorship, a mixed economy, and relative pluralism in society, artistic experimentation, and utopian experiment. The revolutionary storm that blew over Russia and shook away portions of the old order also irrigated the intellectual landscape of early Soviet life, from which sprouted a lush crop of futurology. Physicists, geologists, astronomers, and biologists vied with journalists and literary figures to speculate in the press about the coming character of a world made perfect under communism. Scientists of the New World spoke of prolonging human life, while mystical philosophers with roots in the Old World insisted on the possibility—indeed the moral need for—abolishing death altogether and revivifying all of deceased humanity. Bolshevik worship of the machine, cults of Ford and Taylor, and a frenetic technolotry suffused the arts, the theater, and the military, and shaped all kinds of revolutionary discourse. The tractor became an emblem of modernizing the countryside and fighting religion. Aviation—as a physical binding force in the vast spaces of the USSR and as another symbol of modernity—became a public hobby. Scientists and laymen alike, driven by an intense interest in outer space, saw distant stars as more fortresses for bolshevism to storm along the way to cosmic liberation. [2]

Soviet science fiction in this era, though drawing much from Bogdanov and other prerevolutionary writers, was ruled by the atmosphere of optimistic and limitless change—a collective eulogy to Prometheus and Marx. The revolutionary age, 1917–1930, witnessed a mounting wave of utopian vision and experiment: culture building, egalitarianism, collective creativity, moral brotherhood, sexual revolution, and urbanization inspired thousands of intellectuals, students, and workers to point demonstratively to the glimmering daybreak of their tomorrow. The grim, prosaic realities of daily existence for most Soviet citizens of the period endowed this vision with exceptional pathos. On a time scale, utopianism could be divided into present, near future, and distant times to come. In the first category were the social experimenters of the time, in particular the hundreds of city people and country folk of the communal movement—like-minded comrades of both sexes who lived together, pooled their meager earnings, divided the labor, shared all goods, and tried to cement friendship in a familylike setting. In the near-future category were architects, city planners, economists, and sociologists, who drew blueprints for new schemes of structures, space, and population density—deurbanizing dispersal across the land, conglomerations of towering house communes, fantastic cities and anticities to replace the decaying metropolises of Old Russia. And in the category of further speculation, the science fiction writers took on the depiction of an entirely new world to come. [3]

On peering into the future worlds of the most popular and character-istic of the science fiction works of this period, we see in a flash both the tremendous euphoria that the Revolution had breathed into popular writ-ers (most of them non-Party fellow travelers sympathetic to commu-nism) and the endless horizons it opened to them through the device of blending Marxism with "Marsism." The plots and situations were trivial and remarkably similar. One can speak confidently of an almost common scenario of a world environment, an eco-technical system of governance, and human relations. The dominant motif is the world city, or world-as-a-city: an urbanized planet shaped by technology, with mountains moved and seas rearranged, with megacities and a citified globe. In one example the planet is so completely urbanized that one can travel from Australia to Central Russia by walking or riding along city streets. London and Paris are conjoined; Moscograd is an urban blanket of concrete, steel, and glass. Highways repose upon pillars which themselves house thou-sands of dwellers. Tracks encircle the earth like musculature; an endless parade of aerostats, ethoronephs, airbuses, and skycars transport citi-zens to the far ends of the world. Nature (if it is permitted at all) is a vestigial adornment of the urbs or a setting for children's colonies.[4]

Government and economy as such do not exist in the communist future world. Production, processing, and extraction are performed by machines and their tasks coordinated by other machines. Computers cal-culate all production and distribution of the world's resources in proto-cybernetic systems akin to that on Bogdanov's Mars. In one novel me-chanically harvested and packed foods are pulled through underground vacuum tubes from agrocenters and delivered to urban dwellers. No one need work very much, yet everyone does. As the quintessential utopian host puts it in *The Coming World* (1923), "You will want to work."[5] Labor is pleasant and rotational. All the contradictions lamented by Marx are gone: the manual-mental, the male-female, the city-country, and the private-public. Who rules? Either no one at all or a faceless set of rotated leaders or elders. In the first case a kind of "machine politics" in the literal sense manages the only struggle left to humanity, that between man and nature—in other words, production, feeding, and supply. The noneconomic sector is conflict-free. In some cases a Council of a Hun-dred or some such shadowy body—never elucidated—makes general policy and oversees the big machines. Both politics and economics have disappeared. The usual array of fancy technology is on full display (processed food, personal flight, and so on), with special emphasis on transport, aviation, space flight, and—bowing to Lenin's compulsive dream—total electrification. The world sparkles with light and enlightenment.

Peopling this wondrous universe are the expected paragons of communist virtue. They do not compete, hoard, covet, engage in social climbing, steal, rape, kill, sell their bodies or souls, or make war. The familiar static harmony of most classical utopias ensures serenity and peace. Vernean pairs of protagonists are often employed to examine and narrate the new ways: an arid professor for the technology and a down-to-earth younger companion to inspect life and love. Equality is palpable here: unisex clothing and haircuts, androgynous shapes, absence of deference and differential reward, egoless friendships, comradely cooperation—all of these drawn not only from the utopian corpus and various stages of socialist ideology, but also from schemes and practices in the everyday Russia of the 1920s. No politics, no conflict, no struggle for existence, no apparent anxieties. What, then, is the meaning of life? The authors give no direct answer, but they seem to share two assumptions. The first is that the absence of violence and struggle is happiness enough. One might pause a moment to recall what most of these writers had recently witnessed in their country in the years 1914 to 1922 and the conditions of their present life. The second—a standard reply—is that there is drama in the continuous struggle against nature or the exploration of space in order to see (or liberate) other worlds. But the authors limit their explication of utopian happiness to stock descriptions of rituals and spectacles of the most banal order: the disturbingly familiar radiant faces, the collective song, and the joyful reverence toward one's own world, one's own existence.[6]

The pendant to utopian communist heaven in this literature is the capitalist hell. The road to utopia is hardly ever analyzed, but in most cases it arises in the aftermath of an apocalyptic war between the forces of emergent socialism (often specified as an "Eastern" federation) and the resistant evil empires of the capitalist West—Germany, England, or the United States. These are dictatorships, fascoid despotisms, industrial hells, slave states, plutocracies, or feudal and militaristic fossils. As in Jack London's *Iron Heel* (1907), the inspirer of them all, the workers are brutalized or drugged with mass culture and religion, their leaders provoked into vain uprisings and then massacred. A whole genre of "war scare" fantasy grew up, rooted in the era of fear that preceded the First World War. Atomic explosions, skies blackened by bombers, death rays, mad scientists, and supervillains who would rather blow up the world than hand over their property menace the peace of Russia and her allies. Extraordinary scenes of battle, star wars launched from satellites, underwater armadas, proletarian mutinies, and armies of liberation mark the pages of Soviet fantasy. The most popular of them, Aleksandr Beliaev's *Struggle in the Atmosphere,* depicts a war with America so grim and comprehensive that the U.S. Air Force had it translated and published in the 1960s. In the end, the capitalists are beaten and their subjugated peoples led into the light of humanitarian justice.[7]

The "dialogue of dreams" that began before the Revolution could not be conducted openly, and so the ratio of hope and fear was reversed in the 1920s. Pessimistic novels and alternate routes to happiness were rare and often masked. The most influential of these was Eugene Zamiatin's *We,* a political dystopia about a nightmare collectivist state so famous that it scarcely needs discussion. Until 1989 it had never been published in the Soviet Union.[8] Two observations need to be made about it. First, it was not an answer to Soviet science fiction, because it preceded all the works I have just described; indeed, some of those novels may have been hopeful answers to Zamiatin. But *We* was an answer to something: partly to the Bolshevik visions of modernization, machine worship, and authoritarianism, and partly to the general wave of scientific technicism that Zamiatin (an engineer who had lived in England) saw as the principal menace of the twentieth century. *We* was a classic warning and a supreme work of art. At a lower level of expression were three counterutopias that offered a pastoral vision of the future. Apollon Karelin's *Russia in 1930* (1920) was an anarchist projection of a stateless world fictionalized and set a decade ahead. Aleksandr Chaianov's *Journey* (1920) was a peasant utopia of the year 1984 which combined social justice, private farm property, technology, and deurbanization. Finally, the nostalgic *Beyond the Thistle,* written by the Cossack general Peter Krasnov, in Berlin in 1921, chronicles the return of the Romanov dynasty to Russia in the mid-twentieth century to preside over a resuscitated Muscovite, Slavophile, and faintly fascist religious state.[9]

The dialogue of dreams ended in 1931—as did most other kinds of dialogue. The anticommunist utopians and dystopians had been almost voiceless anyway. But the officially tolerated and encouraged genre of communist science fiction utopia was also silenced. The last of the genre, Ian Larri's *Land of the Happy* (1931), had satirized Stalin and Kaganovich (Larri and some other science fiction writers were actually purged).[10]

But the reasons went much deeper than this. Mere fellow travelers were held in suspicion under the new censorious literary culture of socialist realism. Strict controls were imposed on all writers—especially popular ones. The private press and journals that had printed much of the early science fiction were now nationalized. After Larri—and until a few years after Stalin's death—Soviet science fiction avoided the proscribed topics of far projection, utopia, and pictures of social justice and stateless equality, and dealt instead almost exclusively in near futures, new weapons, production norms, espionage, and wrecking. The Stalinist "single state" (the name of Zamiatin's antiutopia) could not tolerate blueprints of a bright future that was easily contrasted with some of the harsh realities of the 1930s, or even detailed treatments of capitalist abuse of workers. It was now forbidden to glorify equality and scorn deference, hierarchy, and authoritarian ruling styles. The whole corpus of science fiction fell into the dreary doldrums of conformity and instrumentalism for a quarter century.

Revival

After Stalin's death in 1953, some of the more notorious terror captains were shot; millions were eventually released from camps; and, in the years 1956–1962, a very large and visible anti-Stalinist campaign was conducted. This partially cleared the literary air of the fear that had hung over it for thirty years. Also, a number of novel developments in Soviet scientific-technological achievement and awareness came to maturity in the mid-1950s: the launching of *Sputnik* in 1957, the revival of public discussion of systems thinking,[11] and the reinstatement of freer modes of scientific discourse in general. In the midst of this euphoric time, Ivan Efremov, a paleontologist, published *The Mists of Andromeda* (1957), called by Darko Suvin "the first utopia in world literature which successfully shows new characters creating and being created by a new society, that is the personal working out of a collective utopia." Efremov depicts a refined race of the distant future, a unified planet with a single language, possessing neutral names (as opposed to the "revolutionary" names often used in the 1920s such as Youth, Revo, Spark, Joy, Will), and eager to explore outer space. Its many innovative elements include a knock at Stalin through the character of Bet Lon (probably referring to the Russian word *beton:* concrete, or hard), and the prominence of women characters as equals to men. It was the first communist utopia in thirty years. Although opposed by some in the Stalinist literary establishment, it was a huge success, enjoying twenty-four editions and translation into twenty-three languages before the author's death in 1972.[12]

Andromeda opened the floodgates for a deluge of new Soviet science fiction, called by an American critic the "second great age" in the history of this genre, and by a Soviet writer "our own 'Campbellian Golden Age,'" thus comparing it to American science fiction of the 1930s and 1940s. Closet science fiction writers, popularizers and writers of other fiction, and—most important—dozens of professional scientists, researchers, and engineers "crossed over" into a now legitimized field of speculation. Efremov replaced Bogdanov as the new Gogol, laying out for an entire generation previously taboo themes and endless possibilities for refinement and development of his ideas and suggestions. Efremov himself remained the leading figure until his death. Science fiction volumes now began to appear in circulations of one hundred thousand copies and were sold out instantly—a pattern that would continue with periodic crackdowns and falloffs until the 1980s. The first such crackdown came in the late 1960s—after the fall of Khrushchev and the inauguration of the stricter Brezhnev regime—in connection with the works of the Strugatskii brothers (whom I discuss in more detail later in this chapter).[13]

By the mid-1970s Efremov was gone, and the atmosphere of political control had led to a shift into more personal and apolitical themes and a vague sense of malaise among writers. The genre not only survived, however, but increased in output and variety. By the early 1980s a kind of tacit compromise had been reached between the purveyors of fantasy and the censors, represented in the works of Kirill Bulychev: science fiction would not be subjected to the rules of "social command" (purely pro-Soviet works of realism and optimism); and the writers would refrain from wandering into the territory of alternative political utopian vision. And so it largely stood until the era of glasnost.[14]

Who are the writers? Who are the readers? An estimate from the mid-1970s suggests that about six hundred writers have produced works of science fiction—fifty or so as full-time science fiction authors.[15] A dozen or more of these have achieved high visibility in the West in translation: the Strugatskii brothers, Dmitrii Bilenkin, Bulychev, Mikhail Emtsev and Eremei Parnov, Vladimir Savchenko, Anatolii Dneprov, Il'ia Varshavskii, Olga Larionova, Gennadii Gor, E. Voiskunskii and Isai Lukodianov, Vladimir Gakov, and several others. Notable is the dominance of male writers (no one of the stature of Ursula LeGuin has emerged among females) and the frequency of writing partnerships. A few of these authors were born before the Revolution of 1917 (Gor, Varshavskii, Efremov), a few more in the Civil War period and the 1920s (Dneprov and Voiskunskii, for example). All these men were deeply involved in the Second World War. The rest were mostly children of the 1930s who came to maturity and received their education after the war. A rather large percentage did not begin writing science fiction until the Efremov thaw of the late 1950s. In contrast to American and many other communities of science fiction writers, the great majority of the Soviet writers have a serious scientific background: geology, paleontology, chemistry, physics, mathematics, astronomy, ethnography, oceanography. And most of the others are also "scientists" in the European sense—that is, scholars of the humanities or social sciences, including a Japanologist (Arkady Strugatskii) and a specialist on Burmese history (Bulygin).

It is harder to speak with assurance about the readers' market for science fiction, but it is certainly possible to make some broad statements. Most fiction of this sort reaches the public through anthologies such as *Fantastika* (published on an irregular schedule since 1962), books and journals published by Detlit (children's literature), Znanie (Knowledge), and Molodaia Gvardiia (Young Guard, the Komsomol outlet). In the 1960s the circulation for the average book was about 100,000, in the 1970s 200,000 to 300,000. As of the 1970s almost two-fifths of the roughly 2,500 to 3,000 titles of Soviet science fiction published since 1917 had appeared since 1957. In addition, between 1917 and 1970 some 1,500 foreign works had been printed in various Soviet languages, most of them since 1957. Among the most popular authors are Isaac Asimov, Ray Bradbury, Kurt Vonnegut, Robert Heinlein, and the classics of Wells and Verne. Censors attached to every publishing house, at least until recently, scrutinized Soviet science fiction works to prevent the leakage of specific technical and scientific details, discoveries, and secrets. This was done because of the heavy crossover from active scientific work to fiction. Most readers, apparently, are young, male, urban skilled workers, members of the technical intelligentsia—scientists and engineers. Half are of the age group fifteen to twenty-five—pupils and students. This pattern and the estimated number of readers (2 to 3 percent of the population) is similar to that in the United States, except that there are rather more older readers in the USSR. Inquiries into the motivation for reading such works have yielded the following: curiosity about the future, interest in science and technology, and moral content. The genre is known to be a magnet for drawing the young into scientific careers. [16]

Contemporary Science Fiction

Before examining some of the major themes of Soviet science fiction, I want to examine how it compares with popular science futurology. These have a common purpose: to make science accessible, a particularly acute problem in less highly developed countries. It seems clear that science fiction does a much better job. In two examples of scientific futurology, one from 1957 and the other from 1976, what we see is mostly descriptive technologizing. The book *Russian Science in the Twenty-first Century* contains portraits of future foods, cities, transport, amenities, resources, industrialized agriculture, electric power, and oceanic exploitation. Even in the chapters about the "good life" of the coming age—on schools, homes, town life—there is not a single speculation about the nature of man, his reshaping, his relationship to the environment, or the special nature of socialism or communism in that environment. L. E. Etingen's *Mankind in the Future: Its Appearance, Structure, and Form* gives a timetable of changes in the human organism from 1967 to 2030, by which time humanity is seen to be experiencing organ transplants on a regular basis, along with implantation of artificial organs and electronic mechanisms, personality and intelligence modifications through medicine, biomechanical stimulation of the growth of organs, a symbiosis of man and machine, direct communication between persons and between persons and computers, chemical control of aging, and the prolonging of life. [17]

In these and similar studies the technical is divorced from life—emotions, relationships, family, taste, political problems of power, individuality, cooperation, love, and all the rest. These questions are, of course, taken up separately in works on sociology and Communist Party program commentary, but rarely are the two fitted together. Soviet science fiction attempts to do this in a variety of ways. All the problems just enumerated have been treated in science fiction in the last twenty-five years, but they are usually accompanied by an examination of their effects on people and their society, by ironic twists, unexpected outcomes, anxiety, failure, reversals, or details about the role of technology in everyday life. With few exceptions, the problems are treated individually, in stories. The deficiency in Soviet science fiction futurology is the working of all these themes together into a grand mosaic, a master blueprint of what science and society will look like and how they will interact in a better world.

Nevertheless, the genre is rich in insights into Soviet mentalities. One Soviet critic sees science fiction as a bridge between different cultures (she seems to be including communist and capitalist as well as national cultures), and between scientists and the masses within society.[18] Science fiction writers pose necessary questions: How are we changing? What will come of it? How does technological "improvement" impinge on society, on men and women, on the biological structure of the species? What is the role of science in a world full of mutual hostilities and marked by the ambivalence of most people toward science? Science fiction themes in the USSR—and I deal here only with the Russian-language corpus, and not with that of other nationalities—include very broad-based and variegated alternative worlds and cultures, space travel, extrapolation into the unknown, technological fantasy, and ecological awareness, and they are usually sensitive as well to problems of social welfare and security, aesthetics, and the meaning of happiness. These themes reflect a whole range of opinions about Soviet mores and values, about the capitalist West, about the past, and about the future. In comparing this body of work to the science fiction of the revolutionary period, we see that modern works are much more scientifically literate than the older ones. It is also apparent that the genre has gone beyond the bounds of socialist realism and possesses a sharp critical edge. Some of it is good literature. Knowledgeable people claim that it is also equal in merit to Western science fiction.

I have arbitrarily organized my discussion around the following themes: the "problem" of science; the use of technology; Soviet values; anticapitalism; social criticism; and antimodernism. The last flows naturally into the realm of dystopia and into the puzzling works of the Strugatskiis.

To the layman in the modern world, the problem of science is the contradictions it creates: the growing rift between "two cultures"; the explosion of unabsorbable knowledge; the creative urge to invent and develop a thing to its outer limits, including "things" that can destroy the entire planet through a trivial accident. In this matter, Soviet science fiction offers nothing particularly new, although the expression of anxiety may have a peculiarly Russian or Soviet character. The Dr. Frankenstein with a test tube, the mad scientist, the doomsday weapon, and the death ray were common elements of 1920s science fiction, and always attributed to evil capitalists. Contemporary science fiction has come a long way; it recognizes the danger to be universal and not class bound. Vladimir Savchenko's hero in *Self-Discovery* (1967) muses over the familiar repertoire: poison gas in the First World War, the death camps of Nazi Germany and their technology of extermination, Hiroshima, and the present balance of terror. No direct allusion to a possible Soviet-made disaster is apparent; but, as in most discussions of this question, it can be inferred.[19] Dmitrii Bilenkin, in "The Ban," has the world's most respected scientific figure falsify his evidence and suppress a discovery in order to save mankind from its evil potential. Sudden global holocaust is not the only perceived danger. Bilenkin specializes in warnings. In one of his stories prospective parents are required to pass a battery of examinations in moral values to determine their fitness for reproduction. In another space explorers unwittingly blind an entire planet with their search beams. In still another a painter has found the last piece of undeveloped nature and settles down to capture it on canvas when a "servo-robot" appears from nowhere to prepare his food and a storm of green rain descends to kill the mosquitoes that are annoying him.[20]

Some writers believe that the way out of this threat of mindless and untrammeled scientific growth is to create more smart people—"smart" being identified with wise and humane in almost all positive Soviet characters. Professor Tsesevich in Pavel Amnuel's "Today, Tomorrow, and Always" wishes to multiply the number of geniuses and abolish "mental inequality." Savchenko in *Self-Discovery* has the hero manufacture new people one by one by means of a computer. Echoing some motifs of the "nihilist" science-worshiper Dmitrii Pisarev, uttered a century earlier, he laments the inertia and egoism of the masses and sets out to make new people brimming with energy and brains. The production of such people would fill the gap in competence and integrity—so obviously lacking in the negative characters around him—give science a broader social base, link the elite more closely with the untutored, and finally better control the increasingly uncontrollable aspects of science and technology. Savchenko rounds out his decision by facing the moral dilemmas inherent in it: the denial to man-made people of childhood and family; the danger of the "processing" of created humans as voters, prostitutes, soldiers, or commercial commodities.[21]

The scientists in the Soviet science fiction of our time are a varied and complicated lot—not always the strong and virtuous Bolshevik heroes of the early days who did no wrong and made no mistakes. They are fallible, weak, puzzled, energetic, brave, devious, ambitious, even lazy. The heroes are better than the average, to be sure: they try. And the reality of their condition is far more convincing than the equivalent in older works of socialist realism. One gets the feeling in almost all of this fiction that the general public, while perhaps ignorant about science, holds science and its practitioners in high regard. The scenes of scientific conferences are held in resplendent palaces of learning and other dignified settings. Within the scientific community fossilized types abound, sometimes bearing indicative names of the sort used by Griboedov in the nineteenth century. A perfect example is Professor Mesozoiskii in one of Savchenko's tales, whose prehistoric views match his name. Heroes and heroines battle against "mesozoic" attitudes all the time.[22]

Internationalism and cosmopolitanism in science, in the most literal sense, is a major background motif in the years since Stalin's death—and quite unthinkable, of course, during his regime. Soviet scientists in fiction, whether in future or present scenarios, are constantly flying off to foreign capitals for conferences and consultations. Limitless travel, a world community of science, permanent mutual communications—sometimes instantaneous—learned friendships and collaboration, central information depositories, a sense of international equality and a single spirit of open inquiry are almost mandatory in the longer novels that deal with global crises. Internationalism in science fiction contains an element of pathos, given the actual restrictions on scientific cooperation in some quarters, but it clearly reflects the broad opinion of Soviet scientists themselves and forms a permanent buffer to the occasional anticapitalist stories and themes. Historical references in the works celebrate intellectual community in the same way: the names of Norbert Wiener, Einstein, Bohr, and many others are constantly invoked and properly credited with the scrupulousness of a scholar with a footnote. Nothing could be further from Zhdanov's intellectual pogrom of the late 1940s, when Russians were flaunting their "own" science and crediting themselves with every sort of discovery and invention.[23]

The treatment of new technologies is not really separable from scientific progress as such, but it operates fictionally on a more concrete level and therefore deserves an analysis of its own. Soviet science fiction is modernist and Promethean as a rule. Some authors are still playing the Stalinist-period game of merely projecting replicas of what the government is doing now—as in a short piece by Mikhail Greshnev, "Sagan-Dalin" (1984), which features the dream of two scientists of a future city, another BAM (Baikal-Amur Mainline), another giant power station, another vision of a streamlined, populated, industrial Siberia, a dream originally sketched in a striking prose poem, "Express" (1916), by the proletarian poet and apostle of Taylorism Alexei Gastev.[24]

Then there is the usual assortment of plausible toys for the future: airbus, videophone, robot, time machine—all of them familiar to readers of nineteenth-century fantasy, even in Russia. Far more interesting, though of course also not novel, is the notion of computerizing all of life and work. Savchenko gives us a computer factory the size of a television set which receives orders for computer systems, calculates the requirements, plots the correct circuitry, and prints out a thin plate in twenty seconds to be mailed to a client (factory, hospital, traffic control headquarters) and plugged in. In another story he suggests the application of cybernetics to personal affairs and daily life, to manage independently all kinetic and organizational tasks and allow the unburdened human brain to deal with creativity, love, friendship, pleasure, and other presumably nonorganizable matters.[25]

Dmitrii Bilenkin and Vladlen Bakhnov take up where Savchenko leaves off: scientizing emotions, dreams, and psyche. In "What Never Was," Bilenkin's main character falls into a voluptuous reverie on love, nature, and serenity so real that he thinks he is living it. But the dream has been manufactured and fed to him by a "biowave" method used to prevent depressives from committing suicide. In "Personality Probe" Bilenkin uses collective cognition to allow students in an experiment to think together without voicing their thoughts aloud, with periodic and automatic feedback through phonoclips of solutions arrived at. The wittiest of these devices are in Bakhnov's "Cheap Sale": a "mood battery" to regulate the emotions and store up joy for later use in moments of adversity; reified emotions that can be "put on" and taken off; a love potion to inspire creativity (not procreation); and "fond memories" for the enjoyment of selective nostalgia.[26]

The most imaginative invention I have seen in this literature is Savchenko's "computer womb" for the generation of new and brainy scientists. In *Self-Discovery*, which draws on the myth of the homunculus and cybernetic research, the hero, Valentin Krivoshein, builds three new people by combining his own organism and thoughts with a computer. Biology, systemology, bionics, electronics, and chemistry are enlisted in an elaborate system for feeding impulses into the machine, which then follows the metabolic control patterns of Krivoshein through his brain waves and creates not a clone but a replica with desired variants of size, personality, looks, and brainpower. A remarkable episode in this story, written twenty years before Chernobyl, is the vivid description of an atomic reactor explosion, the burning of graphite rods, the deliberate self-radiation of Krivoshein, and his self-cure through immediate protection of bone marrow and marathon showers.[27]

The darker side of technology does not escape the scrutiny of modern science fiction writers. Bilenkin offers novelty with a twist: the "time bank," in which one can save up useless time (time spent standing in line, for example) and later retrieve it for serious purposes. It is the Taylorist equivalent of the "mood battery." Taylorism in everyday life and the battle against wasted time (and motion) had a lively history in the 1920s when the NOT (Nauchnaia Organizatsiia Truda, or Scientific Organization of Labor) movement tried to mechanize and robotize the rustic Soviet labor recruit, and when the League of Time tried to schedule the lives of all people by the clock and by a rigid division of the day into useful time morsels. In Bilenkin's story it all falls apart, just as it did in real life fifty years ago. Soviet censorship and control are mildly ridiculed (or so it seems) by Il'ia Varshavskii's "phenotype" and Bakhnov's "Ah-meter." The first is a sensor beam which locates students on any part of a huge campus to summon them in for a talk. The second is a machine that measures the exclamations of art appreciators in order to determine the authenticity and artistic worth of a painting—an unequivocal sermon against gauging the value of art, as in the traditional Soviet way, by the number of people who can understand it.[28]

One more illustration is perhaps worth a note: Viktor Pronin's story "The Power of the Word," which deals with ESP and telekinetic energy, the former having a long history in Russian science fiction although banned under Stalin as a subject of speculation. This tale shows how a "simple" cleaning woman can move objects with the power of her mind. The implication is made clear: in addition to the social application of this power that can lighten the burdens of everyday life, its military application would make it so dangerous that it would soon render nuclear weapons obsolete.[29]

In the realm of Soviet values and their relationship to realities, ethnicity and gender are dealt with in Soviet science fiction in the same way they are dealt with in general fiction (as far as I can judge from wide but spotty reading in the latter). Ethnicity is treated according to the official position, whereas gender is treated in accord with reality. The topic of ethnicity in literature is always a delicate matter, particularly in cultures where a myth of complete equality and harmony is officially held. In stories with a Soviet setting, the main characters are almost invariably Russian. But characters with Georgian, Armenian, Ukrainian, Tatar, or Central Asian names are common and are often situated in high places in the academic, scientific, or administrative community. Aside from passing references to swarthy skin, black eyes, or "unpronounceable" names, no issue is made of their national origins. They form in a sense the minority backdrop for the action. Crews of spaceships in futurist stories are multinational. Only a few identifiably Jewish characters appear: one, Erik Erdman of Odessa, is a prominent coprotagonist of *World Soul* (one of whose coauthors is Jewish). I have found no anti-Semitism, no tension over ethnic interaction, no false exaltation of Jews. One may read this collective text (together with all other official statements on Soviet nationality policy) either as a hypocritical device masking the actual ethnic problems of the USSR or as a commendable way to foster tolerance of non-Russians who conform to the general values of Soviet society. [30]

The general treatment of gender—specifically of women—in a genre that is male dominated is closer to the realities of Soviet society. That reality includes a relatively high level of opportunity in science, technology, and professional life (compared to other societies and to Russia's past), overbalanced by vertical and horizontal segregation in the workplace, lower wages and prestige, relative powerlessness in all walks of life, and the burden of the "double shift"—full-time work as well as the responsibility for home management, domestic work, and child care. Some writers, such as Efremov and Sergei Snegov, have tried to upgrade women into major figures in their fiction. Others, especially the Strugatskiis, have opened themselves to charges of overt sexual chauvinism, particularly in *The Snail on the Slope,* in which females play negative roles and the female gender even symbolizes a kind of evil force. In the mainstream of Soviet works the treatment of women characters is more realistic. They are always a minority on expeditions or scientific teams, are intelligent and well educated but distinctly overshadowed by the men, are often subordinate to theoretical equals, are nurturing and inspiring rather than dynamic and active. Whether scientifically competent or not, the women seem to be more materialistic, more down to earth, more skeptical of free flights of imagination—more "womanly" in the stereotypes that have dominated Soviet imagery since the 1930s. Soviet science fiction is clearly a genre written by men and for men. It differs from its Western counterpart in the absence of overtly sexual scenes and the coarser kind of male attitudes toward women.[31]

Although, in a general sense, women seem less enamored of technology in the stories I have read, there is nothing like a "feminist" science fiction or specifically feminist utopia. American feminist utopias of the 1970s tackled the matter of technology head on, displaying hostility to violence and ecological destruction and repudiating the vision of technology as synonymous with human progress. In these works women are aligned with nature; and the male use of technology to make war against nature is equated with male dominance over females. The fact that this perspective has not emerged among Soviet science fiction writers, male or female, says something about the level of a feminist consciousness in the Soviet Union.[32]

Attitudes toward the state and authority do not emerge very sharply, except in the bolder dystopic works, which I discuss later in this chapter. Most science fiction is very vague on the ideal state, or nonstate, of the future. Like the novels of the 1920s, recent ones allude to shadowy councils and galactic committees of wise people, usually men, with an occasional essay into machine-run politics and fuzzy references to consensus or computer calculation. It is on this issue apparently that the limits of political alternatives bump up against censorship and self-censorship. Fundamental questions of individual freedom and social or intellectual rebellion are hardly addressed. The most interesting characters are usually those who fight against bureaucratic forces, mindless conformity, and careerism—but they do so in the name of values that are fully accepted in official Soviet life, and in this respect are almost identical with themes in general literature since the 1950s. The texture of everyday life is decidedly urban. There are few peasants or collective farms. Soviet town life is familiarly depicted—restaurants, tram rides to the institute and university, lecture halls, laboratories, academic councils and dissertation defenses, *komandirovkas* (official trips), "ordinary" sexism, but not yet much on street violence, alcoholism, drugs and *fartsovchiks* (dealers), blue jeans, and rock bands.[33]

Anti-Western polemics in the 1920s helped to focus anxieties about the world without producing ambivalence in the reader. The format was strictly we-they/good-evil. This element is rare in modern Soviet science fiction, as compared to the earlier period, and especially the Stalinist years, when it was reduced to the crudest forms, very much resembling the treatment of Germans and Japanese in American popular culture in the prewar and wartime years. Matters are more complicated nowadays. One finds various kinds of jibes against the West in recent stories. For example, when unexpected disaster ensues after two Soviet scientists develop a "world soul," the American press accuses the Soviet government of spreading an epidemic. In the same story the Soviet hero is disgusted with a social system (capitalism) in an unspecified South American country because it produces prostitution. Spartak Akhmetov in "Shock" (1984) is still deploying the old clichés of soulless decadence in a modern Western city—a bleak cityscape, faceless people, leaden skies, derelicts lying on the streets ignored by the heartless crowd, suicides, and a culture of neurotic haste—fairly reminiscent of one of the earliest Russian fictional assaults on European industrial society, "A Journey to Mars" (1901). In Liudmila Sveshnikova's "How to Outwit Pain" (1984) a lonely American widow who has lost her husband and son in some distant imperialist war retreats into reveries and loses touch with the world. Aleksandr Potupa's "Effect of a Lucky Man" (1984) tries to show the extreme mutual distrust that pervades mendacious Western society. These are all rather formulaic and pallid critiques, without depth; most of them could be interpreted as Aesopian critiques of the Soviet scene, which possesses its own quotient of mistrust, indifference to suffering, prostitution, and a distant war in which the Joe Sr. and the Joe Jr. of Sveshnikova's story could easily be replaced by the fathers and sons who died in Afghanistan.[34]

Somewhat sharper in their bite are the scenarios in *World Soul* whose main premise is the obliteration of private thought and secrecy. Aside from the many other complications this creates, it forces capitalist businessmen in the West to disclose their profits and shady practices and to tell the truth about their products—thus eliminating advertisement and forcing firms into bankruptcy. Amnuel's recent "Innocent" traces the career of a Harvard-educated scientist who is so intent on pursuing his experiments to the very end—even though he knows that they will endanger the species, the environment, and the entire planet—in the interest of "truth" and to satisfy his intellectual ego that he brings the world to the brink of disaster and is arrested and charged with irresponsible crimes toward humanity. And a 1985 fantasy describes in great detail how American corporations and government agencies develop new modes of electronic surveillance, including bee-shaped sensors that fly among the citizens, pick up their conversations, and try to apprehend the mood of the people.[35]

When we attempt to isolate and categorize varieties of social criticism in modern Soviet science fiction, we face two problems. The first is epistemological and has to do with determining the line between anticapitalist and antisocialist critiques, the latter easily produced through Aesopian modes of expression. Further, one must search for the equally elusive boundary between legitimate social criticism and political dissidence. The moral dilemma for the Western student is the danger of attributing dissidence to works in which it was never intended.

Most of the critical material in these stories refers to Stalinism— either its fearful past or its lingering presence in Soviet life. Straight-forward references to the purges can sometimes be found in the biographies and memories of heroes, as with Savchenko's character who muses over his father, a Red Cossack in the Civil War, who was arrested in 1937. No more than those last three words is offered; and for all Soviet readers, no more is needed. With Il'ia Varshavskii and others, we are left to wonder whether or not the dreadful gulag-like camps, the slave labor, and the brutal guards (such as the "stupids" in Bulychev's "Half a Life") are only Kafkaesque fantasy and spaceship conventions or allusions to the real thing, which flourished in Stalin's Russia longer than it did in Hitler's Germany.[36]

Contemporary careerism and opportunism, especially in high places, offers the most fertile field for alerting readers about the remnants of the mentality of Stalin's ruling class. The extraordinarily entertaining story "Success Algorithm" by Savchenko creates a computerized program for success in life by feeding in the clichés of amoral upward mobility: "You scratch my back and I'll scratch yours"; "An eye for an eye, a tooth for a tooth"; "Don't rock the boat"; "Dog eat dog"; and "Every man for himself." The completed program even teaches the user the facial language of hierarchy and deference.[37] An unforgettably obnoxious figure in *Self-Discovery* is a perfect illustration: Harry Haritonovich is a fluffy, vain, false, well-turned-out toady, bootlicker, backbiter, and conformist who manages to push through a wholly worthless dissertation because the board is too lethargic and routinized to stop a man with connections. An angry outburst from one of the positive characters sums up the author's view: "And *don't* we [in the Soviet Union] have people who are ready to use everything from the ideas of communism to false radio reports, from their work situation to quotes from the classics [of Marxism] in order to become wealthy and have a good position and then to get more and more for themselves, at no matter what cost?"[38]

But beyond the banalities of ordinary careerism lies what we might call "reptilian careerism," familiar to all who have studied the history of Soviet science and the arts. It has to do with advancing or holding one's privileged position through character assassination of prominent but vulnerable people. A masterful rendition of this is Bilenkin's "Personality Probe," in which a famous nineteenth-century writer, editor, and publicist (who also wrote science fiction, by the way), Faddei Bulgarin, is brought back to life by a classroom of computer students. Bulgarin, the king of the "reptile" press in the reign of Tsar Nicholas I (1825–1855), made a specialty of slandering his rivals and spreading poisonous gossip about them in high places. His most famous victim was the poet Aleksandr Pushkin. The twentieth-century students reconstruct Bulgarin in a "phantomatic" holograph resembling ectoplasm or a three-dimensional apparition and then subject him to a withering interrogation about the immortality of his career—a perfect vehicle for exposing to the knowledgeable reader the shameful, not-so-distant history of many Soviet intellectuals.[39]

Another target is the "masses," sometimes disguised in other national identities, sometimes not. Many stories contain contemptuous allusions to consumerism, materialism, philistinism, mindlessness, conformism, and the herdlike shape of the unspecified mass—an inert mediocrity. This has been a favorite target of abuse in the West from Nietzsche and Ortega through H. L. Mencken to Ayn Rand, often a rather easy target for cheap shots. Soviet writers must be circumspect; they can hardly be permitted to attack "the working class" as such, for it is still the mythical hegemon of Soviet culture. But they do it nonetheless, lamenting the palpable power of mob inertia against the creative or innovative spirit of scientists and the gap between elitist science and mass ignorance. In the many instances of bribery through gadgets, the facile purveying of alcohol to oil the way, and corruption at various levels, one can even see glimmers of the massive "second economy" that is visible almost everywhere in Soviet urban life (and even in some rural areas) but could not until recently be openly discussed as a pervasive phenomenon.[40]

Conservatism and resistance to healthy change appear everywhere. Often they are embodied in the stuffy blimps who inhabit research centers, institutes, and academies—such as Professor Voltamperov in *Self-Discovery*. A different twist is provided in Amnuel's "Higher than the

Clouds, Higher than the Mountains, Higher than the Sky" (1984). The protagonist, Log (from the Greek *logos,* for "word" or "reason"), lives in a village somewhere in time and space but wishes to travel widely and discover other worlds (perhaps also a mild reference to the travel restrictions to the West placed on Soviet citizens). The village elder tells him: "A sensible man will not travel from one settlement to another when life is difficult and the same everywhere. Doesn't the law teach us this?" But Log persists in wanting to know what is out there, higher than clouds, mountains, and sky. Imprisoned by the community, he escapes to pursue the "beautiful sickness of mind"—the urge to know and to master. [41]

Social control and overpowering mediocrity flow into the prominent theme of antimodernism, familiar to readers of Soviet literature in the last two decades as expressed in village prose, or the exaltation of traditional Russian rural, national, and even religious values. Science fiction writers are more oblique than the *derevenshchiki* and focus on the harmfulness of useless machinery and excessive technology. In Iurii Medvedev's "Love for Paganini," a zealous scientist on Mars decides to transform nature beyond recognition in his society. Mars is already a highly technical utopia, and the inhabitants resent his compulsion to change the climate on Neptune, move mountains and seas, and invent an antimelancholy device, an atomizer of moonlight, and an audible replicator of ocean waves. In the end, the Martians send him to earth and allow themselves to keep in touch with the beauties of the past—revering Paganini as well as Euripides. [42]

Aleksandr Petrin, in "Vasil Fomich and the EVM," unveils a mild but firm variety of Luddism in the modern workplace. The EVM, or Electronic Calculator, is introduced into an office and greeted with skepticism by the bookkeeper Vasil Fomich (a very archaic name). He believes that personal labor and human psychology (and, presumably, the abacus) are better able than an insentient machine to do his job and meet the plan. He wins this duel: the employees produce their quota, the plan is met, the bonuses are meted out, the machine is expelled, and Vasil Fomich is the hero. It is tempting to read in this the coming wave of resistance by the Soviet office force (overwhelmingly female at the lower ranks) to personal computers and other forms of machinery. But Luddism, fictional or real, has never long delayed the march of technology in modern societies. Liudmila Ovsiannikova's "Machine of Happiness" is another expression of the malaise over the rapid development of a machine culture. She may be voicing the Soviet woman's perspective (not, however, a Western feminist one) when she identifies machines of happiness with appliances to ease the weight of everyday life and questions the utility of machines that carry cosmonauts to the moon. Is this the voice also of grass-roots men and women who long for consumer goods and who wonder at the awesome expenditures for space programs?[43]

Unlike utopia, dystopia has not been subjected to endless searches for typology. I do not wish to violate this tradition, but I do suggest here at least a basic dividing line between the dystopias of entropy and those of energy. The former is the more familiar and finds its classic expression in the One State of *We,* a frozen world of symmetrical regularities, uniform appearance and motion, cyclic existence, and mathematized ritual. Ideally, nothing happens in the United States; it is locked in entropy. The second type releases an evil force, a wave of destructive energy. Around the time of the Revolution of 1905, Valerii Briusov—a well-known poet of the day—had written several stories of both genres. In "Earth" the planetary city-state lies underground and the hero fights to let in the air—and then dies. In "Republic of the Southern Cross," an antarctic megalopolis of high technology and surface welfarism explodes with a strange disease that reverses the behavior of all the inhabitants and wreaks disaster on society.[44]

In modern Soviet science fiction we have Varshavskii's miniature tale "The Violet." In it the earth has become a ball of continental cities from pole to equator, two miles deep into the earth and thirty miles above; no nature, no soil, no verdure exists. A small child, taken to the bowels of the earth in order to visit a "field" in the museum of antiquities, weeps for the death of a flower. In *World Soul* a polymer growth from a fragment of seaweed blossoms turns into a biotosis, or "world soul," a universal telepathic computer that links all minds together, generates empathy, and opens communications. Its coinventor argues that the biotosis simplifies and automates the world, makes government truly democratic, abolishes deception, and eliminates violence (since each person suffers when any individual does). "Life is simpler and happier," he says; "the scope of [people's] feeling has grown from the single 'I' to the 4 billion 'we.'" Critics in the novel point out that the biotosis causes transformations, accidents, tragedies, pseudo resurrections, insane visions, and death; and that even if its chaotic energy is corrected, it will mark the end of human will and the ascendance of a "supercybernetic" collectivist machine that will destroy individuality.[45]

The path of the Strugatskii brothers—the Soviet science fiction writers best known to Western readers—has been called by one critic "paradise lost," or a journey from utopia to dystopia, by another as an agonizing struggle between their utopian visions and the dystopian forces around them. Whatever the case, it is apparent that their vision has developed, opened out, and become desimplified. An example from each of the two basic periods of their production can serve as illustrations. Their first major science fiction work, following directly in the tradition of *Andromeda*, was *Homecoming* (1962), a utopia of the twenty-second century, inhabited by physically and spiritually superior people and run by a world council. Russia has become the first communist society and the nucleus of planetary perfection. There is a moving flashback to the siege of Leningrad (1941–1944) and then a pause at a statue of Lenin: "Lenin stretched out his hand over this city, over this world. Because this is his world; he envisioned it this way—radiant and lovely—two hundred years ago." The Strugatskiis have retrieved all the romanticism of the 1920s, all the Promethean force of the *chelovek vsemogushchii* (omnipotent man), the serenity and humaneness of the revolutionary vision.[46]

Ten years later, in *The Ugly Swans* (1972), we have a dictatorship deploying its security forces, the Legion of Freedom, or "woodpeckers," against the "slimies"—outcasts and rebels whose status is openly compared to that of Negroes and Jews—who crave books and ideas, maintain a high moral standard, and vacillate between utopian and dystopian visions (a fairly close description of the Soviet underground intellectual community, including but not coterminous with the dissidents). Between these two novels the Strugatskiis obviously underwent an intellectual and emotional change, in addition to a broadening of their perspectives through exposure to the writings of Kafka and the Polish science fiction writer Stanislaw Lem. Their stories became more problematical, more complex, more cryptic—perhaps a way of warding off the measures taken against their work by censors in the late 1960s and 1970s, but more important, a sign that Soviet science fiction has passed far beyond the rosy simplicities of the revolutionary age. The work of the Strugatskiis is a major fictional corpus; it deserves (and has received) lengthy analysis in its own right. It is still impossible to gauge how representative it is of Soviet science fiction as a whole or how reflective of the values of science fiction audiences.[47]

In Retrospect

It would not be a gross oversimplification to say that Soviet science fiction has seen three main ages: the Age of Bogdanov (1908–1931); the Age of Stalin, who created much fantasy but wrote no science fiction (1931–1956); and the Age of Efremov (1957 to about 1970). The first age was dominated by sweeping optimistic utopianism in the midst of a deplorably backward society; the second by tight control of fantasy and imagination and a sharp narrowing of thematics; the last by a renewed flash of utopian hope fueled by a de-Stalinizing thaw and by a maturing culture of scientific and technical achievement. What has occurred in about the last two decades is not so easy to characterize because our perspective is too short. This recent period resembles very much the world around us—full of patternless confusion, fluidity, blends of bright hope and dark fear, new dialogues of dream and nightmare.

Thus the history of science fiction in the Soviet Union seems to show a natural fit between certain periods of the genre as a whole and the historical backdrop against which it appears. Some might see this simply as a result of "command literature," a reflection of new styles and programs of Party leaders. But it is surely more than that. Although it is impossible to track the history of social values directly by reading science fiction, it does give a feel for some of the most significant attitudes that writers—and thus many of their readers—hold about the outer world, technology, progress, scientific morality, professionalism, and a dozen other elements of the collective, educated Soviet mentality.

What is the promise of the new period that is opening up before our eyes, this age of glasnost and perestroika? It is, of course, very difficult to speculate about an era that is in such rapid flux. But some straws are in the wind. First, there are some indications that readership of science fiction may be growing. No data are available, but cultural critics have been bewailing the decline in "serious" reading among the citizenry and a rise in "low" reading tastes—that is, for the detective story and science fiction. This is a perennial complaint, to be sure, but one that gathers force almost daily.[48] With the extraordinary lessening of censorship, science fiction may join the enormous trend in popular culture toward exploring hitherto unmentionable dimensions of life—including sex, social abuses on the grandest scale, and democratic criticism of the regime itself. And it may lead to the more experimental modes of literary expression that are emerging rapidly in the arts, especially in film. Thus far, however, I have seen no clear-cut examples of open eroticism or of avant-garde aesthetic forms in the genre of science fiction. Yet neither is there any sign of the racism that has crept into the writings of some belletristic writers—particularly Viktor Astaf'ev.

Similarly, it is hard to discern any major shift in the political thrust of Soviet science fiction, but there are signs that some of the most influential writers are squarely in the camp of Gorbachev-style reform. In March 1987 the Council on Adventure, Entertainment, and Science Fiction of the powerful Union of Writers convened for the purpose of discussing the role of this genre of writing in combating reactionary propaganda—one of the code words for antiperestroika thinking. Among its members were Eremei Parnov, A. Kazantsev, the Strugatskiis, Dmitrii Bilenkin, G. Gurevich, Olga Larionova, E. Voiskunskii, and V. Revich—in other words, the crowned heads of Soviet science fiction of the last two decades.[49]

Old books have a way of inspiring new ones, especially old forbidden books of explosive content that have been long suppressed. Masses of Soviet citizens have at long last had the opportunity of reading Evgenii Zamiatin's classic dystopian science fiction thriller *We,* written in 1920 but not published in the Soviet Union until 1989. In his introduction to the first Soviet edition, the critic V. Lakshin pulls no punches, relating the novel not only to its revolutionary context—particularly War Communism and the visions of Gastev, Bogdanov, and Taylorism—but also to the barbarities and hypocrisies of high Stalinism. The literary echoes to this book—and there are many—may have the effect of drawing science fiction into further political speculation and fantasy. In any case, it is hard to imagine Soviet science fiction remaining untouched by this cultural bombshell from out of the past (particularly with the publication in Russian of Aldous Huxley's *Brave New World* and George Orwell's *1984*).[50]

Taking Huxley for inspiration in his article of July 1988, "O, Brave New World," Eremei Parnov, a dean of science fiction letters, cast a passionate glance at the Stalinist past, which he described as a lapse into medievalism, the recurrence of darkness that seems to happen every century. Parnov's target is not only the cruel repressions and stupendous errors of Stalin but the legacy of twisted truth, the cult of Big Brother (his words), and the lies that are the main foe of human reason, sanity, and global survival. Prejudice, racism, and noxious mythologies were the fruits of that ugly time; and at the intellectual core of the Stalinist deformation, says Parnov, was the death of science—all science—and the bigoted exclusion of new fields such as cybernetics. Parnov's cry of collective conscience is striking not only because of his acknowledged stature in the community of science fiction, but also because it appeared prominently in the mass publication *Soviet Culture*.[51]

Equally moving was the last essay of Dmitrii Bilenkin, who died soon after its publication. It was his farewell address to the readers of the *Science Fantasy Annual*. Bilenkin wished to broaden the range of science fiction, to widen its content, and, by implication, raise it to the same level as "literature"; to spread its appeal, as happens in advanced industrial societies that are leaders in science and technology, by the spontaneous formation of nationwide science fiction fan clubs. In the coded idiom of perestroika, Bilenkin told his readers that in a rapidly changing world, the people must prepare themselves spiritually and intellectually for adaptation to the unheard-of novelty of the coming age. The failure to make such a mental adjustment, he warned, leads to psychological disorientation, a flight from reality, spiritual paralysis, and a "hatred of progress."[52]

Thus some major figures in this exciting genre are plainly on the side of reform, of change, of a betterment of Soviet society, and closer links with those nations in which science is free and dynamic. None of this tells us, however, what the coming fate of Soviet science fiction will be. Its traditional role, in times of relative freedom, has been to popularize technology and science, to indulge in healthy fantasy, to promote civic virtues, to tell a good story, and—at times—to offer oblique social and even political criticism. Works performing the last-named function may now be drowned in the flood of what Andrei Bitov has called "half-food" or docunovels[53]—that is, books by village prose writers who want to protect ancient Russian rural values as well as natural resources and historical-political exposés such as *Children of Arbat* by Anatolii Rybakov. To put it more bluntly, social criticism is now so prominent in the daily press that fiction writers may lose interest.

But the current growth of belletristic fantasy in Russia can also compete with science fiction since major Soviet writers, like those of the "magic realism" school in Latin America, are treating large philosophical themes in a setting of lyrical fantasy. In other words, lines are blurring between mainstream and genre science fiction writing. The likely outcome is that new writers of science fiction will turn more to highly scientific and technical subjects and exploit the leading edge of world scientific experiment, research, and speculation.[54] Whatever happens, it is certain that Soviet readers—and perhaps outsiders too—can look forward to many surprises, twists, and rich insights into how the scientific community and science itself are going to function in the heady years ahead. It is highly doubtful that even a reactionary antiperestroika turn could stop the flood that is now released.

· 11 ·

Bridging the Two Cultures:
The Emergence of Scientific Prose

Mark Kuchment

In Soviet cultural life the genre of scientific prose (*nauchno-khudo-zhestvennaia literatura*) emerged in the early 1960s. Though not as prominent in comparison with other types of Soviet literature such as poetry and fiction, it has navigated the changing winds of Soviet cultural policies. What is of special interest is that Soviet scientific prose became a language that could be shared by a variety of intellectuals, professional Soviet scientists, science writers, historians of science, historians of culture, novelists, and essayists. Using the same idiom, they effectively transcended the boundary between the two cultures: the humanistic culture and the culture of science.

What follows is a brief history from the emergence of scientific prose in the 1960s to the time in the 1970s when its appeal to the Soviet public reached its peak. Scientific prose originated with a literary manifesto, which I shall discuss along with some of the best-known and most typical works of scientific prose. I shall describe how, in the last twenty-five years, the Soviet public came to read scientific prose despite stiff competition from Soviet poets and novelists such as Solzhenitsyn, Evtushenko, and Voznesenskii. I shall conclude by examining scientific prose in its broader social, political, and historical perspectives, showing that its emergence, flourishing, and decline were affected by Soviet political history. Its appearance in the immediate post-Stalin period was a natural response to the enormous thirst for rationality that Soviet society experienced after decades of the irrational cult of Stalin. By the late 1970s the reading public had become more interested in nationalistic and religious messages carried in Soviet essays and also in novels and short stories, and scientific prose suffered a temporary decline.

The Literary Manifesto

The manifesto of this movement was published in 1960 by Daniil Danin, one of the best science writers in the Soviet Union. Entitled "The Thirst for Clarity," it was published in *Novyi mir* and began with the paradoxical statement that popular science cannot teach anything. Danin proposed that scientific knowledge is not a trim garden, growing behind a high brick wall: there is no easy way to achieve scientific knowledge. One can enter the garden by breaking through the wall, but the problem, according to Danin, is that knowledge is not like a garden but more like a maze, a labyrinth, and by breaking through the walls of the labyrinth, one destroys it and knowledge disappears. That is why no popularization of science, however skillful, can serve as a substitute for direct knowledge.[1]

This definition of a new type of literature was introduced, perhaps for the first time, by Maxim Gorky in the 1930s. At that time Gorky spoke about "images of scientific prose."[2] That was a period when the Russian term *nauchno-khudozhestvennaia literatura* was coined and the first successful books written. Danin cites as early examples of Soviet scientific prose *Sunny Substance*,[3] written by a fine Soviet physicist, Matvei Bronstein, and books by Mikhail Il'in, author of the now famous *Story of the Great Plan*.[4]

At that time *Sunny Substance* had only recently been rediscovered; along with its author it had been a victim of the Stalinist purges of the 1930s. Bronstein, a prominent Soviet theoretical physicist who was married to the young writer Lidiia Chukovskaia—by now a well-known dissident—was arrested by the secret police during the Great Purge and shot.[5] His book was banned and destroyed in all libraries in the Soviet Union. Only after his posthumous rehabilitation was it republished, with an introduction by his friend, the Soviet Nobel prizewinner Lev Landau.

Sunny Substance is the story of the discovery of helium. The book begins: "I will tell you about a substance which people found first on the sun and only later here on earth." As Danin points out, with this first phrase Bronstein hooks the reader, making it impossible to put the book down until the very last sentence: "Helium had an unusual fate indeed."[6]

Being a story about scientific discovery, *Sunny Substance* takes as its theme not so much personalities and circumstances as the details of the process. For example, Bronstein never attempts to portray the scientists who participated in the discovery of helium. The single exception is

Henry Cavendish, about whom Bronstein provides a few words of description mainly to explain why his major discoveries remained unknown to his contemporaries.

The economy of approach, the clarity, and the concentration on a chemical and not on a character exemplify Soviet scientific prose. But Danin takes one further step to mark the border separating scientific prose from ordinary prose, and specifically ordinary prose whose heroes are scientists. Danin uses as an example Chekhov's *Dull Story,* which made him famous and inspired an article by Thomas Mann. The hero of Chekhov's story is an old professor of pathology at Moscow University. In what became a tradition of scientific prose, the scientist is neither protagonist nor hero. Indeed, the story is not about him but about something else. To support his thesis, Danin quotes Albert Einstein's statement to Leopold Infeld concerning the plan for their book *Evolution of Physics:* "It should be a drama—a drama of ideas."[7] According to Danin, the hero of scientific prose is the scientific pursuit of knowledge, a kind of no-man's-land lacking history or personalities. Yet Danin maintains that in this no-man's-land the natural language of science can be translated into the language of everyday life through the human element of the storyteller.[8]

Objectivity and impartiality are stressed, however. Danin held the idealistic belief that the story of scientific research could be reduced to the drama of pure ideas, assuming on the part of the reader both sophistication and patience. This theory, which originated in the USSR, was remarkably free of standard Marxist-Leninist phraseology and ideas. Could such theoretical prescriptions lead to the creation of a good book? The answer was a definite yes, because just such a book, enjoying enormous success in the first half of the 1960s, had been written by Daniil Danin himself.

The book in question, called *Inevitability of the Strange World,*[9] in five years went through three editions, each edition selling 100,000 copies. Its protagonist was the drama of relativity theory and quantum mechanics. Danin's achievement was to write the book as if it were translated from the language of science into everyday Russian, so the circle of its readers was practically unlimited. In my review of the book in *Novyi mir,* published in 1962, I wrote that Danin's book made the theory of relativity and quantum mechanics clearer and more logical than the ideas of classical physics, which embodied logic itself, and that these new theories were not only inevitable but simple in their own way.[10]

Danin describes relativity and quantum mechanics by taking as an example two elementary particles, the photon and the electron. He extracts the main logic of the special theory of relativity from two facts: that the speed of light is the highest speed attainable by a particle of matter, and that the mass at rest of a photon equals zero. From these two facts he explains the thought experiment at the center of the theory of relativity: What would happen if an observer tried to fly alongside a photon?

With the electron, the author develops the idea of the duality of wave and particle. One of his examples became popular: "If you have a tile with a curved surface," he wrote, "then you could feel its duality in a direct way. If it strikes you on the head, you wouldn't care about its wavy surface, but you would feel very keenly its mass and its energy. If a boy running in the street barefoot stepped on such a tile, he would immediately feel its curved, wavy surface without thinking about its mass or kinetic energy."[11] Such allegories and examples, and the continuous thread of logical conclusions, enable the author to bring the readers along and involve them in a story at the summit of human intellect.

The book begins with a trip to an astrophysical observatory high in the mountains of Armenia, a trip symbolic of the intellectual effort the reader has to make. It ends with an assertion that if we make the effort of reading the book, if we let our reason be our guide, we will be able to overcome our instincts and accept a new and at first uncomfortable and paradoxical picture of the world. On this difficult journey the readers following the adventures of the electron and the photon are guided by Daniil Danin. That is why the guide's personality and background really matter.

Danin, a Russian Soviet writer and a longtime member of the Soviet Communist Party, began to write professionally in 1938.[12] Before the Second World War Danin attended Moscow State University, where he studied first chemistry and then physics for four years before the war began. He also became interested in literature, and started to write, mostly as a poetry critic. At the beginning of the war Danin worked for the army press but was arrested by military authorities, ostensibly for making an unauthorized trip. At that time he could have been shot for lack of discipline, but he had the good fortune to speak with Il'ia Ehrenburg[13] as he was being led by sentries to trial. Ehrenburg, an influential Soviet journalist who knew Danin personally, intervened and helped him return to duty.[14]

Danin first joined the Communist Party during the Battle of Stalingrad. After returning from the war, he became a well-known poetry critic. In this capacity Danin had a tough encounter with the new generation of young, conservative Soviet poets, such as Gribachev and Sofronov. When the fight against "cosmopolitanism"[15] began in 1947–48, Danin was among the leading cosmopolitans; his name was frequently used in the Soviet press in this connection, on several occasions in the plural, with a lower-case *d (daninyi),* as the epitome of cosmopolitanism. Born a Jew, he was a natural target for such a campaign.[16]

All of this developed while the secretary of the Union of Soviet Writers, Aleksandr Fadeev, who knew and liked Danin, was on an extended tour abroad. In Fadeev's absence Danin was promptly expelled from the Party. When Fadeev returned, he could contain only a fraction of the damage by repeatedly postponing the meeting at which Danin was to have been expelled from the union. Danin was not arrested, as had happened to many cosmopolitans; he just disappeared from Moscow for a long time, working as a geologist on expeditions. During those lean years he gradually shifted his interests, first to popular science, and then to the field of scientific prose, of which he was the founder.

In the post-1956 period, when Danin was readmitted into the Party and returned in full force to literature, he wrote on many topics: the peaceful use of atomic energy; changes in downtown Moscow, including the opening of the big department store GUM; modern Soviet architecture. The essay on architecture, "Material and Style," is of special interest.[17] In it Danin came out strongly in support of the new industrially constructed apartment buildings made from prefabricated concrete; he spoke in support of Bauhaus-style architecture, and was one of the first Soviet journalists at the time to quote Le Corbusier approvingly: "The city generates joy or despair, pride or revolt, neutral attitude or disgust, increases your forces or makes you tired—all depending on the choice of proper form."[18] The former cosmopolitan did not reject his past.

Paths to the Unknown

In his early article on Party organization and literature, Lenin declared that a newspaper is not only a collective propagandist and collective agitator but also a collective organizer.[19] Indeed, it has not been just books that have had a lasting impact on Soviet literature; literary magazines such as *Krasnaya nov'*, *Novyi mir*, and *Iunost'* have opened the way to new Soviet fiction and poetry. *Iunost'*, started in the post-Stalinist era by Valentin Kataev, has continued under a succession of editors to the present.

Another periodical, a literary almanac called *Literaturnaia Moskva* (Literary Moscow), was started shortly after the Twentieth Party Congress. The moving force behind its publication was Emmanuel Kazakevich, a close friend of Daniil Danin. This almanac published a famous story, "The Levers," by Aleksandr Iashin, and poems by Pasternak and Tsvetaeva. It attracted considerable attention and sharp criticism and was terminated after the second issue, around 1958.

The experience of this periodical influenced Danin's own plans. He hoped that by starting a scientific periodical, he could influence the whole field; he also realized that in moving too rapidly or being too radical, he could wreck the periodical's future. Still, he collaborated on starting a new publication, by the name of *Puti v neznaemoe* (Paths to the unknown). [20] First published in the Khrushchev era, this periodical has continued into the present without encountering any severe government criticism, or without compromising its original high standards.

Each issue of approximately five hundred to six hundred pages includes the work of about a dozen different authors. It is published by the Union of Soviet Writers, the Moscow writers' organization. From the time the first issue was published in 1960, the Union of Soviet Writers made a significant decision: to treat those works that would be published in *Puti v neznaemoe* as qualifying for the highest honoraria in the Soviet Union. In the 1960s, as today, the honorarium for about twenty pages of typed, double-spaced text is 600 rubles; the usual article in scientific prose runs about a hundred pages, which earns the author roughly 3,000 rubles—very high by Soviet standards. At the time I published my first piece in *Paths to the Unknown*, my salary as a college professor was 150 rubles per month; my honorarium was roughly equal to two years' salary. Participation in such a magazine therefore carried a strong financial motivation. To embark on such a writing project was risky, however. In the 1960s *Paths to the Unknown* was practically the only buyer of scientific prose. If one wrote a hundred pages of scientific prose, one would not be able to publish it in any of the other Soviet popular science magazines (*Science and Life, Knowledge Is Power*, and so on) because length constraints would not allow it. Despite the risk, the prospects of financial reward and of working with the finest authors in the field were attractive.

Although Danin played a major role in the organization of *Paths to the Unknown,* and even directed the whole enterprise for a number of years, he did not become its first editor in chief, nor even its second. People with more reliable public records became the editors of the publication, and it is worth describing them briefly.

The first editor, Oleg Nikolaevich Pisarzhevskii, primarily a popular science writer, had many connections in the scientific community. From 1936 to 1946 he had been an assistant to the famous physicist Peter Kapitsa. He combined his work for Kapitsa with popular science writings, primarily about chemistry. When T. D. Lysenko was in power, Pisarzhevskii—perhaps keeping his true thoughts to himself—published articles in support of the official view of Lysenko as an embodiment of the new Soviet biology.[21] When, under Khrushchev, Lysenko's position began to crumble, Pisarzhevskii criticized Lysenko, but not in writing. When Lysenko was finally overthrown at the beginning of 1965, Pisarzhevskii went public and became the first journalist in the Soviet Union Soviet Union to write a critical anti-Lysenko article, which he published in *Literaturnaia gazeta.*[22] Pisarzhevskii was extremely excited that this article was finally being published, an excitement that may have led, on the day after the newspaper went on sale, to his death from a sudden heart attack.[23]

Boris Nikolaevich Agapov, the second editor in chief of *Puti v neznaemoe,* was a very different sort of person, and one of the most prominent essayists in the Soviet Union.[24] Agapov belonged to the first generation of Soviet writers, having begun his successful career under the tutelage of Gorky himself. Gorky noted Agapov's first publication and frequently his name in his own articles, which helped Agapov in many ways.[25] Gorky even played a role in obtaining for him his own apartment in downtown Moscow, one block from the Kremlin, an enormous luxury for a writer in the Moscow of the 1930s.[26]

Agapov was a self-educated man, without any formal background in science and technology, but with fine intuition, good literary style, and an enormous capacity for mastering tasks. Making the most of his career from the 1930s through the postwar era, he tried as much as possible under the circumstances to protect his integrity,[27] and he was known for his skepticism toward the concept of heroism, one of the main topics of official Soviet literature. "It's very good that, when learning about a heroic deed, you say: 'Glory to the hero.' And you are very lucky if you don't have to add to it: 'But who is guilty, if you need a real hero to fix the shortcoming?'"[28]

Agapov was a bridge connecting the younger generation of science writers to the tradition of Russian prerevolutionary literature and to the time when Soviet literature was created. In private conversations with Agapov, we could see that he kept alive as best he could the moral stance of Russian and Soviet literature; he also valued detachment from the problems of everyday life, objectivity, coolness, and the ability to be rational in an irrational epoch. I once ran into Agapov at the Union of Soviet Writers, and he suddenly stopped me and said, "Look, I have been rereading Marx for the last week or so. I haven't done it for a long time, and you know, it occurred to me that it's fine journalism. Very fine journalism indeed. But is it philosophy? Is it mature political science which forms the unshakable foundation of our state? I don't know," he said. "And I doubt it."[29] It is difficult to overestimate the impact on younger writers of such informal and frequent encounters with one of the founders of Soviet literature.

After Agapov's death in 1973, Daniil Danin became editor in chief of *Puti v neznaemoe* and gave new opportunities to the younger generation of writers. More recently, every issue is put together by an acting editor, or *sostavitel'* (composer), a system that offers the kind of collective leadership, always friendly to its authors, that has enough authority to push works through the bureaucracy of the Soviet Writer, the journal's publishing house, and, of course, past the censors.

The contributors to *Puti v neznaemoe* can be divided into four groups. First are the science writers, like Danin,[30] Medvedev,[31] and Volodin,[32] who have written books on the central ideas of physics; biographies of scientists such as Rutherford, Bohr, and Mendel; and essays about Soviet nuclear physicists, medical researchers, and inventors. The second group includes professional writers interested in problems of science and technology, or of science and society. Among them are the Soviet poet Margarita Aliger, who wrote an essay about anthropology in Chile;[33] or Iuri Davydov,[34] the author of fine novels and essays on life in imperial Russia in the epoch of Alexander II; and Natan Eidelman,[35] historian of the revolutionary movement in Russia and a specialist on Pushkin. The third group is composed of historians of science who write for the general public, such as Viktor Frenkel.[36] The fourth group includes scientists with writing ability; such scientists who publish in *Puti* are Bruno Pontecorvo,[37] a defector from the West and a theoretical physicist working on exotic elementary particles such as the neutrino, and Elevter Andronikashvili[38] from Georgia, one of the leading researchers on low-temperature physics.

The thirst for lucidity and the ability to write about science were combined with a strange mix of escapism and activism to bring the two cultures together. For a science writer a favorite topic might be the biography of a prominent scientist, or ideas about relativity and quantum mechanics; a historian of science might explore an emerging scientific discipline such as mathematical linguistics, mathematical economy, or unusual biographical details of a well-known figure (for example, Einstein's work at the patent office in Switzerland). But the division is never precise.

Sometimes a professional scientist will raise broad issues concerning the link between science and society. A good example is an essay by Lev Artsimovich, "Physicist in Our Time," published in 1967, although not in *Puti v neznaemoe*.[39] Artsimovich poses several problems that an experimental physicist might meet in his career. For instance, as soon as the scientist becomes successful at his physics, he is rapidly promoted and becomes the leader of a group of researchers, making it impossible for him to continue his experimental work. What attracted special attention to this essay, however, was the discussion of competition between the United States and the Soviet Union. He presented the competition in an abstract form, as between "Country A" and "Country B." Scientists from Country B visit Country A. Upon their return, they report that Country B is ideologically superior to Country A, and its theoretical research is on a higher scale. Unfortunately, however, Country A is building an experimental device more powerful than anything currently available in Country B; to catch up, Country B must build a device even more powerful than the one in Country A. Then scientists from Country A visit Country B and return with a similar report. The result is that financing is arranged for both scientific communities at the expense of their national economies (or, as Artsimovich puts it, "Scientific research in our time is the satisfaction of one's own curiosity at the expense of one's government").[40] It seems at first that Artsimovich is simply describing the "accelerator race" for more powerful elementary particle accelerators. But in retrospect, taking into account that Artsimovich was a participant in the Pugwash movement, and that the Soviets had been debating the creation of a national antiballistic missile defense system, we can surmise that his model has a broader reference, not only concerning scientific research and its costly experiments, but also the military-technological race of the late 1960s.

Scientific Prose, Present and Future

The rising public interest in scientific prose is evidenced by the fact that *Inevitability of the Strange World* went through three editions of 100,000 copies each in a matter of five years, and that *Puti v neznaemoe* began as a publication of 75,000 copies and has maintained this high distribution. The reasons for this enormous success are more profound than a general interest in science or the Russian tradition of merging the two cultures. There were significant historical reasons why, in the post-Stalinist era, the reading public turned so readily to literature on science in general and to scientific prose in particular.

In the early 1920s many Russian intellectuals felt that the Russian Revolution of 1917 had been full of paradoxes in that the emerging social order had little room for rationalism, humanism, or internationalism, the most attractive slogans of the victorious Revolution. Osip Mandelstam, one of the best twentieth-century Russian poets and a fine essayist, wrote in his article "Nineteenth Century":

> In the veins of our twentieth century flows the heavy blood of very remote and monumental cultures, such as Egyptian and Assyrian ones. The task of those survivors of the nineteenth century who were shipwrecked on the new historical continent is to humanize and Europeanize the twentieth century, and in this work it is much easier for them to rely, not on yesterday, but on the day before yesterday . . . The elementary formulas, the concepts of the eighteenth century, could once again be of great use. The spirit of the French encyclopedia, the spirit of the law, idea of social contract, naive materialism, which was ridiculed in the nineteenth century, schematic intellect, the spirit of purposefulness can still be of use to humanity. Now is not the time to be afraid of rationalism. The irrational root of the coming epoch, as if the stone temple of an alien idol, casts its shadow upon us. In such days the intellect of the French encyclopedists is for us the sacred flame of Prometheus. [41]

Mandelstam's essay, first published in 1922, by a strange quirk of fate reappeared in 1928 at the beginning of the era of Stalinism, which was a feast of irrationalism in Soviet public life. Only after 1953, at the end of this period, did Mandelstam's assessment become relevant and the return to an eighteenth-century type of rationalism become possible.[42] A strong reaction to Stalinist irrationalism emerged and elevated the scientist—the embodiment of rationality—to hero and idol. Suddenly the everyday occupation of scientist acquired strong moral qualities. This view was crystallized by the Soviet mathematician Aleksandr Khinchin, who wrote in glowing and idealistic terms about the pursuit of mathematical truth, which from his point of view is absolutely selfless and devoid of any self-interest.[43] Khinchin maintained that a scientist's attitude toward his own research could spill over into his day-to-day behavior and his relation to society—a public mood expressed most clearly in a section on the so-called physicists and lyricists.[44] The main outcome of Khinchin's proposition was the conviction that in the 1960s, science and technology (physicists) started to play a much greater role in the intellectual life of Soviet society than did literature and the arts (lyricists).[45] As a result, scientific prose flourished, sharing the spotlight with traditionally modest genres of Russian literature such as essays and memoirs. Still, at the same time, the poetry of Evtushenko and Voznesenskii, relying on the tradition of Pasternak and Akhmatova, was not only holding its ground but was actually starting to expand its influence.

After the publication of *One Day in the Life of Ivan Denisovich* in 1962, which gave an enormous boost to Russian prose, the equilibrium between prose, poetry, and the more subordinate areas of Soviet literature, such as scientific writing, started to shift in favor of traditional genres. The influence of scientific prose waned because of the moral compromises it had to make in order to survive. To see this clearly, we have to take into account not only those works published in *Puti* but also those that were not. In the mid-1960s, while visiting an editor of *Puti*, I bumped into a strange, high-strung man who was on his way out, obviously after a very heated conversation. I learned that he was Zhores Medvedev, who had sought to have his book on Lysenko published in *Puti*. I do not know how serious the editorial board was about printing the book, but when the editors met with resistance from higher authorities, they did not persevere. After the manuscript appeared in samizdat and was smuggled abroad, they ceased their efforts.

In a similar situation, the editors never made the slightest attempt to publish the memoirs of G. Ozerov, entitled *Tupolevskaia sharaga*, about Soviet aviation designers and engineers who worked under Tupolev's direction while serving as prisoners of the KGB.[46] One could add more examples, but the general point is that while Soviet scientific prose in the

1960s and 1970s survived intact, it carried more interesting, more analytical, more critical material about Western than Soviet science and technology. Materials about the Soviet research and development community were mostly informative, descriptive, and laudatory.

This situation was summed up brilliantly in 1977 by a prominent sociologist, Vladimir Shubkin, in his article "The Limits."[47] Shubkin spoke mostly about the relation between Soviet prose and modern Soviet sociology. He pointed out that the pretensions of modern Soviet sociology to pronounce judgment on Soviet literature, or even to try to win territory in its domain, are without any serious foundation. Here lies the natural limit of the rational approach in the formulation of a collective consciousness of Soviet society. Shubkin takes Danin's argument—that in parallel to the problem of protecting the environment the Soviets have the problem of protecting the inner, independent, spiritual world of man—and shows that if, in solving the first problem, the role of science is most significant, then, in solving the second problem, the role of art is paramount.[48]

Continuing this analysis Shubkin becomes critical of the type of Soviet intellectual whom he calls a rational man. Shubkin feels that within modern Soviet society such a rational man can play a destructive role. He points out several features of the Soviet rationalist who firmly believes that a human being is the product of this social environment and thus has a moral excuse; therefore, he is not responsible for his deeds. If he does not violate certain rules, it is only because he fears punishment. His response to the pressure of society is to develop his own counterculture. He has no conscience but only his own self-interest. He always weighs his choices thus: "This thing is advantageous to me, and that thing is not so advantageous." He could even use a computer to make the choices; it would not change the nature of his behavior. He tries not to develop ideas or produce them but rather to trade socially useful information. His usual attitude is to get more than he gives, or at least to even the score. Such a rational man is very much attached to the material world and its rewards. He has a highly developed sense of self-preservation. He is a cosmopolitan. He is usually cool to any feeling for his native land, but his feeling of belonging to an interest group is intense. He is not brave, but he is flexible; he is always ready to compromise. He lacks spiritual, creative fire. He is a good executor and a fine critic, but he is not imaginative. Neither is he religious, but he is superstitious; he remains in the grip of a nonreligious form of religious consciousness, as Marx called it.

It is obvious, then, that the rational man is not perfect, and that the task of art is to study and reform him, using the force of art to create a spiritual man, a man with conscience. Such a spiritual man is not against rational knowledge, but he feels that such knowledge is limited, that there are many things in the world about which our sages never dreamed.[49] That is why Shubkin feels that this movement from social to spiritual man puts certain limits on the use of the tools of modern social knowledge, while the use of intuition, imagination, and the traditional tools of prose is taking new opportunities.

A severe critic will find in Shubkin's approach the redressed arguments of a Russian Slavophile or even protofascist. But the arguments of the sociologist cannot be dismissed that easily, especially when one considers the source of his analysis: modern Soviet literature. Opening at random *Nash sovremennik,* the main publication of the Russian nationalists in the USSR, one can easily find one or two stories about members of the Soviet technical intelligentsia, and one would be amazed at the directness of the approach. One example, selected at random, is a story by Anatolii Chernoisov entitled "Business and an Extra."[50]

It tells of an assistant professor who teaches rigidity of materials at a technical college. It is well known that in Soviet engineering colleges, rigidity of materials is one of the basic courses. (After you pass an exam in rigidity of materials, says a student proverb, you can get married.) Because the poor assistant professor does not have a Ph.D. (it is very difficult to get a doctorate in rigidity of materials before age forty, because of the rigidity of the subject itself), he finds himself in desperate financial straits. His wife pushes him to take odd jobs; she tells him that other assistant professors are moonlighting by covering the entrance doors to apartments with synthetic leather, by repairing plumbing or electrical wiring, making a ruble on the side. The man finally decides to go work as a carrier at a meat packing plant. When he arrives, he finds himself among an odd crowd: drunkards, nincompoops, and enterprising young men, mostly students. It happens that the leader of his brigade turns out to be one of his own students. Before they begin working, the leader tells the men frankly that the wage they will be paid is very low; each will receive only three rubles at the end of a long and strenuous day. He can do nothing about it, but he can reward them in a different way: he can give them the product itself. They can take home as much meat as they can load on themselves, and although it is illegal, the guard will not pay any attention. The poor professor is deeply humiliated, but still he takes the meat and then has a long discussion with his student, who explains to him that there is no way around it. Of course if everybody steals, then the whole society will collapse; but still he feels keenly that his teacher is severely underpaid by the college. The assistant professor does all the work while the full professor, the department chairman, makes four times as much money for doing almost nothing. The poor assistant professor buys this argument, and what is more, when the student is required to pass an exam in rigidity of materials, the professor gives him a C instead of the D he deserves. The reader gets the strong impression that Shubkin is right, and that such a story says more to and about Soviet society than an entire issue of *Puti*. But to be fair, let us take one final step.

About the same time Shubkin published his piece, Danin wrote a short memoir about a famous Russian constructivist, Vladimir Tatlin (1885–1953).[51] Tatlin was a modernist who combined the humanistic and scientific features of Soviet culture. He was known as a fine painter, a stage designer, and, above all, the creator of a model for a monument to the Third International. Tatlin envisioned a building 400 meters tall, the outside surface of which was to be shaped like a double helix. Inside this double helix would be four colossal structures made from steel and glass—a glass pyramid, a large glass cylinder, a small glass cylinder, and a glass hemisphere—all rotating with different periods, one every 365 days, another every month, a third every 24 hours. The figure 400 meters also had a meaning: it represents one hundred-thousandth part of the meridian of the earth. The axis of the helix was not straight, but met the surface of the earth at an angle equal to the angle between the axis of the earth and the plane of its orbit. This monumental project, which could not be realized when it was conceived in the 1920s for technical reasons, but which is now quite practicable, is at the foundation of modern architecture. We can find traces of it in different places: in Moscow's television tower, in Brasilia, in Rockefeller Plaza in Albany, New York. Tatlin was also known for his fantastic ornithopter, a plane he called *Letatlin*, a combination of his own last name and the Russian verb *letat'*, to fly. He built a life-size nonworking model of it.

Danin did not deal with the scientific, technical, or cultural significance of Tatlin's work, however. Instead he wrote a highly personal memoir about his own acquaintance with Tatlin, telling of one evening late in 1931 when a special meeting took place dedicated to the memory of Vladimir Maiakovskii, at which the leading figures of 1920s Soviet modernism were gathered. Pasternak was reciting his poetry; on the podium was Meyerhold; and from the ceiling of the old Moscow mansion where part of *War and Peace* was set, *Letatlin* hung like a chandelier. The real roots of Soviet scientific prose could be found here in this hall, through the influence of those on the podium. This memoir also shows that the roots of Soviet scientific writing are not in the modern West nor in eighteenth-century France but mostly in the Soviet Union of the 1920s. If ever there is a return to the mood and creativity of this fabulous period of Russian and Soviet history, Soviet scientific prose may have something interesting to say, because it works as part of the mighty, irrepressible tradition of Russian and Soviet culture and literature.

Why Scientific Prose?

The emergence of scientific prose can be explained as a reaction to Stalinist irrationality. After Stalin's death, Soviet society strongly recoiled from the irrationality he had spawned and reinstated a type of eighteenth-century French rationalism. This return to a schematic intellect, to the idea of purposeful materialism and objective truth, makes the extreme moral claim of Aleksandr Khinchin understandable and in a way natural. Those circumstances also turned Soviet scientists—the embodiment of rationality—into heroes of their time. Having to their credit the atomic and hydrogen bombs, as well as *Sputnik,* Soviet scientists also became heroes in the eyes of their own government—members of the establishment who had to be reckoned with and who enjoyed certain liberties, including the liberty to go public.

Yet Soviet literature, which had only begun to recover from the Stalinist terror, badly needed an ally and defender. The language of scientific prose gave both sides—writers and scientists—the opportunity to join forces. The reading public, which wanted its heroes to be sophisticated, was also delighted. From the desires of those three groups Soviet scientific prose was born.

Soviet scientific writing has three striking features. First, it bears the imprint of the hierarchical structure of Soviet society. Born in Moscow, the center of political, economic, and intellectual power, at the center for the Union of Soviet Writers, it was initiated almost singlehandedly by three or four prominent figures in Soviet literature and science. This cult of personality in science and art is one of the peculiarities of modern

Soviet life, which could sometimes reduce a merger of two cultures to the personal relationships among a few participants.

Second, Soviet scientific prose is concerned more with presenting science in a favorable light than critically assessing scientific phenomena. This peculiarity is the natural consequence of the circumstances under which Soviet scientific prose emerged.

Finally, this prose, being part of the Soviet cultural establishment, for a long time did not encounter serious public criticism. The situation started to change in the second half of the 1970s, when Soviet sociologists, under the substantial influence of Russian nationalist fiction, began to cast suspicious glances at the protagonist of Soviet scientific prose—the rational man, a person they saw as not only unattractive but also alien, even Western.

Soviet scientific prose rose to this challenge with a well-argued claim that it was firmly rooted in Soviet soil, in the fabulous 1920s. This response implies that if the political situation in the USSR becomes favorable for a new burst of revolutionary creativity, then scientific prose will be able to realize its full potential, asserting its independence from dogmatic Marxist ideology and from Soviet science itself.

One of the problems of fruitful interaction between modern Soviet and American society is their asymmetry. A Russian dissident arriving in the United States from Siberian exile receives a most cordial welcome from American conservative politicians but is met with suspicion and aloofness by his American colleagues. An American specialist in science policy, arriving in the USSR in search of fine Soviet research, will be cool to an offer to visit the "Soviet Harvard"—Moscow State University. He would by far prefer one of the institutes belonging to the Academy of Sciences. Following this line, one may say that if an American researcher wants to study the two-culture problem in the USSR, it would be better to skip the discussion about physicists and lyricists; to abandon the symposiums on the issue of creativity in science and art, where major Western arguments are rather artlessly repeated; and to turn instead to the modest field of Soviet scientific prose, where those two cultures happily—or uneasily—interact.

· 12 ·

The Response to Science and Technology in the Visual Arts

Peter Nisbet

The year 1917 brought to power in Russia a party professing a fundamental belief in science and technology. Not only did Marxism claim to have uncovered the scientific laws of history; those laws also pointed to industry (and thereby technology) as the inexorable expression (and motor) of progress throughout that history. However much practical circumstances may have allowed the Bolsheviks to express their central commitment to science and technology, the pervasive ideology was inescapable for writers, artists, and intellectuals attempting to come to terms with the new regime and its implications.

Responses naturally were many and varied. For the visual arts, the power and authority of science and technology were often an unsettling challenge, confronting the creative individual not only as insistent new themes but also as systems of behavior and social organization that potentially undermined much of what the visual artist had apparently stood for. Throughout any discussion of the impact of science and technology on the visual arts in the Soviet Union, we will of course find the artist adjusting to this modern subject matter, reflecting the Party's evolving policy, and contributing in a small way to society's debates. We shall also see, however, the recurrent pattern of the artist finding ways to answer the "threat" posed by science and technology to his creative role. Sometimes engaging science and technology on their own terms, sometimes transfiguring them into less challenging metaphors, the visual artist, although perhaps never as deeply embedded in Russian or Soviet cultural life as the writer, has thereby provided a running commentary on the Bolsheviks' massive investment, both ideological and financial, in science and technology.

A comprehensive chronological survey of the treatment of science and technology by visual artists in the Soviet Union between 1917 and 1987 would surely repeat, in broad outline, the findings for literature presented by Katerina Clark in Chapter 9 of this volume and by other analysts elsewhere.[1] Although modern hermeneutic theory has encouraged us to doubt that there are any unambiguous statements of any kind (whether verbal or visual), it is plausible to suggest that Soviet visual art can provide many examples of a relatively unproblematic and untroubled attitude toward science and technology, both pro and con. In visual expressions that tend toward the illustrational, one can find evidence of wholehearted approval, as well as art that emphatically rejects and condemns science, technology, and all their works. Although this chapter will focus on the more challenging and complex middle ground between these two positions, it is valuable to remind ourselves of the twin poles, which can also be said to bracket our subject chronologically, with affirmation characterizing the early years of Soviet cultural life and negation increasingly evident in more recent decades.

Especially in the popular visual culture of printed propaganda in the 1920s, the glorification of a scientistic or technicist position is common. A striking example is provided by such images as V. P. Krinskii's graphic work showing a Byzantine church converted into a machine shop, with saints in the spandrels replaced by communist agitators, whose heads are in turn replaced by belt pulleys set in motion from the "altar," while the "mechanized congregation" looks on.[2] The message of modern engineering emphatically triumphing over traditional values (indeed, adapting them for its own use) could hardly be clearer. It is a message implicit in the proliferation of technological images on items of mass production— everything from postage stamps to matchboxes and textiles.[3] The First Five-Year Plan naturally provided the political and emotional climate most conducive to this tendency, with posters and paintings stridently asserting the unalloyed advantages of industrialization (and, by extension, of the scientific research and technological advances underlying it).

The mirror image of this mood is neatly represented, for the contemporary period, by the work of the beloved artist Il'ia Glazunov. Here is a clear rejection of modernity in all its manifestations, including twentieth-century science and technology. In turning so aggressively to themes from Russia's medieval and mythical past, in concentrating so energetically on the Russian landscape, national traditions, and folk art, Glazunov unambiguously (and somewhat theatrically) turns his back on the modern world, refusing almost entirely to engage seriously with its issues and problems. The modern world is either entirely absent from his paintings, or is simply juxtaposed as a grim counterpoint to the higher truth derived from "unspoiled" history and countryside.[4]

Between these two sociologically interesting but aesthetically disappointing extremes lies a full range of sophisticated and complex responses by visual artists to the imposing status of science and technology in the Soviet Union. This chapter will very selectively touch on a number of examples to highlight the varying strands in the artists' continuing attempts to come to terms with this aspect of reality.

In the early years of the young Soviet state, science and technology were not so much a theme for the visual arts as a model: for a significant number of the Soviet avant-garde, art was—or should be—science or technology. In other words, the artist was to be a scientist or an engineer. This attitude was not, of course, wholly original to these artists; nor did it completely disappear with the eventual demise of this avant-garde in the 1930s. It was played out, however, with a seriousness and consequentiality that make it a revealing episode in the continuing debate between the two cultures.

Furthermore, one particular theme emerged during these early years that became something of an obsession for both the avant-garde and traditional artists, a theme that could fuse art's most poetic longings with the insistent reality of scientific advances and technological imperatives (not to mention political priorities). Flight (flying, airplanes, pilots, and so on) as image, as metaphor, and as subject neatly brought together the utopian aspirations of both the artist and the scientist-engineer.

The theme of flight, however, was but one special instance of a more general approach to technology by artists, an approach that sought to soften or mitigate the potentially alien intrusiveness of technology by embedding it in images of natural harmony (often calling on traditional associations of landscape and the female form). This strategy can be found in paintings of a more or less socialist-realist stripe, as well as in the work of those dissident artists of the 1960s and beyond who took up the issue of science and technology with renewed concern, often drawing on the heritage and achievements of the "heroic" avant-garde of the years around the October Revolution.

Art as Science, Art as Technology

While the achievements of the Russian avant-garde cannot be said to date from the October Revolution, the events of 1917 certainly did confront the advanced artists with a changed situation which demanded a response. Not only were there novel possibilities for working in the administration of cultural life of the young Soviet Union, but also the insistent fact of a new social order made the question of the future role of art much more acute and pressing than it had been in the ten years of feverish experimentation in the visual arts that preceded the Revolution.

Accompanying the focus on finding a suitable social role for art was a turning away from the glorification of subjectivity, intuition, and illogic that can be said to characterize the prerevolutionary avant-garde. In a sense, Kazimir Malevich's famous painting *Black Square* (1915), his manifesto of suprematism, was the culminating extreme of a flight from the modern world as represented by the machine, positivist science, and the bourgeois order that supported them. The nostalgia of the World of Art Group, the mysticism of the Blue Rose artists, the peasant paintings of Goncharova, Larionov, and Malevich, and the willful alogicality of "transrational" art all defined themselves in opposition to the mundanity of the nineteenth century.[5]

Parallel to the Revolution, this avant-garde tradition underwent an astonishing transformation, with several younger artists, no doubt aware of the immense status and authority conferred on the Marxist revolution by its scientific aura, redefining art as the equivalent of science, pursuing similar aims with similar methods. Analogously, art was also defined as engineering, a practical application of creative design skills to the building of the new society. Inherent in the strategic attempt to define the revolution in art as the precursor and equivalent of the revolution in society was the more specific notion that the new art could draw authority from the same rationalist and utilitarian principles as the new social order.

One of their weapons in the generational conflict with their symbolist and irrationalist predecessors was, therefore, the newcomers' claim to be working in a realm of intersubjective certainty. This claim could, of course, if successfully promoted, also assure that new art official status as the art appropriate to the new social order (and therefore guaranteed future hegemony).

In the years 1917–1922, then, abstract art was described by some of its practitioners as a laboratory art, the product of experiments in the objective qualities of material, form, and color. The codification of the characteristics of art was pursued as a scientific inquiry, far removed from individual emotion or intuition. Much as science concerned itself with uncovering and examining the basic elements of nature, so too would painting now devote itself to understanding its own basic, intrinsic elements: color, line, form, and the laws, so called, of their interaction. Sculpture would investigate the nature of materials.[6]

The most pronounced evidence of this trend was to be found in the programs of the various research institutes to which artists gravitated in the postrevolutionary years. Most notably, the Institute of Artistic Culture (INKhUK) had been founded in March 1920 with the goal of researching all aspects of the science of art. "The aim of the work of the Institute of Artistic Culture is Science, the investigation of the analytical and the synthetic basic elements of the separate arts and of art as a whole."[7]

One of the central debates conducted at INKhUK in 1921 concerned the relative merits of composition versus construction as the basis for contemporary aesthetic activity. The more radical avant-garde defended the latter vociferously (it was part of the stance that led them soon afterward to abandon art in favor of practical design). In the course of these analytical discussions, one of the leading members of the "advanced" faction, Aleksandr Rodchenko, wrote a series of slogans for his teaching at the Moscow Higher Technical-Artistic Studios. Inter alia, he asserted that "construction is the arrangement of elements. Construction is the outlook of our age. Like every science, art is a branch of mathematics . . . Consciousness, experiment . . . function, construction, technology, mathematics—these are the brothers of the art of our age."[8]

Alongside the appeal to scientific modes of thought and behavior, technology and engineering were also exercising a powerful fascination over the advanced artists. By 1921–22 many had decided that the usefulness of design activity was more justifiable and appropriate to the social and political situation than even the most rigorously "scientific" explorations of formal elements of art. A mass declaration of late 1921 affirmed that many of these artists would devote their creative energies to practical utilitarian tasks, "going into production" and the factories. The origins and development of this crucial step into constructivism need not concern us here. What is important to note is only the artists' recognition of the central role of the engineer-technologist in the new social order, and their attempt to redefine their own role so that a place (preferably quite a prominent place) could still be found for their talents.

Several of the constructivist artists explored smaller-scale architectural design, working on newspaper kiosks, agitational displays, and the like. By 1922 Rodchenko was wholly committed to graphic design (including advertising, photomontage illustrations, and film captions). His colleague Liubov Popova was pursuing typography, stage design, and work for the textile industry. In a statement of 1922, she neatly encapsulated the wish for the combined power of science and engineering to eradicate obsolete notions about art and aesthetics:

> Why do we still not have a precise formula that . . . would remove all this aesthetic trash from life and transfer it to the jurisdiction of those who protect antique monuments and items of luxury?
>
> Let this formula be infallible, like the formula of a chemical compound, like the calculated tension of the walls of a steam boiler, like the self-confidence of an American advertisement, like $2 \times 2 = 4$ . . .
>
> Ah! *Expediency! Please* be our criterion at least for a moment. Our entire life in the form of sociology, chemistry, physics, mathematics, engineering, technology, etc. . . . dictates a single, integrated approach to evaluating the facts of everyday life.[9]

Popova's language appeals to the authority of science and technology (the chemical formula, the structural analysis) as the foundation for a creative activity devoted to the public good.[10]

Throughout the 1920s this impulse to put one's creative talents at the service of society resulted in many artists' preferring the factories and the workshops to the studios and the museums. Book covers, clothing, furniture, buildings, stage sets, household utensils, posters—these all attracted the attention of artists who sought a reconciliation with technology. The exemplary activity of the creative designer, fusing art and engineering, could, through an intimate involvement with the production process itself (preferably, of course, the process of mass production), contribute to the revolutionary society appropriate to the new Soviet state.

It was not only artists of this narrowly defined constructivist group who effected the important move from pure to applied art (or, as one of the more extreme critics phrased it, "from the easel to the machine").[11] Vladimir Tatlin, known for his prewar abstract reliefs in nontraditional materials (which he described in 1920 as "material models on the laboratory scale"), began in 1919 to work on his Monument to the Third International, a huge spiraling tower that was intended to be not only a homage to the internationalist revolution but also a structure to house deliberative assemblies, a radio station, and other useful activities. He also experimented with a position teaching technical drawing in a factory.

As he wrote in 1920: "An opportunity emerges of uniting purely artistic forms with utilitarian intentions . . . The results of this are models which stimulate us to inventions in our work of creating a new world, and which call upon the producers to exercise control over the forms encountered in our new everyday life."[12]

By the end of the 1920s this commitment had in Tatlin's case taken the form of research into a unique "flying bicycle," an ornithopter to be powered by a single human being.[13] Around 1929–30 Tatlin devoted his energies to the design of this birdlike machine, the *Letatlin* (a verbal pun fusing *letat'*—to fly—with the artist's name). It exemplified, he wrote, "art going out into technology."[14] This joining of the artist's understanding of materials (and his empathy for nature, for the design was based on careful research into the flight of birds and insects) to the development of "an everyday object for the Soviet masses, an ordinary item of use"[15] can be seen as one of the most impressive results of art's attempt to find a common ground with engineering (here of a peculiar, Leonardoesque kind) in the postrevolutionary situation.

Flight: Image and Metaphor

Tatlin's *Letatlin* was of course more than a demonstration of a particular kind of organic technology. Its function expresses the artist's age-old dream of flight. In the Soviet Union this traditional theme could be deployed by artists who wished to interpret modern technology in a manner consistent with more traditional aspirations to freedom, inventiveness, and achievement. As airplanes and flight became ever more impressive examples of Soviet technological achievement, artists could pay homage to these gigantic strides and yet still redirect attention to flight as a metaphor rather than as a triumphant fact.

The prewar avant-garde, much like its counterparts in the rest of Europe, had quickly recognized the potential of flight as an evocative theme.[16] The futurist poet Vasilii Kamenskii was one of the first aviators in Russia. A famous pilot, he also wrote poems on the theme of flight. This theme also figured prominently in the watershed production in 1913 of the futurist opera *Victory Over the Sun*, with libretto by Kruchenykh, score by Matiushin, and sets by Malevich. A character in this alogical story was the "Traveler Through the Centuries," who "had a broadened concept of time, for his airplane allowed him to move freely through the centuries and to examine the values of various ages. With this knowledge, he suggested that the values of the existing world were limited and relative."[17] The last act of the opera introduces the figure of the Aviator, "a man freed from the old conceptions of time and space who participates in the destruction of the old order and accepts the terrible strength of the new age, unlike others who, being frightened, become insane or commit suicide."[18] Soon after his subsequent leap into wholly abstract art, Malevich created a suprematist painting, with its characteristic abstract forms floating in an undifferentiated cosmos, with the title *Aeroplane Flying*, and, indeed, termed an entire phase in the development of his style "Aerial Suprematism."[19]

After the Revolution, the range of symbolic properties that the avant-garde had exploited in the themes of flight—speed, the aerial viewpoint, and especially overcoming gravity (the symbol of all earthly constrictions)—were further developed, primarily in a metaphorical fashion. Malevich encouraged research into floating, aerial architecture employing suprematist forms. Rodchenko's *Hanging Constructions* were not only sculptures that appeared to be modeled demonstrations of geometrical shapes and relationships; they were also floating bodies, apparently free of the earth's pull. El Lissitzky invented his *Proun* forms, constructions deployed in the illusionistic space of the two-dimensional surface in ways that deliberately contradict our expectations of natural, stable order; he himself attempted a rather literal expression of this impulse in a text inscribed on one of his early *Proun* lithographs: "Construction floating in space, propelled together with its spectator beyond the limits of the earth, and in order to complete it, the spectator must turn it and himself around its axis like a planet."[20]

Parallel to these investigations by the avant-garde, the actual technology and business of aviation was proceeding very rapidly. It was surely the achievements of Igor Sikorsky in 1913–14, for example, that encouraged the futurist allusions to the theme of flight, and gave those allusions a special resonance. As the years progressed, and airplanes assumed an increasingly important role in national life, artists were more and more able to engage directly with the fact and implications of this new technology.[21]

Scholars have often noted the "high priority attached to technical progress in aviation" by the bolshevik government.[22] This extended not only to necessary funds being more readily available (and more closely controlled by the highest Party and government organs) but also to a continuing emphasis on social mobilization. "Huge public interest was created in aircraft and aviation both by party and non-party organizations. In 1923 a Society of Friends of the Air Force (ODVF) was established to support the development of the air force, to organise the collection of funds for the construction of aeroplanes and generally to popularise aviation."[23] The founding in 1923 of the air transport organization Dobrolet to serve as the nucleus for a countrywide airline network was equally a part of this process. It "took the form of a joint-stock company, backed both by individuals and by such organizations as Komsomol, the Communist Party's youth organization."[24]

It is not surprising that a leading member of the avant-garde such as Rodchenko, after his transition to a "productivist" stance committing him to practical work in the applied arts, should devote some of his energies to this new and peculiarly modern social reality. In 1923 he actually designed some trademarks and advertisements for Dobrolet, encouraging the public to buy shares in this promising new enterprise.[25] One image presents a slogan asserting that anyone not a shareholder was not a citizen of the USSR, alongside the image of a rapidly climbing monoplane.[26]

It was also in 1923 that Rodchenko designed the cover for a volume of "aeropoems" on themes of flight,[27] and used many images of airplanes in his pioneering photomontages illustrating poetry by Vladimir Maiakovskii and adorning the cover of the radical literary-cultural journal *LEF*. Most evocative of the latter is the design showing a biplane dropping a massive fountain pen on the cowering form of an ape holding a crude arrow (1923). There could be no clearer affirmation of the avant-garde's association of their own creative talents (the pen) with the newest technology in the battle being waged by the Soviet Union against the outdated and primitive past.[28]

As abstract art faded in prominence in the course of the 1920s, these and other representational images of flight came to carry a broad range of attitudes about art and technology, sometimes with a rather disturbing ambivalence unusual in Soviet art. Aleksandr Tyshler was a member of the OST group, a society of artists founded in the mid-1920s to try to preserve something of the formal experimentalism of the earlier avant-garde years, but allied to representational subject matter more in tune with the demands of the New Economic Policy (NEP) period.[29] Quasi-surrealist in approach, several of Tyshler's works deploy the image of an airplane, perhaps a dreamy symbol of release and imaginative freedom, but also part of a possibly threatening technology. *Woman and Aeroplane* of 1926[30] shows a three-quarter-length figure of a young woman, with an absurdly stretched neck and head craning upward at an impossible angle to gaze at the distant form of an airplane flying away.

The indeterminacy of this image is heightened by even more ambiguous representations of technology in Tyshler's oeuvre, such as a work from his so-called Lyrical Series of 1928.[31] This painting shows a straw beach hut, inhabited by several figures (including a boy and a girl conversing next to a bicycle) and a horse. In the distance, through openings in the hut, we see a speeding sports car and an airplane. Within the immediate enigma of such an image we may glimpse the implication of a critical attitude toward the mechanized forms of transportation in favor of the more familiar, safer methods that rely on muscle power, whether of horses or humans, as with the bicycle. Certainly these elements in the painting are directly associated with the enclosed and protected space of the beach hut, where the everyday activities of human life proceed.

Tyshler was repeatedly criticized for this kind of work. Whatever one may feel about its quality, its intensely personal stance made it difficult for and suspect to even the most intelligent of critics in the 1920s. Ia. Tugendkhold, for example, sought to excuse and neutralize the ramifications of this kind of painting by categorizing Tyshler as essentially a stage-set painter.

> And it is precisely in this orientation toward the theater, in his tendency toward the playful, that we find proof that all his "experiments" are perhaps not meant so seriously as it might appear, that everything is just a theatrical mixture of the tragic and the comic, and that the artist can be "cured" of his "childish illness" of left-wing radicalism. However, we can and must say that Tyshler is in danger of becoming decadent, that he has a very one-sided attitude, that he knows only half of the present.[32]

Certainly the elements of the satirical, the exotic, and the grotesque in Tyshler's art cannot be easily accommodated to the optimistic mood that seemed de rigueur, and common both to the most avant-garde and the most traditional of artists in the first two decades of the Soviet Union.

The advent of the First Five-Year Plan, with its attendant upheavals, reinforced this obligatory optimism and reinvigorated the themes of flight and airplane technology in a manner that reflected the expanding role of aeronautical engineering and research of all kinds. This trend then reached its apogee in the 1930s, when Stalin took a very close personal interest in aviation, promoting aeronautical achievement and record setting to an unprecedented degree.

Images of aviation from the 1930s of various kinds can be fairly directly linked to social and policy imperatives of the time. A famous painting such as Samuil Iakovlevich Adlivankin's *Competition of the New Model Airplane Builders* of 1931,[33] showing five boys and girls with their heads raised, not, like Tyshler's woman, to gaze longingly at a receding plane, but rather to hold up and proudly launch their various models, fits perfectly into a period when

> there was also a stress on the need for members of the Komsomol to take an active part in the work of fostering aviation, both through participation in Osoaviakhim [founded in 1927 as the successor to ODVF] and by enrolling in the aviation educational institutes, setting up aeroclubs and working in the aircraft industry. Particular attention was paid to the role of the Komsomol in aviation at the Ninth Congress of VLKSM [All Union Lenin Communist Union of Youth] in 1931. A great burst of activity apparently resulted with the formation of many aeroclubs, gliding groups and aeromodelling organisations.[34]

Moreover, Adlivankin's painting is not simply about the joys of youth celebrating flight, but is explicitly an image of the start of a competition in which the different models—monoplanes and biplanes of various designs—will be judged against one another. The role of the single inventor, pitting his or her wits against rivals, was not inappropriate to the subject, as "another distinguishing feature of design and development in aircraft was the greater than average importance attributable to the individual. The emergence of design teams concentrated around particular designers and the failure of the attempt to establish a large central design organisation in themselves suggest that there was a greater role for the individual in aircraft development than in other parts of the industry."[35]

For the creative artist, the fusion of the themes of youth (traditionally the most energetic, most creative age), the inventor (a parallel to the artist's own inventiveness), and flight (as we have seen, a natural metaphor for the artist's aspirations) must have been irresistible. Indeed, through the 1930s we can find several other images that bring two or more of these aspects together.

Aleksandr Deineka's *Future Pilots* of 1937[36] has a theme similar to Adlivankin's work. Three young boys sit by the water in Sebastopol, looking out onto the vast expanse of ocean. They are admiring three white flying boats low over the water (one appears to be landing). Deineka has obviously accentuated the harmonious integration of these impressive examples of Soviet aviation technology into the natural landscape of peaceful skies, gentle waves, and bright sunshine, enjoyed by the boys in their unself-conscious near-nakedness. There is a plane for each boy, and no implication of competition or danger.[37]

There appears to be a connection between the growing popularity of this theme and the political uses to which aviation exploits were put as part of Stalin's campaign to secure legitimacy for his rule.

> As Stalin and his associates moved from their underground past into the rapid modernization of the 1930s, they increasingly emphasized the "scientific" basis of their hegemony and particularly accomplishments of technology as evidence of their right to rule . . . Aviation was only one form of modern technology which was used in this way by the Stalinists. (Gigantic hydroelectric stations, canals and metallurgical plants were others.) But aviation played a very prominent and dramatic role in this respect at a critical time in the development of Stalinism.[38]

After 1933, when Stalin singled out the aviation industry for special praise as one of the First Five-Year Plan's major accomplishments and instituted the annual celebration of a new festival, Aviation Day (August 18), the attempt and achievement of ever more difficult aviation feats became an increasingly prominent part of the Soviet propaganda effort. "By 1938, Soviet spokesmen claimed to have set some sixty-two world records, including the longest, highest, and fastest flights, the first landing at the North Pole, and the first flight between the Soviet Union and the United States by a polar route."[39]

An artist such as Lissitzky, whose earlier creative activity would, as we have seen, have predisposed him to an interest in the theme of flight, indeed put his design talents in the service of trumpeting these achievements of Stalinist Russia. As one of the chief layout designers in the 1930s for the monthly magazine *USSR in Construction,* he was responsible for issues that dealt, for example, with the world-record ascent of a stratosphere balloon (February 1934) and the epic air rescue of the crew of the icebreaker *Cheliushkin* (October 1934).[40]

Alongside many similar instances in the printed media, a painting such as Vassili Kupzov's *ANT 20 "Maxim Gorky"* of 1934,[41] showing an aerial view of a massive silver-gray airliner over a town, accompanied by a fleet of other smaller planes and a dirigible,[42] can stand as a typical example of the required glorification of Soviet aviation. Created by the dean of Soviet airplane designers, A. N. Tupelov (who was also "the designer most closely associated with long-distance record attempts"),[43] this machine had the added advantage of carrying the name of a great artistic hero, the writer Maxim Gorky, thereby providing the immediate link between achievements in technology and art under socialism.

The plane was "an agitation and propaganda aircraft intended to popularize the regime. This eight-engine craft, which made its first flight in June 1934, could carry eighty passengers at a speed of 280 kph. Weighing forty tons, it had a radio station, printing press, photo laboratory, telephone switchboard, telegraph office, and motion-picture projectors."[44] Such an image is somewhat removed from the "Traveler Through the Centuries" of *Victory Over the Sun*, yet it does embody the continuing romance with flight that appealed so strongly to so many artists in the Soviet Union. By linking art to the modern notions of speed, communication, agitation, and weightlessness, it continues a tradition with deep roots in the experimental avant-garde. Though somewhat crude, it surely forms something of a bridge to later treatments of this unique conglomeration of themes. In the postwar years, one could surely trace the same network of allusions and connotations through the visual artists' response to the glory and excitement of space travel, especially manned space travel.

Harmony with the Natural

Although flight, as a scientific-technological theme, may have held special appeal for artists at various times, and for various reasons, it was, of course, but one such theme among many that were possible. With the increasing tempo of industrialization during the 1920s, as the NEP attempted to put the Russian economy back on its feet, artists struggled to find new ways to record, and indeed encourage, the rise of new industry. Several groups of artists committed to a realist, almost documentary, style emerged in the course of the 1920s, and were attracted to the factories for their political correctness as a subject as well as their picturesque and painterly possibilities. This naturally angered an avant-garde, constructivistically inclined critic such as Boris Arvatov, who complained in 1925 that such artists should not be visiting steel foundries and similar plants to depict them, but should be helping in their construction.[45] Nevertheless, the revival of representational easel painting could not be prevented: powerful political, aesthetic, and economic forces supported its reemergence as the dominant mode.

Throughout the Soviet period there would be little or no direct criticism of science and technology in paint. By and large, the visual artists confined themselves to affirmations of the value of science and technology, with one particular slant that is perhaps the most interesting constant thread running through the official images. In one form or another, artists sought to evoke a state of harmony between nature and industry, often compositionally subordinating the latter to the former in a way that suggests a rather measured enthusiasm for science and technology, appropriate for artists presumably committed to the significance of the handmade object expressing human and even spiritual values.

Quite early in the history of images of science and technology, this posited harmony between industry and nature finds particular expression in the incorporation of female figures, more or less supple and sensual, into the industrial landscape. This is strikingly noticeable in two canvases of the mid-1920s by Deineka.

His *Constructing New Factories* of 1926[46] shows two monumental women in rhythmic arrangement in the foreground, one pulling a coal cart, against a distant high-vantage-point view of a skeletal scaffolding shed with rail lines. The rippling, curvaceous strength of the women's bodies is forcefully underlined. Similarly, *Textile Workers* of 1927[47] presents three young women, dressed modestly but in tight and skimpy clothes, attending to various aspects of the cotton spinning factory's production process. In both cases a mild, suppressed eroticism seems to be a factor in the representation of industry. Women do not contradict engineering but rather complement it.[48]

Certainly this feature is not unique to images of industry in Soviet art. It is not hard to find other works of the period with gently sexual overtones, particularly paintings of sports subjects, and images of motherhood and joyous youth.[49] But the softening presence of young (if not adolescent) women as emblems of the continuing humanity of man's technological encroachments on nature reappears regularly in Soviet painting. One postwar example is a work from the 1970s, Sarkis Mambreevich Muradian's *Under Peaceful Skies* of 1972,[50] in which a girl in a translucent shift balances on an aqueduct bringing water down a gentle slope to a previously barren farming area. As one Soviet critic has written, "The slim girl walking alongside [sic] the irrigation pipe seems to personify the bright future and the new flourishing of this land."[51]

If the prominently placed women in these images are emblems of the natural and the creative, then they are a special case of the more general tendency in Soviet paintings of industry and technology for the natural landscape (often fertile) to be balanced compositionally against the man-made contribution. Typical in this regard are three paintings by the leading artist G. G. Nisskii. His *Autumn: Signal Masts* of 1932[52] has a horizontal composition of railway signal masts, looking much like Monet poplars, against a low, broad horizon. *Above the Snows* of 1959–60[53] shows a sleek plane flying low above a broad expanse of snow with two birch trees to the left. A Soviet description of a third painting from 1957 can stand, in essence, for the spirit of all three: *In the Suburbs of Moscow: Autumn 1957*[54]

> vividly illustrates Nisskii's artistic principles, his romantic perception of life and nature . . . The pure untouched snow, the slender fir trees, the forest on the horizon combine well with the geometrically clear-cut line of the highway and the electric wires etched against the vast dome of the sky. The distinctive marks of the time and the world of nature are organically fused together. In this lies the main attraction of the picture and of its emotional message, consonant with the feelings experienced by many who live in the age of scientific and technical progress.[55]

This lyrical view of technology predominates in Soviet art.[56] To be sure, there have been periods when a more aggressive, celebratory tone has been adopted, but it has always been a priority of the Soviet artist to relate technology as manageably as possible to the human realm. As early as 1928 the critic Tugendkhold had summed up this constant in Soviet art when he wrote: "Representation of sport, of radio, of flight, of the automobile is naturally a contemporary and urgent task. But they also have that in the bourgeois West, and that does not yet reflect a new art, which is connected to the development of a *new man*."[57] In most acceptable painting, at least, this "new man" and the natural landscape of which he was an integral part, were rarely if ever overwhelmed by the effects of science and technology.

Around 1960 three connected developments in Soviet cultural life converged to reanimate the question of the role of science and technology in the visual arts. Most publicly, official policy embraced the scientific and technological revolution as part of the Party program in 1961. Second, a debate on the relative value of art and science erupted in 1959, encompassing several different points of view about the importance of artistic creativity in society and its adequacy to the new scientific-technological realities.[58] The issues in the so-called physicists' and lyricists' debate, named after a poem by Boris Slutsky, were summarized by *Literaturnaia gazeta:* "The question of the relationship between contemporary art and the scientific revolution, the mutual influence of scientific and artistic creativity, today arouses widespread interest. Will they become antagonistic? Or will they peacefully share out 'spheres of influence'? Or perhaps somewhere the boundaries between them will disappear?"[59]

It was also in these years that the relative liberalization following the death of Stalin began to bear fruit in a lively "alternative" art scene. Inspired to some extent by exposure to Western models, young artists began to explore themes and styles beyond official socialist realism. To the extent that any generalizations about "dissident" art can be made at all, it is probably fair to say that the issue of science and technology was not, as such, a pressing priority for these innovators. Their concerns were more introspective, lyrical, and often religious,[60] and it might be said that the very act of ignoring these themes was an implicit comment on their ultimate insignificance.[61]

One tendency within the emerging unofficial art scene, however, can plausibly be understood within the context of the political and cultural debates on the role of science and technology. In 1962 Lev Nusberg, together with a group of like-minded colleagues, founded the group Dvizhenie (Movement). They were committed to the search for an artistic technology focused on "kineticism."

Two programmatic statements by the group help to explain its aims and methods. In his 1965 manifesto "What Is Kineticism?" Nusberg wrote: "We demand the utilization of all potentials and all media, all technological and aesthetic, physical and chemical phenomena, all forms of art, all processes and forms of perception as well as the interrelationship of physical reactions and the action of various human mental impressions as a means of artistic expression."[62] Another statement is a little more specific:

> Goal of the group [is] the common utilization of the most modern technical means (naturally in addition to those already existing) in artistic design of public interiors and exteriors, of festivals, ultimately of entire cities; utilization of these means to achieve aesthetic goals in the theater, cinema, and television; and furthermore for the search and development of new symbolic forms appropriate to modern man's conception of the world and himself.[63]

The work of the Dvizhenie group is, not unexpectedly, very diverse. It ranges over intense studies of symmetry to the design of temporary utopian environments in the guise of industrial exhibition installations for such official Soviet celebrations of technology as Elektro '72 and the pavilion devoted to glass at the Building Materials exhibition of 1971. The group also contributed to the decoration of the city of Leningrad on the occasion of the fiftieth anniversary of the Revolution.

Displaying a fascination for kinetic sculpture (a tradition that goes back to the early Soviet avant-garde),[64] for modern materials, and for organizing quasi-scientific seminars and discussions, the Dvizhenie group, throughout its changing membership, sought to integrate and extend the accomplishments of modern science in aesthetic projects with all but cosmic implications. Reviving the constructivists' more extreme aspirations of the 1920s toward a total renovation of man and society, the Dvizhenie artists pursued research into an "Artificial Bio-Kinetic Environment" which aims at the harmonious integration of man and nature.

In some of their theatrical and experimental performances the group has also incorporated a "paganistic" element that reintroduces nature in the form of nudity and eroticism in a way that has curious echoes of some of the themes of official art. Collective erotic mystery plays such as *Devils' Women,* performed in the Crimea in 1970, bring together elements of science (cybernetics, optics with geometrical mirrors, electronics, and so on) and sexuality.

This goal of harmony between man and nature (or between science and nature) is, as I have made clear, also a concern of the "official" artists. It seems as if the visual artists have never subscribed wholeheartedly to the technological Prometheanism sometimes espoused by the writers and politicians of the Soviet state. The conquest of and absolute control over nature seem antipathetic to the artists' desire to limit the authority and range of science and technology. Often fighting for a more traditional, subjective, even romantic or earthy approach to the world, many of the artists I have discussed have sought to show how the achievements of science and technology either are or should be compatible with nature, very broadly defined as the traditional landscape, as individual creativity, or as the human dimension to life.

At all events, for the better Soviet artists the question of the relationship of their creativity to the impressive and commanding achievements of science and technology remains troubling and unresolved. A 1982 painting by Rein Ennovich Tammik encapsulates the self-conscious thematization of this problem. *In the Studio*[65] shows a studio filled with printed images of modern technology, including an Aeroflot poster advertising the Tu-134, a cover of *Ogonek* showing a jet fighter plane, and a photograph, pinned to an easel, of a nuclear mushroom cloud. The artist, dressed in Rembrandtesque costume (!), is seated in front of an enormous blank canvas on which he has managed to make one curving red line, his (desperate, defiant, or respectful?) personal response to the mechanically mediated reminders of science and technology. Not only is the debate between art and science here still undecided, but the debate itself is also still a worthy subject for the artist.

Notes

Contributors

Index

Notes

Introduction

1. David Joravsky, *Soviet Marxism and Natural Science* (New York: Columbia University Press, 1961).
2. Paul Josephson, "Science and Ideology in the Soviet Union: The Transformation of Science into a Direct Productive Force," *Soviet Union/Union Soviétique* 8, pt. 2 (1981), 159–185. Also see *Program of the CPSU (1961)* (New York: International Publishers, 1963), p. 81.
3. Paul Doty of Harvard University has described to me in conversation how these early contacts between American and Soviet scientists on technical matters created a level of trust that enabled them to proceed subsequently to difficult political subjects such as arms control.
4. See, for example, Karin D. Knorr-Cetina and Michael Mulkay, eds., *Science Observed: Perspectives on the Social Study of Science* (London: Sage Publications, 1983), especially the chapters by Barry Barnes and H. M. Collins. Also see Everett Mendelsohn, Peter Weingart, and Richard Whitley, eds., *The Social Production of Scientific Knowledge* (Dordrecht: D. Reidel, 1977).
5. Herman Ermolaev, *Soviet Literary Theories, 1917–1934: The Genesis of Socialist Realism* (New York: Octagon Books, 1977), pp. 19–26.
6. Kurt Johansson, *Aleksei Gastev: Proletarian Bard of the Machine Age* (Stockholm: Almquist and Wiksell International, 1983).
7. Loren R. Graham and Richard Stites, eds., *Red Star: The First Bolshevik Utopia*, trans. Charles Rougle (Bloomington: Indiana University Press, 1984).
8. Ermolaev, *Soviet Literary Theories;* and René Fülöp-Miller, *Geist und Gesicht des Bolschewismus* (Vienna: Almathea-Verlag, 1926).
9. Boris Schwarz, *Music and Musical Life in Soviet Russia* (Bloomington: Indiana University Press, 1983), esp. pp. 53, 70, 75, 85; also René Fülöp-Miller, *The Mind and Face of Bolshevism: An Examination of Cultural Life in Soviet Russia*, trans. F. S. Flint and D. F. Tait (London: G. P. Putnam's Sons, 1927).
10. Richard Taylor, *The Politics of the Soviet Cinema, 1917–1929* (Cambridge: Cambridge University Press, 1979), p. 125.
11. A Soviet debate in the late 1950s and early 1960s, often called physicists versus

lyricists, was a forerunner of the continuing reevaluation of science and technology in the 1970s and 1980s. See B. Slutskii, "Fiziki i liriki," *Komsomol'skaia pravda,* 13 October 1959; and other articles on the theme appearing in *Komsomol'skaia pravda* and *Literaturnaia gazeta* in late 1959 and early 1960.

12. See Eric Hobsbawm, *Industry and Empire: An Economic History of Britain since 1750* (London: Weidenfeld and Nicholson, 1968); Eric Hobsbawm and George Rudé, *Captain Swing* (London: Lawrence and Wishart, 1969); Maxine Berg, *The Machine Question and the Making of Political Economy, 1815–1848* (Cambridge: Cambridge University Press, 1980); Fritz Stern, *The Politics of Cultural Despair: A Study in the Rise of German Ideology* (Berkeley: University of California Press, 1974); and Leo Marx, *The Machine in the Garden: Technology and the Pastoral Ideal in America* (New York: Oxford University Press, 1964).

13. Valentin Rasputin, *Farewell to Matyora* (New York: Macmillan, 1979).
    Much of the ensuing discussion relies heavily on a separate article of mine, "Adapting to New Technology," in T. Anthony Jones, David Powell, and Walter Connor, eds., *Soviet Social Problems* (Boulder: Westview Press, forthcoming).

14. Theodore Shabad, "Soviet, after Studies, Shelves Plan to Turn Siberian Rivers," *New York Times,* 16 December 1983, 1. Rasputin has also expressed concern about the pollution of Lake Baikal as well as other Siberian environmental problems. See Valentin Rasputin, "Posluzhit' otechestvu Sibir'iu," *Izvestiia,* 3 November 1985, 3.

15. Vladimir Soloukhin and Il'ia Glazunov, *Pisatel' i khudozhnik* (Moscow: Izobrazitel'noe iskusstvo, 1979); *Vystavka proizvedenii Il'i Glazunova: Katalog* (Moscow: Izobrazitel'noe iskusstvo, 1977).

16. *Il'ia Glazunov* (Moscow: Planeta, 1978), esp. pp. 190ff.

17. Olga Carlisle, "From Russia with Scorn," *New York Times,* 29 November 1987, H23, H34. Neither my research assistant Lisa Halustick nor I have been able to locate the woman with the book and the rat in the film; one wonders if she was subsequently deleted.

18. See Richard De George's chapter in this volume and his book *Soviet Ethics and Morality* (Ann Arbor: University of Michigan Press, 1969).

19. Eugene Kamenka, *The Ethical Foundations of Marxism* (London: Routledge and Kegan Paul, 1972); and Philip T. Grier, *Marxist Ethical Theory in the Soviet Union* (Dordrecht: D. Reidel, 1978).

20. "Test Tube Babies," *Hastings Center Report* (October 1978). Also see Loren R. Graham, "Concerns about Science and Pierre Soupart's Proposal," testimony presented to Ethics Advisory Board, Department of Health, Education and Welfare, Boston, 14 October 1978.

21. Loren R. Graham, "Science, Citizens, and the Policy-Making Process: Comparing U.S. and Soviet Experiences," *Environment* 26, no. 7 (September 1984), 6–37.

22. Ibid.

23. "Khristianskii vzgliad na ekologicheskuiu problemu," *Zhurnal moskovskoi patriarkhii,* 4 (1980), 35–39.

24. Frolov is the Soviet coordinator of the project on biomedical ethics for the Sub-commission on the History, Philosophy, and Social Study of Science and Technology, sponsored by the American Council of Learned Societies, the Social Sciences Research Council, and the Academy of Sciences of the USSR. The project includes meetings on biomedical ethics in both countries.

25. *Vremennye pravila bezopasnosti rabot s rekombinantnymi DNK* (Pushchino, 1978). See also my discussion in "Reasons for Studying Soviet Science: The Example of Genetic Engineering," in Linda L. Lubrano and Susan Gross Solomon, eds., *The Social Context of Soviet Science* (Boulder: Westview Press, 1980), pp. 205–240.

26. See Eugene Skolnikoff, "The Technological Factor Shaping East/West Relations," paper prepared for the Institute for East/West Security Studies Conference, "The Impact of Technology on the Future of European Security and Cooperation," Finland, 11–13 June 1987.

27. Lenin often expressed his opinion that wars under capitalism are inevitable. See, for example, V. I. Lenin, *Collected Works,* 2nd rev. ed. (Moscow: Foreign Languages Publishing House, 1960), vol. 8, p. 53; vol. 13, p. 80; vol. 21, p. 39. Less frequently cited is his belief that socialists should nonetheless try to avoid wars. See ibid., vol. 21, p. 299. When Soviet Russia needed a breathing space in 1918, Lenin supported the Brest-Litovsk peace against the advice of his colleagues Bukharin and Trotskii.

28. Quoted in David J. Dallin, *Soviet Foreign Policy after Stalin* (Philadelphia: J. B. Lippincott, 1961), p. 323. John Lewis Gaddis has maintained that the term *cold war* is a misnomer for the period after the Second World War and has emphasized the stability of great-power relations. See his book *The Long Peace: Inquiries into the History of the Cold War* (Oxford: Oxford University Press, 1987).

29. Nikita Khrushchev, *For Victory in Peaceful Competition with Capitalism* (New York: E. P. Dutton, 1960), p. xv.

30. Mikhail Gorbachev, *Perestroika: New Thinking for Our Country and the World* (New York: Harper and Row, 1987), p. 11.

31. "Vystuplenie M. S. Gorbacheva v OON," *Sotsialisticheskaia industriia,* 8 December 1988, 1; and *New York Times,* 8 December 1988, A16.

32. Thomas P. Barnett, "The Concept of Technocracy and the Soviet Politburo," unpublished paper, presented to Department of Government, Harvard University, 1988.

33. T. Anthony Jones, Walter Connor, and David Powell, eds., *Soviet Social Problems* (Boulder: Westview Press, forthcoming).

34. "Debating the Need for River Diversion," *Current Digest of the Soviet Press* 38, no. 7 (19 March 1986), 1. An excellent analysis of the different motives and groups in the controversy over river diversion is Robert G. Darst, Jr., "Environmentalism in the USSR: The Opposition to the River Diversion Projects," *Soviet Economy* 4 (July–September 1988), 223–252.

35. In Soviet newspaper and journal articles since Gorbachev came to power the psychological element has been clearly recognized. A family that rents its own

land and cares for it separately is often praised as the "master of the farm" (*khoziain na ferme*). See, for example, "Khoziain na ferme," *Pravda,* 3 September 1988, 1.

36. Interview with Evgenii Velikhov, Presidium building of the Academy of Sciences, USSR, Moscow, 4 December 1986.

37. The recent cessation of Soviet jamming of Western radio broadcasts fits well with this observation. But controls over personal computers are still in effect. See S. Ushanov, "Sploshnaia komp'iu-terrorizatsiia," *Literaturnaia gazeta,* 27 January 1988, 10.

38. Velikhov has actively pushed for widespread use of personal computers in the Soviet Union; he has a poster on his office wall proclaiming, "Personal computers for everybody!" At the same time, other Soviet officials have insisted that computers be used in the Soviet Union in a different way than in Western countries and have stopped several desk-top publishing enterprises. Even such a promoter of computers as Academician A. Ershov has warned that the introduction of computers into Soviet schools must be done with "full consideration for our social system, its realities, and our cultural and social traditions." A. Ershov, "EVM v klasse," *Pravda,* 6 February 1985, 3.

39. In a talk at Georgetown University on 7 October 1988, Aleksandr Chakovskii, editor in chief of *Literaturnaia gazeta,* said that Glavlit (the governmental censorship bureau) still has a censor posted in his editorial office with a list of items that cannot be printed, officially called the *Perechen'* (informally known as "the Talmud"). According to Chakovskii, most of the items are about military secrets. SOVSET message 1692, 10 October 1988 (electronic bulletin board for Soviet studies).

40. Press conference, American Academy of Arts and Sciences, Cambridge, Massachusetts, 5 November 1988.

41. See Jones, Connor, and Powell, *Soviet Social Problems.* Also see Nick Lampert, "Russia's New Democrats," *Detente,* nos. 9/10 (1987), 10–12.

42. For an early warning, see A. Merkulov, "Trevoga o Baikale," *Pravda,* 28 February 1965, 4. Also see Marshall Goldman, *The Spoils of Progress: Environmental Pollution in the Soviet Union* (Cambridge, Mass.: MIT Press, 1972).

43. See Thane Gustafson, "Environmental Issues Rise to Official Legitimacy," in his *Reform in Soviet Politics: Lessons of Recent Policies on Land and Water* (Cambridge: Cambridge University Press, 1981), pp. 39–52.

44. V. Umnov, "Tsepnaia reaktsiia," *Komsomol'skaia pravda,* 27 January 1988, 2. Also see Bill Keller, "Soviet Scraps a New Atomic Plant in Face of Protest over Chernobyl," *New York Times,* 28 January 1988, 1, A9, and "No Longer Merely Voices in the Russian Wilderness," *New York Times,* 27 December 1987, E14. On the resistance to a biotechnology plant in Kirishi, see "The Ministry vs. the Press," *Moscow News,* no. 30 (24 July 1988), 8. Also see "Protest by the 'Greens' in Irkutsk," and "Yerevan in Trouble: The Chemical Attack Continues," *Glasnost: Information Bulletin,* nos. 7, 8, and 9 (November 1987), 46–47.

45. "In Gorbachev's Words: 'To Preserve the Vitality of Civilization,'" *New York Times,* 8 December 1988, A16.

1. New Communications Technologies and Civil Society

1. S. Frederick Starr, "The Changing Nature of Change in the USSR," in Seweryn Bialer, ed., *Change in the Soviet Union and American Foreign Policy* (Boulder: Westview Press, 1988).

2. "The USSR Confronts the Information Revolution," proceedings of a conference held at Airlie House, Virginia, 12–13 November 1986, U.S. Government, Directorate of Intelligence; "Communications and Control in the USSR," research memorandum, United States Information Agency, 24 November 1986; Hans Heymann, Jr., "Commentary: A Note on the Critical Telecommunications Lag," conference article, 27 April 1987, Hudson Institute, Indianapolis; Richard W. Judy et al., *Soviet Informatics Project Phase I,* draft report, HI-3884-DP, Hudson Institute, 12 February 1987; and Loren Graham, "The Computer Revolution Is Bypassing the Soviet Union," *Washington Post,* 2 April 1984, 24–25.

3. Karl W. Deutsch et al., *Political Community and the North Atlantic Area* (Princeton: Princeton University Press, 1957), p. 54.

4. Lucien W. Pye, ed., *Communications and Political Development* (Princeton: Princeton University Press, 1963).

5. Marshall McLuhan, *Understanding Media: The Extensions of Man* (New York: New American Library, 1964).

6. Oswald H. Ganley and Gladys Ganley, *To Inform or To Control? The New Communications Networks* (New York: McGraw Hill, 1982).

7. Among the exceptions are Marjorie Ferguson, ed., *New Communications Technologies and the Public Interest* (Beverly Hills: Sage Publications, 1986), p. 53. See also Daniel Bell, "The Social Framework of the Information Society," in M. L. Dertouzos and J. Moses, eds., *The Computer Age: A Twenty-Year View* (Cambridge, Mass.: MIT Press, 1979).

8. Daniel Lerner, "Toward a Communications Theory," in Pye, *Communications and Political Development,* p. 328.

9. Deutsch, *Political Community,* p. 51.
   Horizontal communication is that which passes among members of a society at any given level. Vertical communication is that which passes from the authorities to the citizens of a society (top down) or from the citizens to those in charge (bottom up).

10. Marshall McLuhan, *The Gutenberg Galaxy* (New York: Routledge and Kegan Paul, 1962), pp. 141, 220, 240, 246; see also Walter J. Ong, S.J., *The Presence of the Word* (New Haven: Yale University Press, 1967), p. 64.

11. A. S. Zernova, *Nachalo knigopechataniia v Moskve i na Ukraine* (Moscow, 1947), chaps. 1–3; and M. N. Tikhomirov, *Nachalo moskovskogo knigopechataniia* (Moscow, 1947), chaps. 1 and 2.

12. Iurii Ovsiannikov, *Lubok: Russkie narodnye kartinki XVII–XVIII vv.* (Moscow: Sovetskii khudozhnik, 1967), pp. 24–26.

13. I. P. Kozlovskii, *Pervye pochty i pochmeistery v Moskovskom gosudarstve,* 2 vols. (Warsaw, 1913), I, chaps. 1–4.

14. *Kratkii istoricheskii ocherk razvitiia i deiatel'nosti vedomstva Putei soobshcheniia za sto let ego sushchestvovaniia, 1798–1898* (St. Petersburg, 1898), pp. 13–30.

15. Ibid.
16. S. A. Urodkov, *Peterburgo-moskovskaia zheleznaia doroga: Istoriia stroitel'stva, 1842–1851* (Leningrad: Izd-vo Leningradskogo universiteta, 1951), pp. 34–35.
17. William L. Blackwell, *The Beginnings of Industrialization* (Princeton: Princeton University Press, 1968), pp. 273–274.
18. Ibid., pp. 283, 294.
19. Ibid., p. 269.
20. D. D. Blagoi, *Istoriia russkoi literatury XVIII veka*, 3rd ed. (Moscow: Academy of Sciences, 1955), pp. 212–215.
21. A. A. Sidorov, ed., *400 let knigopechataniia*, 2 vols. (Moscow: Lenin Library, 1964), I, chap. 4.
22. S. Frederick Starr, *Decentralization and Self-Government in Russia, 1830–1870* (Princeton: Princeton University Press, 1972), pp. 333–334.
23. Michael T. Florinsky, *Russia: A History and an Interpretation*, 2 vols. (New York: Macmillan, 1968), II, 937.
24. Ibid., p. 789.
25. Ibid., p. 937.
26. *Ministerstvo vnutrennykh del za sto let* (St. Petersburg, 1901).
27. Donald W. Treadgold, *The Great Siberian Migration: Government and Peasant in Resettlement from Emancipation to the First World War* (Princeton: Princeton University Press, 1957).
28. M. Lemke, *Epokha tsenzurnykh reform, 1859–1865* (St. Petersburg: Gerol'd, 1904); also see Charles A. Ruud, "The Russian Censorship, 1855–1865: A Study in the Formation of Policy" (Ph.D. diss., University of California, Berkeley, 1966).
29. Charles A. Ruud, "Russian Entrepreneur: The Publisher Ivan Sytin of Moscow, 1851–1934," unpublished ms.; also see Charles A. Ruud, *Fighting Words* (Toronto: University of Toronto Press, 1982), pp. 203–205, 217–218.
30. Jeffrey Brooks, *When Russia Learned to Read: Literacy and Popular Literature, 1861–1917* (Princeton: Princeton University Press, 1985).
31. Starr, *Decentralization*, pp. 333–334. Franking privileges were suspended in this situation as well.
32. On Russian inventors of this era, see V. S. Virginskii, *Tvortsy novoi tekhniki v krepostnoi Rossii* (Moscow: State Pedagogical Institute, 1962), pp. 298–317.
33. A. M. Prokhorov, ed., *Bol'shaia sovetskaia entsiklopediia*, vol. 28 (Moscow: Sovetskaia entsiklopediia, 1974), p. 451.
34. N. D. Psurtsev, ed., *Razvitie sviazi, v SSSR* (Moscow: Sviaz', 1967), p. 26.
35. F. A. Brokgauz and I. A. Efron, *Entsiklopedicheskii slovar'*, vol. 32 (St. Petersburg, 1901), p. 793.
36. Ibid.
37. Psurtsev, *Razvitie sviazi v SSSR*, p. 31.
38. Brokgauz and Efron, *Entsiklopedicheskii slovar'*, vol. 32, pp. 814–815.
39. Psurtsev, *Razvitie sviazi v SSSR*, p. 26.
40. Ibid., p. 33.
41. Marc Ferro, *The Russian Revolution of February 1917*, trans. J. L. Richards (Englewood Cliffs, N.J.: Prentice-Hall, 1972), pp. 87–91.

42. V. V. Uchenova, *Partiino-sovetskaia pechat': Vosstanovitel'nogo perioda* (Moscow: Moscow University Press, 1964), p. 5.

43. V. I. Lenin, *Polnoe sobranie sochinenii*, 55 vols., 5th ed. (Moscow: State Press for Political Literature, 1967–70), V, 39–41.

44. Cf. Richard R. Fagen, *Politics and Communications* (Boston: Little, Brown, 1966), p. 34.

45. Peter Kenez, *The Birth of the Propaganda State: Soviet Methods of Mass Mobilization, 1917–1929* (Cambridge: Cambridge University Press, 1985), p. 254.

46. Psurtsev, *Razvitie sviazi v SSSR*, p. 82. For a stimulating analysis of Russian telephones in the 1920s, see Steven L. Solnick, "Soviet Telephones, 1917–1927: An Early Case Study in Modernization and Economic Reform," unpublished ms., SSRC Summer Workshop on Soviet Domestic Politics, University of Toronto, 1988.

47. Jeffrey Brooks, "The Breakdown in Production and Distribution of Printed Material, 1917–1927," in Abbott Gleason, Peter Kenez, and Richard Stites, eds., *Bolshevik Culture: Experiment and Order in the Russian Revolution* (Bloomington: Indiana University Press, 1985), pp. 151–154.

48. Kenez, *Birth of the Propaganda State*, p. 101.

49. Brooks, "Breakdown," pp. 155–166.

50. Kenez, *Birth of the Propaganda State*, p. 104.

51. Psurtsev, *Razvitie sviazi v SSSR*, p. 178.

52. Ibid., p. 66.

53. Kenez, *Birth of the Propaganda State*, pp. 105–107.

54. Psurtsev, *Razvitie sviazi v SSSR*, pp. 38, 188.

55. Lilian-Dorette Rimmele, *Der Rundfunk in Norddeutschland, 1933–1945: Ein Beitrag zur Nationale Organizations-Personal-und Kulturpolitik* (Hamburg: H. Ludke, 1977); and Franco Monteleone, *La radio italiana nel periodo fascista* (Venice: Marsilio, 1976).

56. Psurtsev, *Razvitie sviazi v SSSR*, p. 221.

57. Ibid., pp. 222–227.

58. Ibid., pp. 222–223.

59. *Narodnoe khoziaistvo v 1962 godu* (Moscow, 1963), p. 422.

60. Kenez, *Birth of the Propaganda State*, p. 252.

61. Gayle Durham Hollander, "Political Communication and Dissent in the Soviet Union," in Rudolf L. Tokes, ed., *Dissent in the USSR: Politics, Ideology, and People* (Baltimore: Johns Hopkins University Press, 1975), p. 251.

62. Ellen Proffer Mickiewicz, *Soviet Political Schools: The Communist Party Adult Instruction System* (New Haven: Yale University Press, 1967), pp. 8–10.

63. *Ustav sviaz SSSR* (Moscow, 1954).

64. Prokhorov, *Bol'shaia sovetskaia entsiklopediia*, vol. 23 (1976), p. 94.

65. Psurtsev, *Razvitie sviazi v SSSR*, p. 269, table 26.

66. R. Volkova, "And Along Came the Postman," *Pravda*, 18 March 1985, 7, reported in *Current Digest of the Soviet Press* (hereafter cited as *CDSP*) 37, no. 11 (1985), 20.

67. Prokhorov, *Bol'shaia sovetskaia entsiklopediia*, vol. 23 (1976), p. 93; also see Psurtsev, *Razvitie sviazi v SSSR*, pp. 268–272.

68. "Sviaz' sluzhit vsem," *Pravda,* 28 May 1984, 1.

69. International Telecommunications Union, *Yearbook of Common Carrier Telecommunications Statistics,* cited in Heymann, "Commentary," p. 5.

70. Gerald Stanton Smith, *Songs to Seven Strings: Russia's Guitar Poetry and Soviet "Mass Song"* (Bloomington: Indiana University Press, 1984), p. 94.

71. Psurtsev, *Razvitie sviazi v SSSR,* p. 269, table 26.

72. Ibid., p. 331.

73. Ibid., p. 273.

74. *Narodnoe khoziaistvo SSSR v 1962 godu* (Moscow, 1963), p. 422.

75. *Narodnoe khoziaistvo SSSR v 1968 godu* (Moscow, 1969), p. 506.

76. Lewis Feuer, "The Intelligentsia in Opposition," *Problems of Communism* 19, no. 6 (November–December 1970), 1–16.

77. *Narodnoe khoziaistvo SSSR v 1968 godu,* p. 506.

78. Psurtsev, *Razvitie sviazi v SSSR,* pp. 94–97.

79. Aleksandr Petrov, "Communications Is a Key Service," *Izvestiia,* 3 September 1985, 2, reported in *CDSP* 37, no. 35 (1985), 26.

80. Psurtsev, *Razvitie sviazi v SSSR,* pp. 94–98.

81. "Russian Cars," *Economist,* 3 December 1983, 79.

82. Ye. Shestinskii, "In the USSR State Prices Committee," *Izvestiia,* 10 January 1985, 2, reported in *CDSP* 37, no. 2 (1985), 19.

83. N. Tolstova, "We're Looking for a Gasoline Pump," *Izvestiia,* 1 August 1984, 3, reported in *CDSP* 36, no. 31 (1984), 23.

84. Reply to a letter to the editor, *Sovetskaia Rossiia,* 9 April 1985, 3.

85. Shestinskii, "In the USSR State Prices Committee," p. 2.

86. *Sotsialisticheskaia industriia,* 14 January 1985, 4, photographs and accompanying captions.

87. S. E. Goodman and Alan Ross Stapleton, "Microcomputing in the Soviet Union and Eastern Europe," *Abacus* 3, no. 1 (1985), 6–22.

88. Ivan Selin, "Trip Report," unpublished ms., 2 April 1984, 2; also see "Update," unpublished ms., 6 January 1986.

89. Ganley and Ganley, *To Inform or to Control?* p. 85.

90. E. Jakubitis, "Po puti tekhnicheskogo progressa," *Trud,* 21 June 1986, 2.

91. S. Frederick Starr, *Red and Hot: The Fate of Jazz in the Soviet Union* (New York: Oxford University Press, 1983), pp. 263, 278.

92. K. Glukhov, "Fotografiia kak sposob reproduktsii," *Svobodnaia mysl',* 20 December 1971.

93. Quoted by Timothy Ryback in *Rock Around the Bloc* (New York: Oxford University Press, 1989), p. 32.

94. *Narodnoe khoziaistvo* (Moscow, 1970), p. 251.

95. "Talks by Anatole Kuznetsov," *Radio Liberty,* no. 17, 10–11 March 1973. Also see Gene Sosin, "Magnitizdat: Uncensored Songs of Dissent," in Rudolf Tokes, ed., *Dissent in the USSR* (Baltimore: Johns Hopkins University Press, 1975), chap. 8.

96. The best accounts of this phenomenon are in F. Gayle Durham, *Amateur Radio Operation in the Soviet Union* (Cambridge, Mass.: MIT, Center for International Studies, 1965). Also see Hollander, "Political Communication and Dissent," pp. 262–263.

97. Ibid. Also see Smith, *Songs to Seven Strings,* p. 95; and "Radiozdat," *Russkaia mysl',* 6 February 1975, 5.

98. N. Kishchik and Ye. Vostrukhov, "Video Lessons," *Izvestiia,* 15 October 1985, 30, reported in *CDSP* 37, no. 41 (1986), 25.

99. D. Pilipenko, "Videocassette Dealer," *Komsomolskaia pravda,* 20 September 1985, 4, reported in *CDSP* 37, no. 4 (1985), 23.

100. Viktor Yasmann, "Video in the Soviet Union: Trouble with a Capricious Step-Child," *Radio Liberty,* no. 129/86, 21 March 1986; Viktor Yasmann, "The Collectivization of Videos?" *Radio Liberty,* no. 355/86, 22 September 1986.

101. Cf. Chuck Anderson, *Video Power: Grass Roots Television* (New York: Praeger, 1975).

102. "Communications and Control in the USSR," p. 3.

103. K. Abaiev, "Coming Next on Video," *Izvestiia,* 23 June 1985, 6, reported in *CDSP* 37, no. 25 (1985), 24.

104. Kishchik and Vostrukhov, "Video Lessons," p. 25.

105. *Sovetskaia kul'tura,* 10 June 1986, quoted by Yasmann in "The Collectivization of Videos?" p. 2.

106. Aleksandr Petrov, "Medlennyi progress," *Izvestiia,* 3 September 1985, 2.

107. Ibid.

108. International Telecommunications Union, *Yearbook,* quoted in Heymann, "Commentary," p. 5.

109. "Russian Cars," *Economist,* 3 December 1983, 79.

110. V. M. Chebrikov's address on the 110th anniversary of the birth of F. E. Dzerzhinskii, *Pravda,* 11 September 1987, 3.

111. M. Vulfson, "More Decisive Action Was Not Necessary," *Sovetskaia Latviia,* 18 June 1987, 3, reported in *CDSP* 39, no. 27 (1987), 1.

112. Mark D'Anastasio, "Soviets Are Preparing Measures to Stop Expansion of Independent Publishers," *Wall Street Journal,* 9 September 1987, 29.

113. Vladimir Simonov, "Amerikanets i kompiuter," *Literaturnaia gazeta,* 24 June 1987, 14.

114. Deutsch, *Political Community,* p. 41.

115. E. Zlain, "Contraband Songs Can Damage Young People's Education," *Komsomol'skaia pravda,* 18 October 1985, 4, reported in *CDSP* 37, no. 44 (1985), 31.

116. Liudmila Kazymova, "Behind the Film Distributor's Poster," *Sovetskaia Rossiia,* 18 July 1985, 2, reported in *CDSP* 37, no. 40 (1985), 18.

117. Marquis de Custine, *Russia* (New York: D. Appleton and Company, 1854), p. 83.

118. Dimitry Liubosvetrov, "Time on the Screen," *Pravda,* 19 May 1986, 3, reported in *CDSP* 38, no. 20 (1986), 9.

119. Cf. Fagen, *Politics in Communications,* p. 39.

120. Bill Keller, "Soviet Political Clubs on Unofficial Stage," *New York Times,* 9 October 1987, 4.

121. Chebrikov, address, p. 3; also see Jonathan Steele, "Moscow Opens the Door to Reform Groups," *Guardian,* 12 September 1987, 1.

122. Chebrikov, address, p. 3.

123. U.S.-Helsinki Watch Committee, *Reinventing Civil Society: Poland's Quiet Revolution, 1981–1986* (New York, 1986), pp. 43, 53–59, 71–78.

124. "Dolg kazhdogo grazhdanina," *Pravda,* 14 September 1987, 1.
125. Feliks Kuznetsov, "Kul'tura: Narodnost' i massovost'," *Literaturnaia gazeta,* 5 January 1983, 3.
126. Vladimir Simonov, "Khaiping i ego iznanki," *Literaturnaia gazeta,* 23 June 1982, 15.
127. Vladimir Soloukhin, "Skazki mogut i umeret," *Literaturnaia gazeta,* 22 September 1982, 3.
128. These views are conveniently summarized in Stanislav Kuniaev, "Ot velikogo do smeshnogo," *Literaturnaia gazeta,* 9 June 1982, 3.
129. Andrei Bitov, "Net! Nikogda ia zavisti ne znal," *Literaturnaia gazeta,* 7 July 1982, 3.
130. See Jerry F. Hough, *The Soviet Union and Social Science Theory* (Cambridge, Mass.: Harvard University Press, 1977), p. 234.

2. Information Technologies and the Citizen

1. For our purposes, the information technologies (IT) will include computing, telecommunications, and some consumer and commercial electronics technologies (for example, VCRs and photocopying machines).
2. For a more extensive general discussion of these and other sources of stability, see Seweryn Bialer, *The Soviet Paradox: External Expansion, Internal Decline* (New York: Knopf, 1986), pp. 19–40.
3. These four goals are taken from a more comprehensive model of a "Soviet-style information society" presented in S. E. Goodman, "The Information Technologies and Soviet Society: Problems and Prospects," *IEEE Transactions on Systems, Man, and Cybernetics* SMC-17, no. 4 (July–August 1987), 529–552 (esp. pp. 539–542). This model was developed in early 1986 from an analysis of Soviet capabilities, driving forces, systemic conditions, policies, and prospects. To some extent this model might be viewed as a perception of the thrust and setting of perestroika for the purposes of this chapter.

   For selected extensive recent assessments of Soviet progress in computing, see S. E. Goodman, "Technology Transfer and the Development of the Soviet Computer Industry," in Bruce Parrott, ed., *Trade, Technology, and Soviet-American Relations* (Bloomington: Indiana University Press, 1985), pp. 117–140; C. Hammer et al., *Soviet Computer Science Research,* FASAC-TAR-2020 (McLean, Va.: SAIC, July 1984); Ross A. Stapleton and Seymour E. Goodman, "Microcomputing in the Soviet Union and Eastern Europe," *Abacus* 3, no. 1 (Fall 1985), 6–22; S. E. Goodman and W. K. McHenry, "Computing in the USSR: Recent Progress and Policies," *Soviet Economy* 2, no. 4 (October–December 1986), 327–354; Jack Baranson, ed., *Soviet Automation: Perspectives and Prospects* (Mount Airy, Md.: Lomond, 1987); Richard F. Staar, ed., *The Future Information Revolution in the USSR* (New York: Crane Russak, 1988); Richard W. Judy and Jane M. Lommel, "Soviet Educational Computing," in Staar, *Future Information Revolution;* William K. McHenry, "Computer Networks and the Soviet-Style Information Society," in Staar, *Future Information Revolution,* pp. 85–113; Ross A. Stapleton and Seymour E. Goodman, "The

Soviet Union and the Personal Computer Revolution," in Staar, *Future Information Revolution*, pp. 61–83; Ross A. Stapleton, "Personal Computing in the CEMA Community: A Study of International Technology and Development" (Ph.D. diss., University of Arizona, 1988); Peter Wolcott and Seymour E. Goodman, "High-Speed Computers of the Soviet Union," *Computer* (IEEE) 21, no. 9 (September 1988), 32–41; Joel M. Snyder, "Pact Countries Clone U.S. Computers," *Signal* 43, no. 4 (December 1988), 55–62; William K. McHenry, "Computing Technology in the Soviet Union and Other CMEA Countries," in *Global Trends in Computer Technology and Their Impact on Export Control*, Report of the Committee to Study International Developments in Computer Science and Technology, National Research Council (Washington, D.C.: National Academy Press, 1988); and P. Wegner et al., " System Software for Soviet Computers," FASAC Technical Assessment Report 4080 (McLean, Va.: SAIC, 1989). Also see note 7.

4. Bialer, *Soviet Paradox*, p. 23.
5. Mikhail Heller and Aleksandr Nekrich, *Utopia in Power: The History of the Soviet Union from 1917 to the Present* (New York: Summit, 1986), pp. 731–732.
6. Much work on this and related subjects has been done by Erik P. Hoffmann. See his articles "The 'Scientific Management' of Soviet Society," *Problems of Communism* (May–June 1977), 59–67; "Soviet Views of the 'Scientific-Technological Revolution,'" *World Politics* 4 (July 1978), 615–644; and, with Robbin F. Laird, *Technocratic Socialism: The Soviet Union in the Advanced Industrial Era* (Durham, N.C.: Duke University Press, 1985).
7. William K. McHenry has written extensive and detailed studies of computing at the enterprise level in the Soviet economy. See his *"Absorption of Computerized Management Information Systems in Soviet Enterprises"* (Ph.D. diss., University of Arizona, 1985); with S. E. Goodman, "MIS in Soviet Industrial Enterprises: The Limits of Reform from Above," *Communications of the Association of Computing Machinery* 29, no. 11 (November 1986), 1034–43; "Application of Computer Aided Design in Soviet Enterprises: An Overview," in Baranson, *Soviet Automation*, pp. 57–76; and "Enterprise Level Computing in the Soviet Economy," CIA SOV C 87-10043 (August 1987).
8. Petur Ivanov, "A Mental Barrier against Scientific and Technical Innovations," *Politicheskaia agitatsiia* 14 (1985), 36–40.
9. Julian Cooper, "The Application of Industrial Robots in the Soviet Engineering Industry," *Omega: The International Journal of Management Science* 12, no. 3 (1984), 291–298; and John M. Dolan, "The Soviet Robotics Program," in Baranson, *Soviet Automation*, pp. 27–56.
10. For an example of informatization, the existence and exercise of restrictions on the ownership and use of "traditional" products for the dissemination of information—printing presses, telephones, broadcast and receiving equipment—have long been characteristics of Soviet society.

Strong controls have also been imposed on more modern technologies. For example, photocopying machines are produced in the USSR and imported from abroad in unknown numbers. There is now even a Soviet-made desk-top model. "Electrophotographic Copying Device ER-121," *Pribory i sistemy upravleniia,* no. 9 (September 1986), back cover. These machines are kept under tight wraps, sometimes in rooms that are literally sealed at night. In spite of rhetoric—and permission given to the American firm AlphaGraphics to open a couple of stores in Moscow—there had through 1988 been no widespread decontrol of photocopying machines under the Gorbachev administration. It is clear that the Soviets have decided that these machines present certain problems. Control of photocopying machine technology is fairly straightforward: simply control access to the room and, if necessary, to the paper. The need for this technology is generally restricted to institutions. It is not a major form of entertainment for the Soviet population (although it is interesting to note that some Soviet visitors to the West will go to the nearest photocopying shop and, deeply mesmerized, copy large sections of their host's library). Apparently, Soviets have decided that the economic opportunity and productivity lost through the lack of photocopying machines are acceptable in order to maintain control over them. One expects nonetheless that some decontrol of photocopying machines is inevitable.

The control of photocopying machines is a good example of the balancing decisions the Soviets have had to make, and one that has been made on the side of strong control. So far this control has not been a problem because there are few of the pressures for access to photocopying machines and little of the leakage of information about them that have characterized other products such as VCRs. The decision to impose strong controls on VCRs has proven somewhat unstable because of popular pressures; see Goodman, "Information Technologies," pp. 536–537; and Viktor Yasmann, *Radio Liberty Research Bulletin,* RL 129/86 (21 March 1986), and RL 355/86 (22 September 1986). For an assessment of the worldwide impact of VCRs see Gladys D. Ganley and Oswald H. Ganley, *Global Political Fallout: The VCR's First Decade* (Norwood, N.J.: Ablex, 1987).

For examples concerning democratization, see Andrei Sakharov, "Meeting Report" (Washington, D.C.: The Wilson Center, 14 November 1988). Sakharov argued that current attempts to carry out "democratic reforms through undemocratic means" would create a "nonviable antidemocratic structure" within the USSR.

11. So far, attempts (or at least those that we have seen) by Soviet analysts to define something along the lines of a Soviet-style information society have not been especially complete or bold. See, for example, A. I. Rakitov, "Questions of Theory: The Introduction of Information Science into Society and the Strategy of Acceleration," *Pravda,* 23 January 1987, 2–3; and A. B. Vengerov, "Law on Display," *NTR: Problemy i resheniia,* 20 January–2 February 1987, 5. Interest in the subject is growing, however, and at least five groups in the Soviet Academy of Sciences are seriously considering the subject. Not surprisingly, they are working to develop a long-term plan for the informatization of Soviet society, which they hope to present to the state and Party. A notable part of this effort is concerned with the development and implementation of laws dealing with information.

A case can be made that the emerging "Hungarian-style information society" shows weak signs of being a hybrid of the USSR and U.S. models discussed in Goodman, "Information Technologies." Nevertheless, the Hungarians are having serious problems with developing and applying the information technologies and may be approaching something of a dead end if technological and economic relations with the West are not strengthened. There remains a large gap between Hungary's position and that of most of Western Europe.

12. Bialer, *The Soviet Paradox*, p. 21.

13. Several examples might provide a useful sense of Soviet limitations and problems in this area.

I estimate that in mid-1988 there were 100,000 to 150,000 personal computers in the USSR, against 30 to 40 million in the United States, with a similar ratio for VCRs. Furthermore, the quality of the products in the USSR is generally much lower; it is much more difficult to acquire such goods; and the supporting infrastructure (related products and services such as peripherals and software for PCs, repair services, and an information dissemination framework for advertising) is much weaker.

To date, the practical availability of Soviet personal computer models to the general public or even on a broad institutional basis, such as for the much-publicized computer-literacy program for the public schools or the computer clubs that are being formed in major cities, ranges from minimal to nonexistent. Apparently "it is only on the pages of the Soviet press that personal computers have become 'heroes of the day' . . . In real life, they continue to be exotic rarities, access to which is rigidly restricted in one way or another. In this respect, another Soviet novelty—the word processor—provides an interesting parallel. The author of an article on the subject in *Pravda* wrote of his visit to the "Schetmash" factory in Kursk, which turned out the first five Iskra-226SOT editing sets in the country. The proud factory director was careful, however, to qualify the good news: 'I would merely like to add the proviso that we are sending out these sets according to a strict allocation list. So I do not recommend anyone dispatching "expeditors." It will be a waste of time and money.' Viktor Yasmann, "Home Computers Have Gone on Sale," *Radio Liberty Research Report*, RL 407/85 (6 December 1985).

The Soviets entered 1988 without the mass production of any respectable personal computer configuration. Here "mass production" is taken to mean hundreds of thousands per year; and "respectable configurations" include reliable equipment (based on at least sixteen-bit architectures) that can be expected to function without problems for months at a time, quality monitors, inexpensive floppy and Winchester disks and printers, and a modest range of applications software that should include the elementary programming languages, good I/O handling capabilities, a general-purpose text editor, and so on. By current world standards these are modest criteria.

Well-publicized plans call for the production of 1.1 million unspecified microcomputers with unspecified configurations by the end of the Twelfth Five-Year Plan (1990). About half of these are intended for use in the well-publicized Soviet national computer-literacy program in the secondary schools. Current Soviet inventories have been supplemented by a few thousand weak Japanese microcomputers, and a deal is apparently being completed to import 100,000 Peruvian IBM-PC compatibles over five years. Barbara Durr, "USSR to Buy Peruvian PCs," *Journal of Commerce* (8 January 1987). Several joint ventures for the production of microcomputers are also under discussion with Western firms. Even if all of this should come to pass—and it would be somewhat surprising if the Soviets could not somehow collect 1.1 million microcomputers by 1990—this number would be minimal by current world standards.

Finally, a few words on technical services for VCRs and other personal electronic products: "Scattered around Moscow are workshops with familiar decals in their small (and usually dirty) windows—Panasonic, Sony, Hitachi. These 'lavki' repair televisions and other electronics, but in fact are part of a nationwide 'electronics mafia' which remains busy around the clock designing and manufacturing television system converters which adapt European and American TVs and VCRs to Soviet standards, build radar detectors, and 'soup up' shortwave receivers.

"This enterprise is tolerated (encouraged?) since major clients are the nomenklatura itself. One London electronics shop told me that it was shipping more than eighty VCRs a week to Moscow via Soviet diplomats—but not the televisions. Those were being provided via modifications to Soviet models by the electronics mafia.

"This is a parallel activity to black market money exchanges, which are tolerated because the prime users of the system are the KGB case officers and diplomats themselves, who are limited in the amount of hard currency they can legally take out for their own purposes." J. E. Barrie, private communication to author, VAXmail, 2 February 1987.

14. George Miller, "News in Brief: Automated Unemployment," *Soviet Labour Review* (London) 4, no. 4 (December 1986), 8.

15. Barrie, private communication.

16. "While the US emphasizes the rights of the individual, such as freedom of speech and religion, the Soviets stress the notion that individual rights are contingent upon the rights of the collective. They regard full employment, housing and comprehensive health care as the fulfillment of basic human rights; less tangible rights are barely considered." *Time*, 25 March 1985, 31, quoted in Aaron Trehub, "Social and Economic Rights in the Soviet Union: Work, Health Care, Social Security, and Housing," *Radio Liberty Research Bulletin*, RLS-3/86 (29 December 1986), 1–51. Trehub examines Soviet human rights performance on its own terms and finds it falling short of the image of achievement the Soviets would like to project. So far the information technologies have played almost no role in improving Soviet human or civil rights performance by any definition, and the little we have seen of Soviet discussions in this context is fairly insipid. See, for example, L. K. Tereshchenko and I. B. Yermilin, "Review of Book *Civil Rights and ASU*," *Sovetskoe gosudarstvo i pravo*, 2 February 1985, 141–143.

3. Prometheus Rechained

1. For an expanded treatment of this theme through the 1930s, see Douglas R. Weiner, *Models of Nature: Ecology, Conservation, and Cultural Revolution in Soviet Russia* (Bloomington: Indiana University Press, 1988).
2. Similar to but not identical with our term *ecosystem*.
3. Actually the idea of the *etalon* can be traced to the soil scientist V. V. Dokuchaev and even back to the 1870s. See Douglas R. Weiner, "The History of the Conservation Movement in Russia and the USSR from Its Origins to the Stalin Period" (Ph.D. diss., Columbia University, 1984), p. 40; and G. I. Dokhman, *Istoriia geobotaniki v Rossii* (Moscow: Nauka, 1973), passim.
4. S. I. Medvedev and N. T. Nechaeva, "Pamiati V. V. Stanchinskogo," *Biulleten' MOIP, Otdel biologicheskii* 82, no. 6 (1977), 109–117, esp. p. 113.
5. "Otchet VOOP o rabote obshchestva za 1939 god," Central Governmental Archive of the October Revolution (TsGAOR), collection 494, list 1, sheet 1.
6. L. G. Ramensky, "Basic Regularities of Vegetation Cover and Their Study (on the Basis of Geobotanic Researches in Voronezh Province)," trans. John L. Brooks, *Bulletin of the Ecological Society of America* 64, no. 1 (March 1983), 12.
7. L. G. Ramenskii, "Klassifikatsiia zemel' po rastitel'nomu pokrovu," *Problemy botaniki* 1 (1950), 484.
8. Interview with Iakov Mikhailovich Gall, Leningrad, 20 May 1986. Sukachev's observation will appear in a volume of his works now being prepared for publication.
9. N. F. Reimers and F. R. Shtil'mark, *Osobo okhraniaemye prirodnye territorii* (Moscow: Mysl', 1978), p. 39.
10. A. V. Malinovskii (1899–1981) graduated from the Petrograd Forestry Institute and worked as a forester, later organizing plantations in various regions. He started working in Narkomles SSSR in 1934, and then served in the Glavlesokhrana of the USSR Council of Ministers. During the war he directed the Briansk Forestry-Economics Institute and then led the State Forestry Inspectorate, assuming leadership of the Main Administration for Zapovedniki in 1950. The late A. A. Nasimovich relayed the rumor that Malinovskii was tied to Beria.
11. A. V. Malinovskii, "Tovarishchu Merkulovu, V. N.: Spravka o motivakh ostavlenii i sokrashchenii ploshchadi riada zapovednikov, peredavaemykh Glavnomu upravleniiu po zapovednikam pri Sovete ministrov SSSR," collection 200, Moscow State University Archives. Evgenii Shvarts, of the USSR Academy of Sciences' Institute of Geography, has insightfully suggested that the 1951 liquidation was the consequence not of a renewed political offensive against the reserves but rather of a desperate search by an inefficient economy for opportunities for further "extensive" development. The reserves were appealing precisely because they already contained infrastructures: staff housing, buildings, vehicles, and of course exploitable resources.
12. Klim Voroshilov and the aviator-explorer Papanin actually approached Stalin on

the matter at the behest of a delegation of activists. Conversation with Andrei Aleksandrovich Nasimovich, 10 April 1980.

13. "Polozhenie o Glavnom upravlenii po zapovednikam pri Sovete ministrov SSSR," Postanovlenie no. 3192, Sovet Ministrov SSSR, 1951.

14. I. F. Barishpol and V. G. Larina, *U prirody druzei milliony* (Moscow: Lesnaia promyshlennost', 1984), p. 61.

15. Speech of V. N. Makarov (c. mid-1953) for first general convocation of Plenum of Commission on Zapovedniki of the Academy of Sciences, collection 200, Moscow State University Archives.

16. Barishpol and Larina, *U prirody druzei milliony*, p. 61.

17. Ibid., p. 65.

18. "Khronika," *Botanicheskii zhurnal*, no. 5 (1955), 773–774.

19. "Postanovlenie Biuro otdeleniia biologicheskikh nauk Akademii nauk SSSR ot 18 iunia 1957 g., Protokol no. 21, p. 1, 'O ratsional'noi seti zapovednikov SSSR,'" "Khronika," *OPZD*, no. 3 (1958), 112–113.

20. Malinovskii's original proposal called for sparing only twenty-eight.

21. Cited in Philip R. Pryde, *Conservation in the Soviet Union* (Cambridge: Cambridge University Press, 1972), p. 51.

22. See, for instance, I. V. Zharkov, *Prosteishie nabliudeniia v prirode (posobie dlia nabliudatelei zapovednikov)*, 2nd ed. (Moscow: Minsel'khoz SSSR, 1956); and V. A. Stepanov, "Zapovednoe delo v Kazakhstane," in *Zapovedniki Kazakhstana: Ocherki*, 2nd ed. (Alma-ata: Kazakh gos. izd. and Upravlenie okhotnich'em khoziaistvom pri Minsel'khoze Kaz SSR, 1963), pp. 4–6.

23. G. P. Dement'ev, "Deiatel'nost' Komissii po okhrane prirody AN SSSR za pervyi god ee sushchestvovaniia," *OPZD*, no. 2 (1956), 11.

24. "Soveshchanie po okhrane prirody," *Zoologicheskii zhurnal* 9 (1956).

25. S. S. Shvarts, "Voprosy akklimatizatsii mlekopitaiushchikh na Urale," *Trudy instituta biologii Ural'skogo filiala AN SSSR* 18 (1959), 3–22, esp. p. 10. He noted that acclimatization always referred to the successful formation of a population working from the existing genotypes of the introduced specimens, and not from the falsely held power of the environment to work directed, adaptive hereditary changes in the introduced individuals.

26. One Soviet informant, a close collaborator of Timofeev-Resovskii's, described Shvarts as "a classic yes-man."

27. Quoted in Vladimir Nikolaevich Bol'shakov, *Ekologicheskie osnovy okhrany prirody (V pomoshch' lektoru)* (Moscow: Znanie RSFSR, 1981), p. 5.

28. Ibid., pp. 3–4.

29. S. S. Shvarts, *Tekhnicheskii progress i okhrana prirody: Lektsiia* (Sverdlovsk, 1974), pp. 1, 12–13.

30. Ibid., p. 15.

31. Ibid., p. 14.

32. *Dialog o prirode* (Sverdlovsk: Sredne-ural'skoe knizhnoe izdatel'stvo, 1977).

33. Ibid., p. 127.

34. Ibid., pp. 37–38.

35. Ibid., p. 49.

36. A. Borodin, "Usilit' bor'bu s volkami," *OOKh*, no. 7 (1979), 4.

37. O. Gusev, "Protiv idealizatsii prirody," ibid., no. 11 (1978), 26.

38. *Dialog o prirode,* p. 43.

39. Ibid., pp. 125, 24, 22.

40. See discussion of this in V. L. Rashek, N. G. Vasil'ev, and A. V. Chumakova, "Okhrana soobshchestv v zapovednikakh," in *Issledovaniia v oblasti zapovednogo dela: Sbornik nauchnykh trudov* (Moscow: Minsel'khoz SSSR, 1984), pp. 3–21. Also see A. M. Krasnitskii, *Problemy zapovednogo dela* (Moscow: Lesnaia promyshlennost', 1983), p. 111.

41. Krasnitskii, *Problemy zapovednogo dela,* p. 112.

42. See M. Gilyarov, "Agrocenology—An Important Field of Modern Biogeocenology," in *Man and the Biosphere* (Moscow: Nauka, 1984), pp. 18–25.

43. N. Fedorenko and N. Reimers, "Nature Conservation: Growing Proximity of Economic and Ecological Goals," in *Man and the Biosphere,* p. 87.

44. Ibid.

45. N. F. Reimers, "Bez prava na oshibku," *Chelovek i priroda,* no. 10 (1980), 17.

46. Ibid., p. 29.

47. See E. A. Kotliarov, *Geografiia otdykha i turizma* (Moscow: Mysl', 1978), pp. 29–30; and V. P. Chizhova, *Rekreatsionnye nagruzki v zonakh otdykha* (Moscow: Lesnaia promyshlennost', 1977).

48. The Academy of Sciences' Institute of Geography, Department of Biogeography, has a section, led by N. A. Kazanskaia, that studies the ecological impact of recreation on landscapes.

49. On *natsional'nye parki,* see A. G. Nikolaevskii, *Natsional'nye parki* (Moscow: Agropromizdat, 1985); and N. Filippovskii, ed., *Natsional'nyi park: Problema sozdaniia,* Seriia "Chelovek i priroda," no. 6 (Moscow: Znanie, 1979).

50. I. T. Frolov and V. A. Los', "Filosofskie osnovaniia sovremennoi ekologii," in *Ekologicheskaia propaganda v SSSR* (Moscow: Nauka, 1984), pp. 5–26.

51. Ibid., pp. 16, 21.

52. Ibid., p. 15.

53. Ibid., p. 22.

54. Reimers and Shtil'mark, *Osobo okhraniaemye prirodnye territorii,* pp. 179, 161, and passim.

55. N. F. Reimers and F. R. Shtil'mark, "Etalony prirody," *Chelovek i priroda,* no. 3 (1979), 15–16.

56. See my *Models of Nature* for an extended discussion of this, especially pp. 134–140 on *kraevedenie,* the movement for the study of local lore.

57. See V. P. Chizova, "Natsional'nye parki SSSR i puti ikh razvitiia," in *Prirodookhrannoe obrazovanie v universitetakh* (Moscow: MGU, 1985), pp. 112–128.

58. *Matematicheskie modeli v ekologii: Bibliograficheskii ukazatel' otechestvennykh rabot* (Moscow: VINITI, 1981), p. 4.

59. As described by Mary Douglas and Aaron Wildavsky in *Risk and Culture* (Berkeley: University of California Press, 1983).

60. See, for example, Eugene Cittadino, "Ecology and the Professionalization of Botany in America, 1890–1905," *Studies in the History of Biology* 4 (1980), 171–198; Philip D. Lowe, "Amateurs and Professionals: The Institutional

Emergence of British Plant Ecology," *Journal of the Society for the Bibliography of Natural History* 7 (1976), 517–535; and Thomas Söderqvist, *The Ecologists: From Merry Naturalists to Saviours of the Nation* (Stockholm: Almqvist and Wiksell, 1986).

61. See Donald Worster, *Nature's Economy: A History of Ecological Ideas* (Cambridge: Cambridge University Press, 1985); and Ronald Tobey, *Saving the Prairies: The Life History of the Founding School in American Plant Ecology, 1895–1955* (Berkeley: University of California Press, 1982).

62. Dorothy Nelkin, "Scientists and Professional Responsibility: The Experience of American Ecologists," *Social Studies of Science* 7 (1977), 75–95.

4. The Soviet Nature-Nurture Debate

1. Research for this chapter has been supported by the History and Philosophy of Science Division of the National Science Foundation; by the Science, Technology, and Society Program of the National Endowment for the Humanities; and by the Department of the History and Sociology of Science of the University of Pennsylvania.

2. See Loren Graham's discussions of these issues in "Reasons for Studying Soviet Science," in Linda Lubrano and Susan Gross Solomon, eds., *The Social Context of Soviet Science* (Boulder: Westview Press, 1980), pp. 205–240; "Russia's Gene Blues: The Heredity Heresy," *Washington Post*, 22 September 1985, C1, C5; and *Science, Philosophy, and Human Behavior in the Soviet Union* (New York: Columbia University Press, 1987), pp. 220–265.

3. M. E. Lobashev, *Genetika*, 2nd ed. (Leningrad: Izdatel'stvo Leningradskogo universiteta, 1969), pp. 714–718.

4. I. K. Liseev and A. Ia. Sharov, "Genetika cheloveka," *Voprosy filosofii*, 1970, no. 7, 106–115, and no. 8, 125–134.

5. V. Efroimson, "Rodoslovnaia al'truizma (Etika s pozitsii evoliutsionnoi genetiki cheloveka)," *Novyi mir*, 1971, no. 10, 193–213.

6. Iu. Ia. Kerkis, "Nuzhna li kriminologam genetika?" *Priroda*, 1976, no. 7, 148–150.

7. Examples of scientific debates that have been illuminated in this way include those over Aristotelianism, Copernicanism, Mesmerism, Darwinism, spontaneous generation, phrenology, the nebular hypothesis, reductionism, relativity, the uncertainty principle, cladistics, and continental drift. Such historical examples sensitize us to the institutional and disciplinary dimensions of current controversies. For instance, ongoing disputes over dinosaur extinction often pit paleontologists against physical scientists over the question of whether there *was* a single, sudden extinction event; in the meantime, some toxicologists have proposed poisoning by angiosperms, meteorologists have focused on climatic change, and astronomers have favored comets or a "death star" (Nemesis). A more recent example is the skepticism and opposition of nuclear physicists from Harvard, MIT, and Berkeley to the idea of "cold fusion" as set forth by two Utah chemists.

8. On these scientists and their theories, see Kendall Bailes, *Vernadsky and His*

School (Bloomington: Indiana University Press, 1990); Mark B. Adams, "Nikolai Ivanovich Vavilov," in *Dictionary of Scientific Biography*, XV, supp. 1 (New York: Charles Scribner's Sons, 1979), pp. 505–513, and "Aleksandr Ivanovich Oparin," ibid., supp. 2 (1990); and L. V. Belousov, A. A. Gurvich, S. Ia. Zalkind, and N. N. Kannegiser, *Aleksandr Gavrilovich Gurvich, 1874–1954* (Moscow: "Nauka," 1970).

9. First published in 1917, V. M. Bekhterev's *Obshchie osnovy refleksologii* went through three Russian editions in the following decade (2nd ed., 1923; 3rd ed., 1925; 4th ed., 1928). The quotations are taken from the English translation of the fourth edition, *General Principles of Human Reflexology*, trans. Emma and William Murphy (London: Jarrolds, 1933), pp. 33 and 15.

10. Susan G. Solomon, "Social Hygiene and Soviet Public Health, 1921–1930," in S. G. Solomon and J. F. Hutchinson, eds., *Health and Society in Revolutionary Russia* (Bloomington: Indiana University Press, 1990).

11. V. Sukachev, *Rastitel'nye soobshchestva (Vvedenie v fitosotsiologiiu)* (Leningrad: "Kniga," 1928).

12. See, for example, the letters from M. V. Sabashnikov to Iu. A. Filipchenko, Manuscript Division of the Saltykov-Schedrin Public Library, Leningrad, fund 813, dossier 528.

13. I have discussed the Russian eugenics movement in greater detail in "Eugenics in Russia, 1900–1940," in Mark B. Adams, ed., *The Wellborn Science: Eugenics in Germany, France, Brazil, and Russia* (Oxford: Oxford University Press, 1989), pp. 153–216, and in "Eugenics as Social Medicine: Prophets, Patrons, and the Dialectics of Discipline-Building in Revolutionary Russia," in Solomon and Hutchinson, *Health and Society*.

14. See Mark B. Adams, "Iurii Aleksandrovich Filipchenko," in *Dictionary of Scientific Biography*, supp. 2.

15. See Mark B. Adams, "Science, Ideology, and Structure: The Kol'tsov Institute," in Lubrano and Solomon, *Social Context*, pp. 173–204; and "Chetverikov, the Kol'tsov Institute, and the Evolutionary Synthesis," in Ernst Mayr and William Provine, eds., *The Evolutionary Synthesis* (Cambridge, Mass.: Harvard University Press, 1980), pp. 242–278.

16. V. V. Sakharov, "Razbor muzykal'nykh genealogii, sobrannykh na evgenicheskom seminarii professora N. K. Kol'tsova," *Russkii evgenicheskii zhurnal* 2, no. 2/3 (1924), 117–125; and A. S. Serebrovskii and V. V. Sakharov, "Novye mutatsii Drosophila melanogaster," *Zhurnal eksperimental'noi biologii*, ser. A, 1, no. 1/2 (1925), 75–91. On his career, see "Vladimir Vladimirovich Sakharov," *Genetika*, 1969, no. 2, 177–182.

17. Born in 1903, Aleksandra Alekseevna Prokof'eva studied with Filipchenko and graduated from Leningrad University in 1930. She worked in the Institute of Genetics in the mid-1930s with H. J. Muller, jointly authoring several works with him. Following marriage to her Leningrad classmate Mark L. Bel'govskii (1906–1959), who also worked in Muller's laboratory, she adopted the family name Prokof'eva-Bel'govskaia. She died in 1984. See the posthumous collection of her work, A. A. Prokof'eva-Bel'govskaia, *Geterokhromaticheskie raiony khromosom* (Moscow: "Nauka," 1986).

18. See K. A. Lange, *Institut fiziologii imeni I. P. Pavlova* (Leningrad: "Nauka," 1975).

19. A. S. Serebrovskii, "Teoriia nasledstvennosti Morgana i Mendelia i marksisty," *Pod znamenem marksizma,* 1926, no. 3, 98–117.

20. See Mark B. Adams, "From 'Gene Fund' to 'Gene Pool': On the Evolution of Evolutionary Language," in William Coleman and Camille Limoges, eds., *Studies in History of Biology,* vol. 3 (Baltimore: Johns Hopkins University Press, 1979), pp. 241–285.

21. Aleksandr Serebrovskii, "Antropogenetika i evgenika v sotsialisticheskom obshchestve," in S. G. Levit and A. S. Serebrovskii, eds., *Trudy Kabineta nasledstvennosti i konstitutsii cheloveka pri Mediko-biologicheskom institute,* vol. 1 (Moscow: Glavnauka, 1929), pp. 3–19.

22. "Great Break" is one translation of *velikii perelom,* a term used by Stalin to refer to year 1929–30 of the First Five-Year Plan. See, for example, David Joravsky, *Soviet Marxism and Natural Science, 1917–1932* (New York: Columbia University Press, 1961), pp. 233–271.

23. The verb *biologizirovat'* also provided the stem for a new noun, *biologizirovanie* (biologization). The philosopher A. A. Maksimov was one of the first to employ these terms. For a contemporary example of their use in relation to "Menshevizing idealism," see P. P. Bondarenko et al., eds., *Protiv mekhanisticheskogo materializma i men'shevistvuiushchego idealizma v biologii* (Moscow: Medgiz, 1931).

24. Raymond Pearl Papers, Library of the American Philosophical Society, Philadelphia.

25. See the letters from Iu. A. Filipchenko to Theodosius Dobzhansky, Dobzhansky Papers, Library of the American Philosophical Society, Philadelphia.

26. Feodosii Grigor'evich Dobrzhanskii (1900–1975) became internationally known as Theodosius Dobzhansky, Nikolai Vladimirovich Timofeev-Resovskii (1900–1981) as N. W. Timoféeff-Ressovsky.

27. See Mark B. Adams, "Boris L'vovich Astaurov," in *Dictionary of Scientific Biography,* supp. 2.

28. I have discussed developments in the 1930s in greater detail in "The Politics of Human Heredity in the USSR," *Genome* 31, no. 2 (1989). On Levit, see also Mark B. Adams, "Solomon Grigorevich Levit," in *Dictionary of Scientific Biography,* supp. 2.

29. See A. R. Luria, *The Making of Mind: A Personal Account of Soviet Psychology,* ed. Michael Cole and Sheila Cole (Cambridge, Mass.: Harvard University Press, 1979), esp. pp. 81–103.

30. On Muller, see E. A. Carlson, *Genes, Radiation, and Society* (Ithaca: Cornell University Press, 1981).

31. On the session, compare the versions given in *Biulleten' IV sessii VASKhNILa,* which provided daily stenographic accounts, and the version published the next year, edited by O. M. Targul'ian, *Spornye voprosy genetiki i selektsii* (Moscow: VASKhNIL, 1937), which is considerably less reliable.

32. See H. J. Muller's letter to Julian Huxley, 9/11 March 1937, Muller Papers, Lilly Library, Bloomington, Indiana.

33. N. P. Dubinin, *Vechnoe dvizhenie* (Moscow: Politizdat, 1973), p. 71. Compare his account of the meeting with the recent version, based on archives, by T. A. Detlaf, "Institut eksperimental'noi biologii," *Ontogenez* 19, no. 1 (1988), 106–107.

34. S. N. Davidenkov, *Evoliutsionno-geneticheskie problemy v nevropatologii* (Leningrad: GIDUV, 1947); A. A. Malinovskii, "Biologicheskie i sotsial'nye faktory v proiskhozhdenii rasovykh razlichii u cheloveka," *Priroda*, 1947, no. 7, 40–48; and Erwin Schrödinger, *Chto takoe zhizn' s tochki zreniia fiziki*, ed. A. A. Malinovskii (Moscow: Izdatel'stvo inostrannoi literatury, 1947).

35. S. V. Kaftanov, *Za bezrazdel'noe gospodstvo michurinskoi biologicheskoi nauki* (Moscow: "Pravda," 1948), p. 10.

36. A. N. Studitskii, "Mukholiuby-chelovekonenavistniki," *Ogonek*, 1949, no. 11.

37. See Mark B. Adams, "Biology after Stalin: A Case Study," *Survey: A Journal of East/West Studies* 23, no. 1 (Winter, 1977–78), 53–80.

38. M. V. Keldysh, "Vstupitel'naia rech'," *Vestnik Akademii nauk SSSR*, 1965, no. 3, 5–10.

39. *Genetika*, 1965, no. 1, 200–202.

40. Ibid. Luk'ianenko was listed, out of alphabetical order, as Dubinin's first assistant. The council also included such "former" Lysenkoists as Pustovoit, V. N. Stoletov, K. S. Sukhov, N. V. Tsitsin, and N. V. Turbin.

41. Dubinin, *Vechnoe dvizhenie*, p. 435.

42. Ibid., p. 434.

43. Ibid., p. 435.

44. *Genetika*, 1966, no. 8, 2–5.

45. Dubinin, *Vechnoe dvizhenie*, p. 434.

46. V. Polynin, *Mama, papa, i ia* (Moscow: "Sovetskaia Rossiia," 1966; 2nd ed., 1969); and *Prorok v svoem otechestve* (Moscow: "Sovetskaia Rossiia," 1969).

47. Biographical information on Vladimir Pavlovich Efroimson is based on my interview with him (3 June 1988, Moscow) and on the manuscript of a testimonial in honor of his eightieth birthday, kindly made available to me by Nikolai N. Vorontsov.

48. V. P. Efroimson, *Vvedenie meditsinskuiu genetiku* (Moscow: Medgiz, 1963; 1964).

49. *Genetika*, 1967, no. 10, 114–127.

50. V. P. Efroimson, E. F. Davidenkova, E. E. Pogosiants, and A. A. Prokof'eva-Bel'govskaia, "Genetika i meditsina," *Genetika*, 1966, no. 10, 92–101.

51. V. P. Efroimson, *Vvedenie v meditsinskuiu genetiku*, 2nd ed. (Moscow: "Meditsina," 1968); and *Immunogenetika* (Moscow: "Meditsina," 1971); and see note 5.

52. Of course, the two networks were linked. For example, the Leningrader S. N. Davidenkov teamed up with V. P. Efroimson in 1961 to write the article on human inheritance for the *Great Medical Encyclopedia*. S. N. Davidenkov and V. P. Efroimson, "Nasledstvennost' cheloveka," in *Bol'shaia meditsinskaia entsiklopediia*, vol. 19 (1961), pp. 1010–65. Davidenkov was a central figure in a Leningrad network that encompassed biologists, physicians, neurologists, and physiologists. He was closely associated with the Koltushi station of the Pavlov

institute, for example, headed by Orbeli. After Davidenkov's death in 1961, his work was continued by his wife and collaborator, E. F. Davidenkova. In 1971 the geneticist Raissa L. Berg—daughter of the renowned Leningrad University ichthyologist and geographer Lev (Leo) Berg—issued an updated version of their earlier writings as a popular book on human heredity. R. L. Berg and S. N. Davidenkov, *Nasledstvennost' i nasledstvennye bolezni cheloveka* (Leningrad: "Nauka," Leningradskoe otdelenie, 1971).

53. On his career, see *M. E. Lobashev i problemy sovremennoi genetiki* (Leningrad: Izdatel'stvo Leningradskogo universiteta, 1978).

54. M. E. Lobashev, *Genetika* (Leningrad: Izdatel'stvo Leningradskogo universiteta, 1963).

55. M. E. Lobashev, *Genetika*, 2nd ed. (Leningrad: Izd. LGU, 1967; reprinted 1969), p. 716.

56. M. E. Lobashev, K. V. Vatti, and M. M. Tikhomirov, *Genetika s osnovami selektsii* (Moscow: "Prosveshchenie," 1970), p. 381.

57. On Timoféeff's career and achievements, see Mark B. Adams, "Timoféeff-Ressovsky and Modern Biology," in Daniel Granin, *Zubr* (New York: Doubleday, 1989).

58. See *Klassiki sovetskoi genetiki* (Leningrad: "Nauka," 1968); and *Vydaiushchiesia sovetskie genetiki* (Moscow: "Nauka," 1980), which gives extensive biographies of twelve major figures in the history of Soviet genetics but fails to mention that four of them died in prison.

59. See, for example, Mark Popovsky, "1000 dnei akademika Vavilova," *Prostor*, 1966, no. 7, 4–27; no. 8, 98–118.

60. The manuscript was subsequently published abroad in Zhores A. Medvedev, *The Medvedev Papers* (London: Macmillan, 1971), pp. 70–112.

61. Zhores A. Medvedev and Roy A. Medvedev, *A Question of Madness*, trans. Ellen de Kadt (New York: Vintage, 1971), p. 46.

62. Ibid., pp. 175, 98, 162.

63. Personal communication. I was spending the week with Dobzhansky at Mather Camp in California in 1973 when he first read the book.

64. Dubinin, *Vechnoe dvizhenie*, p. 60.

65. Ibid., pp. 433–434.

66. Interviews with B. L. Astaurov (August 1971, Moscow) and V. P. Efroimson (June 1988, Moscow).

67. Raissa L. Berg, "The Life and Research of Boris L. Astaurov," *Quarterly Review of Biology* 54 (1979), 402–403. Also see her provocative memoirs, *Sukhovei: Vospominaniia genetika* (New York: Chalidze Publications, 1983), and the translation by David Lowe, *Acquired Traits: Memoirs of a Geneticist from the Soviet Union* (New York: Viking, 1988).

68. Ibid. Gaisinovich refers to the matter opaquely; Raissa Berg is explicit.

69. B. L. Astaurov and P. F. Rokitskii, *Nikolai Konstantinovich Kol'tsov* (Moscow: "Nauka," 1975).

70. An early Bolshevik, Ol'ga Borisovna Lepeshinskaia (1871–1963) asserted that cells do not come from other cells but can spontaneously form from "living" noncellular material. Her theory was first published in the mid-1930s, but in

1945 Lysenko supported it, and after August 1948 it became official doctrine for several years. See her book *Proiskhozhdenie kletok iz zhivogo veshchestva i rol' zhivogo veshchestva v organizme*, 2nd ed. (Moscow: Izdatel'stvo Akademii meditsinskikh nauk SSSR, 1950), which was awarded the 1949 Stalin Prize, first class. For the most recent discussion of her theory, see Valerii Soifer [Valery N. Soyfer], "Lysenkoisty i ikh sud'by," *Kontinent*, no. 48 (New York: Izdatel'stvo "Kontinent," 1986), pp. 263–297.

71. Interview with A. A. Neifakh, May 1988, Moscow.

72. Interview with D. K. Beliaev, August 1981, Bucharest.

73. Dubinin, *Vechnoe dvizhenie*, pp. 433–434.

74. Ernst Mayr, "Chelovek kak biologicheskii vid," *Priroda*, 1973, no. 12; 1974, no. 2.

75. N. P. Dubinin, "O filosofskoi bor'be v biologii," *Filosofskie nauki*, 1975, no. 6, p. 14.

76. D. K. Beliaev, "Problemy biologii cheloveka: Geneticheskie real'nosti i zadachi sinteza sotsial'nogo i biologicheskogo," *Priroda*, 1976, no. 6, p. 30.

77. N. P. Dubinin and Iu. G. Shevchenko, *Nekotorye voprosy biosotsial'noi prirody cheloveka* (Moscow: "Nauka," 1976).

78. *Vestnik Akademii nauk SSSR*, 1981, no. 6, 42–47; no. 12, 123; 1982, no. 4, 121.

79. Interview with N. P. Bochkov, August 1988, Toronto. See N. P. Bochkov, *Genetika cheloveka: Nasledstvennost' i patologiia* (Moscow: "Meditsina," 1978); L. O. Badalian, ed., *Nasledstvennye bolezni* (Tashkent: "Meditsina" UzSSR, 1980); and N. P. Bochkov, A. F. Zakharov, and V. I. Ivanov, *Meditsinskaia genetika* (Moscow: "Meditsina," 1984).

80. See, for example, S. M. Gershenzon, *Osnovy sovremennoi genetiki* (Kiev: Naukova dumka, 1983); and S. I. Alikhanian, A. P. Akif'ev, and L. S. Chernin, *Obshchaia genetika* (Moscow: Vysshaia shkola, 1985), which makes a passing negative reference to eugenics.

81. See, for example, the following works by Ivan T. Frolov: *Mendelizm i filosofskie problemy sovremennoi genetiki* (Moscow: "Mysl'," 1976), written with S. A. Pastushnyi; "Genes or Culture? A Marxist Perspective on Humankind," *Biology and Philosophy* 1, no. 1 (1985), 89–108; *Chelovek—nauka—gumanizm: Novyi sintez* (Moscow: "Progress," 1986); and *Etika nauki: Problemy i diskussii* (Moscow: Politizdat, 1986), written with B. G. Iudin.

82. See, by N. P. Dubinin with I. I. Karpets and V. N. Kudriavtsev, *Genetika, povedenie, otvetstvennost': O prirode antiobshchestvennykh postupkov i putiakh ikh preduprezhdeniia* (Moscow: Politizdat, 1982); Dubinin, *Chto takoe chelovek* (Moscow: "Mysl'," 1983); Dubinin, *Novoe v sovremennoi genetike* (Moscow: "Nauka," 1986).

83. Graham, *Science, Philosophy, and Human Behavior*, pp. 239–243.

84. Valerii Soifer [Valery N. Soyfer], "Gor'kii plod," *Ogonek*, 1988, no. 1, 26–29; no. 2, 4–7, 31.

85. *Novyi mir*, 1987, no. 1, 19–95; no. 2, 7–92; Daniil Granin, *Zubr: Povest'* (Leningrad: Sovetskii pisatel', 1987; Moscow: "Izvestiia," 1987).

86. Vladimir Dudintsev, *Belye odezhdy* (Moscow: "Knizhnaia palata," 1988). The

title, an allusion to Soviet geneticists, comes from the biblical quotation that opens the book (Rev. 7:13): "What are these which are arrayed in white robes? and whence came they?" The reference is to the 144,000 who have been marked by "the seal of the living God." In the passage that follows (Rev. 7:14), the elder answers his own questions: "These are they which came out of great tribulation, and have washed their robes, and made them white in the blood of the Lamb."

87. N. P. Dubinin, *Genetika—stranitsy istorii* (Kishinev: "Shtiintsa," 1988), pp. 316–322.

88. Neifakh kindly gave me a copy of the poster for his lecture when I met with him in May 1988.

89. A. E. Gaisinovich, *Zarozhdenie i razvitie genetiki* (Moscow: "Nauka," 1988).

90. V. P. Efroimson, "Biosotsial'nye faktory povyshennoi umstvennoi aktivnosti," 2 vols. (1982), on deposit at VINITI, no. 1161, 440 pp.

91. Listed as I. I. Kanaev, *Istoriia genetiki cheloveka v SSSR,* 2 vols. (Leningrad: "Nauka").

5. Engineers

1. This section draws heavily on my doctoral dissertation, "Educating Engineers: Economic Politics and Technical Training in Tsarist Russia" (University of Pennsylvania, 1980), and on "The Russian Engineering Profession," in Harley Balzer, ed., *Professions and Professionalization in Russia at the End of the Old Regime* (Ithaca: Cornell University Press, forthcoming).

2. Beginning in the reform era, many leading Russian engineers received part of their advanced training in Europe. A small number took their degrees in European universities or institutes, but these specialists were required to pass equivalency exams based on the Russian higher school curriculum when they returned home.

3. Edwin T. Layton, *The Revolt of the Engineers: Social Responsibility and the American Engineering Profession* (Baltimore: Johns Hopkins University Press, 1971), p. 4.

4. Donald W. Green, "Industrialization and the Engineering Ascendancy: A Comparative Study of the American and Russian Engineering Elites, 1870–1920" (Ph.D. diss., University of California, 1972), pp. 82–85; and Balzer, "Educating Engineers," pp. 18–19.

5. John McKay, *Pioneers for Profit: Foreign Entrepreneurship and Russian Industrialization, 1855–1913* (Chicago: University of Chicago Press, 1970), pp. 186–187; cf. Fred V. Carstensen, *American Enterprise in Foreign Markets: Singer and International Harvester in Imperial Russia* (Chapel Hill: University of North Carolina Press, 1984), p. 100.

6. Balzer, "Educating Engineers," pp. 134–153.

7. The polytechnical institutes were in Kiev, Warsaw, and Petersburg; the mining institute was in Ekaterinberg; and the transport institute was in Moscow. The focus on the periphery was characteristic of Witte and was an important departure from previous emphases.

8. In 1895 approximately four hundred graduated; in 1904 the number was over one thousand; and by 1914 it exceeded two thousand. See Balzer, "Educating Engineers," pp. 401, 469–470.

9. James C. McClelland, *Autocrats and Academics: Education, Culture, and Society in Tsarist Russia* (Chicago: University of Chicago Press, 1979), pp. 106–107. The "democratic" character of the technical institutes also carried over into the type of revolutionary activity favored by the students. While university students inclined toward the intellectual radicalism of liberal and populist movements, the technical institute revolutionaries were more likely to be "practically oriented" Marxists. See Richard Pipes, *Social Democracy and the St. Petersburg Labor Movement* (Cambridge, Mass.: Harvard University Press, 1963), p. 19.

10. The Technical Society did play an important role in fostering informal networks and providing opportunities for technical personnel to meet with civic and government leaders. For more on its limited capacity to express professional interests see Harley D. Balzer, "The Russian Technical Society," *Modern Encyclopedia of Russian and Soviet History* (Gulf Breeze, Fla.: Academic International Press, 1983), vol. 32, pp. 176–180.

11. *Inzhener*, no. 1/2 (1882), vi; no. 1 (1892), 11.

12. Shmuel Galai, *The Liberation Movement in Russia, 1900–1905* (Cambridge: Cambridge University Press, 1973), pp. 234–235; N. G. Filippov, *Nauchnotekhnicheskie obshchestva Rossii (1866–1917 gg.)* (Moscow, 1966); and L. K. Erman, *Intelligentsiia v pervoi russkoi revoliutsii* (Moscow, 1966), pp. 242, 306–307. Erman is not fully reliable. Also see Jonathan Sanders, "The Union of Unions: Economic, Political, and Human Rights Organizations in the 1905 Russian Revolution" (Ph.D. diss., Columbia University, 1985); and Abraham Ascher, *The Revolution of 1905: Russia in Disarray* (Stanford: Stanford University Press, 1988).

13. L. N. Liubimov, "Iz zhizni inzhenera putei soobshcheniia," *Russkaia starina* 157 (March 1914), 584–585.

14. E. O. Paton, *Vospominaniia* (Moscow, 1956), pp. 51–52. Paton was in Kiev. Neutrality may have been somewhat harder to maintain for engineers in the capitals.

15. Two of the leading figures of the pre-1917 period, M. A. Shatelen and P. S. Osadchii, remained very active after the Bolshevik Revolution. Jonathan Coopersmith is completing a manuscript on electrification and electrical engineers that should fill a sizable gap in our knowledge. On the activism of American electrical engineers, see Layton, *Revolt*, pp. 26, 38–39, 84–86.

16. *Elektrichestvo*, no. 2 (1909), 83. An electrical engineer wrote the first Russian monograph discussing engineering ethics; P. S. Osadchii, *K voprosu o printsipakh professional'noi etiki inzhenerov* (St. Petersburg, 1911). The Fifth All-Russian Electrotechnical Conference in Moscow (1908) approved "in principle" the drafting of a code dealing with an engineer's relationship to lower-level technical personnel.

17. V. N. Ipatieff, *Life of a Chemist* (Stanford: Stanford University Press, 1946), p. 169.

18. Central State Historical Archive (TsGIA), collection 90, list 1, folder 864, sheets 1–2.

19. In addition to a high degree of job security and the guarantee of a pension, government positions also frequently provided other perquisites, such as housing and food allowances. These arrangements, which could vary widely, must be taken into account in any assessment of relative salaries earned by Russian (and Soviet) government employees.

20. Balzer, "Educating Engineers," pp. 374–375.

21. M. A. Pavlov, *Vospominaniia metallurga* (Moscow, 1943), p. 97.

22. Kendall Bailes, *Technology and Society under Lenin and Stalin: Origins of the Soviet Technical Intelligentsia, 1917–1941* (Princeton: Princeton University Press, 1978), pp. 306–309; and Bailes, "Stalin and the Making of a New Elite: A Comment," *Slavic Review* 39, no. 2 (June 1980), 286–289.

23. N. G. Garin-Mikhailovskii, *Studenty* and *Inzhenery* (Moscow: Khudozhestvennaia literatura, 1964). Garin was a graduate of the elite transport institute, and he worked as an engineer on several railroad lines both before and during his literary activity. His work on the first state-financed railroad construction project led to his writing the short story "Variant," published posthumously.

24. See Bailes, *Technology and Society*, p. 24, which includes reference to Ipatieff's statement that he owed his life to the Bolsheviks, from *Life of a Chemist*, p. 257. Bailes's comments about the Bolsheviks' being the party of (relative) law and order are echoed in Boris Kagarlitskii, *The Thinking Reed: Intellectuals and the Soviet State from 1917 to the Present*, trans. Brian Pearce (London: Verso, 1988), pp. 38–41.

25. Ipatieff, *Life of a Chemist*, pp. 256–267; Paton, *Vospominaniia*, p. 66.

26. James McClelland, "The Utopian and the Heroic: Divergent Paths to the Communist Educational Ideal," in Abbott Gleason, Peter Kenez, and Richard Stites, eds., *Bolshevik Culture: Experiment and Order in the Russian Revolution* (Bloomington: Indiana University Press, 1985), pp. 114–130, esp. pp. 115–118; and A. P. Kupaigorodskaia, *Vysshaia shkola Leningrada v pervye gody sovetskoi vlasti (1917–1925)* (Leningrad: Nauka, 1984), pp. 30–33.

27. Central State Archive of the Russian Republic (TsGA, RSFSR), collection 1565, list 4, folder 379, sheet 5. See report on Glavprofobr Department for Higher Technical Education, 1918–1921.

28. TsGA, RSFSR, collection 2306, list 18, folder 307, sheet 5. At the Moscow Higher Technical School, as of 1 January 1919, of 5,351 students officially enrolled, only 1,869 (35 percent) were actually studying or taking part in classes. School reports regularly divided students into those "enrolled" and those "actually studying."

29. The memoir literature is replete with the pathos of the "former people." See, for example, Victor Kravchenko, *I Chose Freedom: The Personal and Political Life of a Soviet Official* (New York: Scribner's, 1946), pp. 66–67.

30. On these workers' faculties, or *rabfak*, see Balzer, "Workers' Faculties and the Development of Science Cadres in the First Decade of Soviet Power," in Stuart Blume et al., eds., *The Social Direction of the Public Sciences, Sociology of the Sciences Yearbook, 1987* (Dordrecht: D. Reidel, 1987), pp. 193–212; Larry Holmes, "Rabfak," *Modern Encyclopedia of Russian and Soviet History*, vol. 30, pp. 124–130; and two books by N. M. Katuntseva, *Rol' rabochikh fakul'tetov v*

formirovanii intelligentsii SSSR (Moscow: Nauka, 1966); and *Opyt SSSR po podgotovke intelligentsii iz rabochikh i krest'ian* (Moscow: Mysl', 1974).

31. TsGA RSFSR, collection 1565, list 4, folder 16, sheet 3.

32. Kupaigorodskaia, *Vysshaia shkola Leningrada*, pp. 142–143. For a different perspective, see James McClelland, "Bolshevik Approaches to Higher Education," *Slavic Review* 3, no. 4 (December 1971), 818–831.

33. I. Khodorovskii, "K proverki sostava uchashchikhsia," *Krasnaia molodezh'*, no. 1 (1924), 112–116; Sheila Fitzpatrick, *Education and Social Mobility in the Soviet Union, 1921–1934* (Cambridge: Cambridge University Press, 1979), pp. 97–102.

34. Central State Archive of the October Revolution (TsGAOR), collection 5548, list 1, folders 9 and 10 contain the stenographic records of the First All-Russian Congress of Engineers, December 19–22, 1922. Folders 21 and 103 contain the records of the activities of the Russian Technical Society during this period.

35. TsGIA, collection 90, list 1, folder 864.

36. A. I. Kardash, "Organizatsiia nauchno-tekhnicheskikh obshchestv v SSSR (1921–1929)" (Kandidat diss., Moscow, Lumumba University, 1968), pp. 140–141; and L. V. Ivanova, *Formirovanie Sovetskoi nauchnoi intelligentsii, 1917–1927* (Moscow: Nauka, 1980), pp. 207–212.

37. TsGAOR, collection 5548, list 1, folder 6, sheet 10.

One of the intriguing questions about these organizations is the leading role of the metallists' ITS. The metallists established the first and most durable ITS and encouraged creation of an engineering center to assist other groups. This is an interesting parallel to the activism of the metalworkers among the proletariat. Did the same factors of education and structure of work also influence the engineers? Were they affected by the activism of the workers in the industry? It is particularly noteworthy that the metallists' ITS opted for a membership exclusively of engineers, rather than including technicians as did some other sections.

38. TsGIA, collection 90, list 1, folder 21, sheets 60–61.

39. Ivanova, *Formirovanie*, pp. 177–206.

40. Nikolai Valentinov, *Novaia ekonomicheskaia politika i krizis partii posle smerti Lenina* (Stanford: Hoover Institution Press, 1971), p. 117.

41. TsGAOR, collection 2556, list 4, folder 34, sheet 1, verso, "Upravlenie upol'-nomochenogo narkomprosa po delam vuzov i rabfakov g. Leningrada."

42. See Katerina Clark's contribution to this volume (Chapter 9); Richard Stites, "Iconoclastic Currents in the Russian Revolution: Destroying and Preserving the Past," in Gleason, Kenez, and Stites, *Bolshevik Culture*, pp. 1–24; and Stites, *Revolutionary Dreams: Utopian Vision and Experimental Life in the Russian Revolution* (New York: Oxford University Press, 1989).

43. See Chapter 9 of this volume.

44. Central State Archive of the National Economy (TsGANKh), collection 3429, list 3, folder 2630, sheets 58–70.

45. Ibid., sheet 72.

46. For example, the mining academy in Moscow was divided into six institutes with "narrow specialties": mining, steel, ferrous metallurgy, geology and explo-

ration, oil, and peat. TsGANKh, collection 7297, list 7, folder 107, sheets 107–108.

47. Fitzpatrick, *Education and Social Mobility,* Chaps. 6 and 7; Bailes, *Technology and Society,* pt. 3; and V. F. Khoteenkov, "Rol' vsesoiuznogo komiteta po vysshemy tekhnicheskomy obrazovaniiu v formirovanii sovetskoi sistemy vysshei tekhnicheskoi shkoly, 1932–1936" (Kandidat diss., Moscow Pedagogical Institute, 1962), chap. 1.
48. TsGAOR, collection 8060, list 2, folder 175, sheet 16.
49. TsGANKh, collection 3429, list 3, folder 2638, sheet 112, notes from protocol of session of Buro of ITS Moscow Oblast GAS, 1 July 1929; report by M. A. Gavrilov and L. E. Solov'ev on reorganization of MVTU electrical faculty.
50. TsGAOR, collection 8060, list 2, folder 175, sheet 2. Emphasis in the original.
51. Ibid., sheets 1–8, VKVTO Methodological Council for Higher Schools, Presidium, "Draft Report on Nomenclature of Specialties for all Commissariats."
52. TsGAOR, collection 8060, list 2, folder 175, sheet 38.
53. *KPSS v rezoliutsiiakh i resheniiakh s"ezdov konferentsii i plenumov* (Moscow: Izdatel'stvo politicheskoi literatury, 1970–), 1928 vol., pt. 2, p. 381.
54. Nikita Khrushchev, *Khrushchev Remembers,* trans. Strobe Talbott (Boston: Little, Brown, 1970), pp. 38–40; and Kravchenko, *I Chose Freedom,* pp. 59–65.
55. Fitzpatrick, *Education and Social Mobility,* pp. 246–249; and Jerry Hough, *Soviet Leadership in Transition* (Washington, D.C.: Brookings Institution, 1980), pp. 45–48.
56. See Chapter 9 of this volume.
57. Sheila Fitzpatrick describes this process as the natural evolution of the new elite. I tend to view it more as a response to the disaster perpetrated in the schools during the Cultural Revolution. See Fitzpatrick, *Education and Social Mobility,* chap. 11; and also Sheila Fitzpatrick, "Stalin and the Making of a New Elite (1928–1939)," *Slavic Review* 38, no. 3 (September 1979).
58. Cf. TsGAOR, Collection 8060, list 1, folder 11, and list 2, folder 175.
59. G. A. Kradinova, "Podgotovka inzhenerno-tekhnicheskikh kadrov v zapadnoi sibiri v gody velikoi otechestvennoi voiny (1941–1945 gg.)" (Kandidat diss., Novosibirsk University, 1972), pp. 52–55.
60. M. R. Kruglianskii, *Vysshaia shkola SSSR v gody velikoi otechestvennoi voiny* (Moscow: Vysshaia shkola, 1970), pp. 140–142; and Kradinova, "Podgotovka," pp. 150–151.
61. E. A. Belaev and N. S. Pyshkova, *Formirovanie i razvitie seti nauchnykh uchrezhdenii SSSR* (Moscow: Nauka, 1979), chap. 3.
62. Kradinova, "Podgotovka," pp. 103–104, 118.
63. TsGAOR, collection 1565, list 4, folder 16, sheets 52–53. At the Moscow Energy Institute and the Moscow Chemical Technology Institute, the assistant director for administrative-economic affairs changed eleven times during 1931. In 1935–36 there were only six individuals working as institute directors who had been in the position in 1933. Only four of eleven current institute directors had a higher education. The result was administrative confusion, although we might infer that the director was concerned with political affairs at the higher

level, and that lower-level administrators and faculty managed to run the schools despite the high turnover rate at the top.

64. Hiroaki Kuromiya, *Stalin's Industrial Revolution: Politics and Workers, 1928–1932* (Cambridge: Cambridge University Press, 1988), pp. 277–278.

65. Lewis H. Siegelbaum, *Stakhanovism and the Politics of Productivity in the USSR, 1935–1941* (Cambridge: Cambridge University Press, 1988), pp. 117–123; 249–252; Kravchenko, *I Chose Freedom*, pp. 187–190.

66. There is a growing literature on workers' relations in industry, including their relationship with managers. See Kuromiya, *Stalin's Industrial Revolution;* Nicholas Lampert, *The Technical Intelligentsia and the Soviet State* (New York: Holmes and Meier, 1979); Donald Filtzer, *Soviet Workers and Stalinist Industrialization: The Formation of Modern Soviet Production Relations, 1928–1941* (Armonk, N.Y.: M. E. Sharpe, 1986); Siegelbaum, *Stakhanovism*; and William J. Chase, *Workers, Society, and the Soviet State* (Urbana: University of Illinois Press, 1987). The classic study of managers is Joseph Berliner, *Factory and Manager in the USSR* (Cambridge, Mass.: Harvard University Press, 1957).

67. Bailes, *Technology and Society,* pp. 153–156, 225.

68. Based on interviews with Soviet émigrés, described in Harley D. Balzer, *Soviet Science on the Edge of Reform* (Boulder: Westview Press, 1989).

69. Kravchenko, *I Chose Freedom,* pp. 187–189; and Siegelbaum, *Stakhanovism,* pp. 104–105, 114–115.

70. Kradinova, "Podgotovka," pp. 37–38. The picture was in fact even less complimentary to the success of programs to train engineers for industry. About 350,000 engineers had been trained in the Soviet Union by 1941 (70,000 in the First FYP, 135,000 in the Second FYP, and 120,000 in the first three years of the Third FYP, plus about 25,000 from before 1928). This means that about one-third of all engineers were not in the industrial sphere at all.

71. Ibid., pp. 43–45.

72. *Nauchno-tekhnicheskie obshchestva SSSR: Istoricheskii ocherk* (Moscow: Profizdat, 1968), pp. 308–309.

73. Kradinova, "Podgotovka," pp. 17, 91.

74. Ibid., p. 118; Kruglianskii, *Vysshaia shkola,* pp. 137–139. To this day Soviet scholars refuse to take female professionals seriously. N. Zakharova, A. Posadskaia, and N. Rimashevskaia, "Kak my reshaem zhenskii vopros," *Kommunist* (March 1989), 56–65.

75. G. A. Dokuchaev, *Sibriskii tyl v velikoi otechestvennoi voine* (Novosibirsk: Nauka SO, 1968), pp. 106–109.

76. A. V. Khrulev, "Stanovlenie strategicheskogo tyla v velikoi otechestvennoi voine," *Voenno-istoricheskii zhurnal,* no. 6 (1961), 64–80.

77. This is not necessarily unique to the Soviet context. Some accounts suggest that left to themselves, without the prodding of military taskmasters, the Manhattan Project team might have taken another decade to produce a working bomb.

78. On political interference in the 1920s see Loren Graham, *The Soviet Academy of Sciences and the Communist Party, 1927–1932* (Princeton: Princeton University Press, 1967), pp. 86–88, 129. When discussing the Academy of Sciences,

one must always keep in mind that *nauka* has a much broader meaning than "science." The Russian term includes the social, behavioral, and humanistic scholarly disciplines, as well as natural science. In fact, the Academy has been more concerned to isolate itself from the technical sciences than to question the place in its ranks of historians or philosophers.

79. The best source on the technical intelligentsia in this period is Bailes, *Technology and Society*. His chapters entitled "The Production Specialist" and "Flight from Production" are particularly valuable. Also see Lampert, *Technical Intelligentsia*.

80. Kardash, "Organizatsiia," p. 129.

81. Stephen Fortescue, *The Communist Party and Soviet Science* (Baltimore: Johns Hopkins University Press, 1986); and Fortescue, *The Academy Reorganized: The R&D Role of the Soviet Academy of Sciences since 1961*, occasional paper no. 17, Department of Political Science, Research School of Social Science (Canberra: Australian National University, 1983).

82. A. Fedoseev, *Zapadnia*, 2nd ed. (Frankfurt: Posev, 1979), pp. 97, 103; and interviews with Soviet émigrés cited in Balzer, *Soviet Science*.

83. V. M. Britov and E. L. Potseltsev, "Sovetskaia istoricheskaia literatura o partiinom rukovodstve tekhnicheskoi intelligentsii v velikoi otechestvennoi voiny" (Ivanovo, 1983) (manuscript deposited at INION).

84. Clark (Chapter 9) makes this point both about Azhaev and in reference to Panova's *Kruzhilikha*.

85. Fedoseev, *Zapadnia*, pp. 120–122.

86. Affirmative action for minorities has not been as popular in technical fields as in the social sciences and humanities. *Sotsial'no-kul'turnyi oblik sovetskikh natsii* (Moscow: Nauka, 1986), pp. 57–65.

87. I. A. Liasnikov, *Podgotovka spetsialistov promyshlennosti SSSR* (Moscow, 1954), pp. 15–18; I. A. Krivoi, "Deiatel'nost' partiinykh organov ukrainy po sovershenstvovaniiu sistemy vysshego tekhnicheskogo obrazovaniia (1959–1965)" (Kandidat diss., Dnepropetrovsk University, 1972), pp. 38–40.

88. Ibid.

89. The data, derived from annual volumes of *Narodnoe khoziastvo SSSR* (Moscow: Finansy i statistika), are discussed in Harley Balzer and Murray Feshbach, *The Soviet Technostructure* (forthcoming).

90. Moshe Lewin, *The Gorbachev Phenomenon* (Berkeley: University of California Press, 1988), pp. 45–46.

91. *Sotsialisticheskaia industriia*, 23 May 1984, p. 3.

92. V. I. Ivanov, "Sotsiologicheskoe obespechenie intensifikatsii ekonomiki," *Sotsiologicheskie issledovaniia*, no. 2 (1986), 10–15; V. F. Sbytov, "Strikhi k portretu sovetskoi nauchno-tekhnicheskoi intelligentsii," *Sotsiologicheskie issledovaniia*, no. 3 (1986), 110–115; and G. A. Lakhtin, "Kadrovyi potentsial nauki v usloviakh intensifikatsii," *Vestnik Akademii Nauk SSSR*, no. 3 (1987), 33–42.

93. The classic study is Joseph S. Berliner, *The Innovation Decision in Soviet Industry* (Cambridge, Mass.: MIT Press, 1976).

94. Boris Paton, interviewed by P. Polozhevets, "Shturmany progressa," in *Uskorenie* (Moscow: Izdatel'stvo "Pravda," 1987), p. 185.

95. There is an interesting parallel to the decline in popularity of engineering in the United States: over half the graduate students at U.S. engineering schools are now foreigners.

96. See the 14 December 1954 TsK decree, "On Scientific Engineering-Technical Societies under VTsSPS, Calling for Mass Organizations by Specialty," in Kardash, "Organizatsiia," p. 130.

97. Fortescue, *The Communist Party,* p. 35; Fortescue, "Party Membership in Soviet Research Institutes," *Soviet Union* 2, no. 2 (1984), 129–156; Cynthia Kaplan, "The Impact of World War II on the Party," in Susan Linz, ed., *The Impact of World War II on the Soviet Union* (Totowa, N.J.: Rowman and Allanheld, 1985), pp. 152–188; and T. H. Rigby, *Communist Party Membership in the USSR, 1917–1967* (Princeton: Princeton University Press, 1968).

98. This is *not* a result of demographic factors. Admissions to higher education have continued to increase, but the attrition rate has been increased as part of an attack on "percentomania." Correspondence divisions admit an increasing proportion of the student body.

99. Every second higher school graduate in industry is not working in a position corresponding to his or her specialty. Ivanov, "Sotsiologicheskoe obespechenie"; Sbytov, "Strikhi."

100. On higher education reform, see Balzer, "The Soviet Scientific-Technical Revolution: Education of Cadres," in C. Sinclair, ed., *The Status of Soviet Civil Science* (Dordrecht: Martinus Nijhoff, 1987), pp. 3–18.

101. This theme is developed in my introduction to Harley Balzer, ed., *Professions in Russia at the End of the Old Regime* (Ithaca: Cornell University Press, forthcoming).

102. For example, see Georgii Kulagin, "Inzhener: Vchera, segodnia, zavtra," *Znamia,* no. 10 (1985), 146–167.

103. Blair Ruble, "The Social Dimensions of Perestroika," *Soviet Economy* 3 (April–June 1987), 171–183.

104. It is striking that higher education has not generally been perceived as a means of *geographic* mobility, except insofar as students use matriculation as a means of getting to urban centers. Graduates' unwillingness to leave major urban centers has been a constant for over a century. It was a problem when the Trans-Siberian Railroad was constructed in the 1890s and in the building of the Baikal-Amur Main Line in the 1980s. Geographic concentration and the preference of the intelligentsia for life in urban centers have consistently complicated planning.

6. Rockets, Reactors, and Soviet Culture

1. A. V. Shchusev, "Moskva budushchego," *Nauka i zhizn'* (hereafter cited as *NiZ*) 9 (1947), 33–36; and K. K. Antonov, "Sovetskaia nauka i vysotnoe stroitel'stvo," *Priroda* 2 (1952), 58–65.

2. A. A. Zvorykin, L. V. Zubkov, "Tekhnicheskoe perevooruzhenie narodnogo khoziaistva SSSR za 30 let," *NiZ* 10 (1947), 22–34. The plates and illustrations

that accompany this article are examples of Stalinist gigantomania in construction.

3. V. V. Mikhailov, "Industrializatsiia stroitel'stva," *NiZ* 4 (1955), 19–23.

4. A. Lysiakov, "Skorostnye lifty," *NiZ* 4 (1951), 38–39.

5. I. I. Artobolevskii, "Zamechatel'noe sooruzhenie stalinskoi epokhi," *NiZ* 5 (1952), 3–5; N. A. Karaulov, "Gidroenergeticheskoe stroitel'stvo v SSSR," *Priroda* 5 (1956), 3–16; and L. A. Shubenko, "Turbiny-giganty," *NiZ* 8 (1957), 5–9.

6. See Erik P. Hoffmann, "Soviet Views of 'The Scientific-Technological Revolution,'" *World Politics* 30, no. 4 (October 1977–July 1978), 615–644, for a discussion of the social, political, and economic context for the development of Soviet attitudes toward this revolution.

7. According to E. Faddeev, space research and development was a vital part of the scientific-technological revolution. He describes the "cosmicization of science as one of the most fundamental characteristics of contemporary scientific progress": "Cosmonautics together with several other factors objectively plays an extremely powerful role as its stimulus and accelerator." See E. Faddeev, "Chelovek i vselennaia," *Kommunist* 3 (1966), 46.

8. See, for example, S. I. Vavilov, "Osnovnye nauchnye problemy AN SSSR v blizhaishei piatiletke," *Vestnik Akademii nauk SSSR* (hereafter cited as *VAN*) 8–9 (1946), 7–16; S. I. Vavilov, "Sovetskaia nauka," *Priroda* 10 (1947), 3–24; A. N. Nesmeianov, "Nekotorye zadachi AN SSSR v svete reshenii XIX s"ezda KPSS," *Priroda* 3 (1953), 7–18; A. N. Nesmeianov, "Nekotorye problemy sovetskoi nauki," *VAN* 5 (1954), 3–25; and "Ob osnovnykh napravleniiakh v rabote AN SSSR," *VAN* 2 (1957), 3–42.

9. L. L. Miasnikov, "Lider sovremennogo estestvoznaniia," *NiZ* 4 (1955), 10–12. Another reductionist view of physics is Oleg Pisarzhevskii, "Fizika i tekhnika," *Fizika v shkole* 6 (1957), 5–15.

10. Mark Adams, Zhores Medvedev, and others have written about the protection afforded some geneticists within physico-mathematical institutes. The role of atomic energy in freeing biologists to study genetics is a considerable one. The study of the effects of radioisotopes and radioactivity on living organisms (for example, in medical treatment, as a cause for mutation, and so on) led to increased interest in the study of genetics and the mechanisms of inheritance. See A. N. Nesmeianov, "O zadachakh AN SSSR v svete XX s"ezda KPSS," *VAN* 6 (1956), 5–6; and N. P. Dubinin, "Problemy i zadachi radiatsionnoi genetiki," *VAN* 8 (1956), 22–33.

11. One author suggested that soon biology would become the leading science. Although physics had been the leading science in the first half of the century, Soviet genetics and biophysics would become vital for progress in agriculture, physiology, and so on. See M. Vasil'ev, "Dorogi vo vselennuiu," *Molodoi kommunist* 3 (1959), 107.

12. *XX S"ezd KPSS: Stenograficheskii otchet* (Moscow: Gosizdatpolit, 1956), I, 595–600. The Soviet effort to build industrial-size reactors and determine parameters for serial production of those reactors as quickly as possible has been a trademark of the atomic energy program since its first days.

13. Ibid.
14. See, for example, V. Orlov, "Sorevnovanie s solntsem," *Ogonek* 21 (1956), 4–6.
15. See *Atomnaia energiia* (hereafter cited as *AE*) 1, no. 3 (1956), 6–7, 66–67, for photographs of the Harwell trip.
16. See N. S. Khrushchev, *Khrushchev Remembers*, trans. Strobe Talbott (Boston: Little, Brown, 1974), pp. 58–71, for a discussion of Khrushchev's relationship with the scientific intelligentsia.
17. For a more complete treatment of the issues raised in this section, see my "Atomic Culture and Atomic Energy in the USSR," in T. Anthony Jones, David Powell, and Walter Connor, eds., *Soviet Social Problems* (Boulder: Westview Press, forthcoming).
18. For a discussion of similar attempts in the United States to develop an atomic culture while avoiding discussion of the potential dangers of nuclear energy, see Daniel Ford, *The Cult of the Atom* (New York: Simon and Schuster, 1982).
19. It was inevitable that the USSR would surpass the West in electrical energy production, one author wrote, since in the West atomic energy served only military purposes, and in the United States, atomic energy lagged because of the oil monopoly's opposition to it. Iurii Rytov, "Atomnaia energetika," *Literaturnaia gazeta*, 2 October 1958, 2.
20. V. S. Emel'ianov, "Mirnyi atom na sluzhbe kommunizma," *Novyi mir* 10 (1961), 37–43.
21. Paul Boyer, *By the Bomb's Early Light* (New York: Pantheon, 1985).
22. A. M. Kuzin, "Mechenye atomy v biologii," *NiZ* 4 (1955), 29–32.
23. E. S. Pertsovskii, "Energiia atoma-pishchevoi promyshlennosti," *NiZ* 8 (1956), 17–20; and B. A. Rubin et al., "Ispol'zovanie gamma-luchei pri dlitel'-nom khranenii kartofel'ia," *Priroda* 7 (1958), 91–94. Under the Reagan administration the FDA endorsed this process for a number of food products.
24. Iu. P. Kurdinovskii, "Luchi, rozhdennye atomom," *NiZ* 3 (1956), 17–21.
25. Ibid.; A. Markin, "Atom na sluzhbe cheloveka," *Smena* 1 (1955), 18–19; and V. A. Mezentsev, "Na poroge atomnogo veka," *NiZ* 1 (1956), 5–14.
26. P. Astashenkov, "Pervye shagi atomnoi energetiki," *Novyi mir* 3 (1955), 181. Another article instrumental in the popularization of atomic culture, V. Emel'ianov, "Rasskazy ob atome," *Novyi mir* 8 (1955), 219–234, appeared later that year.
27. For a discussion of the iconography of Project Plowshares and attempts to allay fears of military uses of the bomb through peaceful nuclear explosions in the United States, see Michael Smith, "Cultures of Procurement: Promotion of Weapons in the Nuclear Age," paper delivered at the annual meetings of the History of Science Society and the Society for the History of Technology, Raleigh, N.C., 30 October 1987.
28. Mezentsev, "Na poroge," p. 7.
29. Markin, "Atom na sluzhbe," p. 18.
30. A. Morozov, "Razrushiteli," *Smena* 9–10 (1946), 11–12.
31. "Prochtite eti lektsii," *Kul'turno-prosvetitel'naia rabota* 2 (1955), 29–30; and

V. Leshkovtsev, "Mirnoe ispol'zovanie atomnoi energii," *Kul'turno-prosvetitel'naia rabota* 8 (1955), 43–48.

32. V. Agranovskii, "Budem iskat' uran," *Komsomol'skaia pravda,* 11 April 1957, 2.

33. G. V. Ermakom (chief engineer of Glavatomenergo), "Pervyi gigant atomnoi energetiki," *Izvestiia,* 19 June 1957, 1.

34. A. N. Komarovskii, "Puti ekonomii stali v reaktorostroenii," *AE* 7, no. 3 (1959), 205–215. This practice of using less steel and less reinforced concrete than in Western reactors persisted until quite recently in the construction of containment facilities for Soviet reactors.

35. P. Anan'ev, "Nekotorye voprosy ekonomiki iadernoi energetiki," *AE* 6, no. 3 (1959), 245–251; and I. Vaisman, "Nekotorye ekonomicheskie problemy atomnoi energetiki," *Voprosy ekonomiki* 5 (1957), 87–94.

36. E. Maksimova, "Put' k Obninsku," *Izvestiia,* 1 October 1967, 5; and A. M. Petros'iants, "Atomnaia nauka i tekhnika," *Trud,* 24 October 1967, 4.

37. A. I. Golovin et al., "Utilization of Floating Nuclear Power Plants in Northern Regions," *Soviet Atomic Energy* (February 1982), 497–501; originally published in *AE* 51, no. 2 (August 1981), 83–87.

38. For a discussion of the politics of the space age by means of a comparison of the different technocratic approaches to space technology in the United States and the USSR, see Walter A. McDougall, . . . *The Heavens and the Earth* (New York: Basic Books, 1985).

39. A. P. Aleksandrov, "Novye vekhi istorii," *Izvestiia,* 18 October 1957, 2.

40. Ibid. Aleksandrov concluded his article by observing that the USSR had more engineering students than the United States—in certain specialties two or three times as many—and by suggesting that training in physics and mathematics was especially important to ensure the USSR's leadership.

41. A. Topchiev, "Vklad sovetskikh uchenykh," *Pravda,* 3 October 1958, 3.

42. L. Sedov, "Kosmicheskie laboratorii," *Pravda,* 3 October 1958, 3. Academician and astronomer V. Ambartsumian also spoke of the leading place of science and technology in a country that had been backward forty years earlier. Soviet scholars "did not make a secret of the fact" that the tempo of growth in science was the result of the "most progressive social order," which fully justified optimistic predictions for its performance in the future. See V. Ambartsumian, "Reshena zadacha ogromnoi vazhnosti," *Pravda,* 24 August 1960, 4.

43. Petr Lebedenko, "Genii millionov," *Don* 10 (1960), 166–167.

44. Ibid.

45. V. Petrov and I. Ovchinnikov, "Pered poletom cheloveka v kosmos," *Novyi mir* 11 (1960), 166–167.

46. V. Petrov, "Amerikanskie sputniki-shpiony," *Komsomol'skaia pravda,* 17 May 1960, 2.

47. M. Kroshkin, "Pervyi kosmicheskii korabl'," *Komsomol'skaia pravda,* 17 May 1960, 2.

48. G. Terent'ev, "'Merkurii'-'dzheminai'-'apollon,'" *Krasnaia zvezda,* 26 March 1963, 3.

49. I. Vereshchagin, "'Dzheminai' i tseli pentagona," *Krasnaia zvezda,* 28 August 1965, 4.
50. Aleksandrov, "Novye vekhi istorii," p. 2.
51. Ibid.
52. B. S. Danilin, "Vtorzhenie v kosmos," *NiZ* 12 (1957), 4.
53. M. Vasil'ev, "Dorogi vo vselennuiu," *Molodoi kommunist* 3 (1959), 104. For a chart of Soviet firsts, or *vekhi,* see G. Titov, "Shturm vselennoi prodolzhaetsia," *Agitator* 5 (1968), 17–19.
54. Lebedenko, "Genii millionov."
55. S. Kondrashov, "SSHA—vtoraia strana, poslavshaia cheloveka v kosmos," *Izvestiia,* 22 February 1962, 2.
56. Nikolai Gribachev, "Na dorogu k zvezdam," *Pravda,* 3 October 1958, 3.
57. K. Mikhailov, "V kosmose net legkikh putei," *Izvestiia,* 31 January 1962, 2.
58. Aleksandr Serbin, "Spotykaias' v kosmose," *Ogonek* 52 (1959), 20–21.
59. Vasil'ev, "Dorogi vo vselennuiu," p. 102.
60. See the interview with A. Nikolaev and P. Popovich, the third and fourth men in space, in "Budushchim zvezdoplavateliam," *Iunost'* 10 (1962), 3–6.
61. See, for example, I. D'iachkov, "Avtografy geroev," *Prostor* 6 (1962), 63–68, concerning a chance meeting between D'iachkov and the cosmonauts G. S. Titov and Iu. A. Gagarin, who were on vacation in the Crimea; and G. Ivanov, "Zdravstvui zemlia," *Prostor* 11 (1962), 87–93, on the "ninety-five hours of worry, joy, and pride for our stellar captains Andrian Nikolaev and Pavel Popovich."
62. S. Borzenko, "Geroi nashego vremeni," *Kul'turno-prosvetitel'naia rabota* 11 (1962), 5–8.
63. Vladimir Komarov, "Zhizn' Vladimira Komarova," *Iunost'* 7 (1967), 65–73.
64. G. Semenikhin, "Komandir korablia 'voskhod,'" *Neva* 1 (1965), 171–174.
65. Iurii Letunov, "Kosmodrom: Iiun' 1963-ii god," *Prostor* 7 (1964), 3.
66. Ibid.
67. Soviet writers have a great deal of reverence for Mother. *Literaturnaia gazeta* reported that Tereshkova responded to the prodding to "name the most dear person in your life" with the reply "Mama." Valerii Agranovskii, "Dela nebesnye, dela zemnye," *Literaturnaia gazeta,* 17 June 1963, 2.
68. Letunov, "Kosmodrom," pp. 9–12. Another example of this genre is N. Mel'nikov, "Kosmodrom i liudi," *Molodaia gvardia* 11 (1963), 102–121.
69. Agranovskii, "Dela nebesnye."
70. A. G. Nikolaev and P. R. Popovich, "My zhili i rabotali v kosmose," *Priroda* 9 (1962), 14.
71. Ibid., pp. 15–16.
72. Semen Kirsanov, "Udivlen'e," *Pravda,* 4 October 1962, 2.
73. Sergei Vasil'ev, "Tol'ko god," *Pravda,* 3 October 1958, 3.
74. Nikolai Aseev, "Zdravitsa," *Literaturnaia gazeta,* 1 January 1960, 1.
75. A. Rogachev, "Do vstrechi, tovarishch luna," *Don* 4 (1961), vi.
76. Anatolii Radygin, "Uidut zvezdolety," *Komsomol'skaia pravda,* 17 May 1960, 1. This poem was published two days after the first Soviet spaceship was launched.

77. M. Sholokhov, "Voskhishchenie i gordost'," *Don* 4 (1961), ii.

78. Vasilii Zhuravlev, "Nepremenno," *Krasnaia zvezda,* 28 October 1959, 3.

79. N. Skrebov, "K zvezdam," *Don* 4 (1961), iii.

80. See Mark Kuchment's contribution to this volume (Chapter 11), which discusses the attempt of science writers to overcome the myth of the two cultures, artistic and scientific.

81. Il'ia Erenburg, "O lune, o zemle, o serdtse," *Literaturnaia gazeta,* 1 January 1960, 3.

82. S. Ostrovskogo, "Pesenka o sputnike," *Kul'turno-prosvetitel'naia rabota* 1 (1958), 30–32.

83. G. Liando, "Nebesnye chastushki," *Kul'turno-prosvetitel'naia rabota* 1 (1958), 34. A *chastushka* is a two- or four-line folk verse, sung in a lively fashion.

84. K. Shistovskii, "Iskusstvennyi sputnik zemli," *Sem'ia i shkola* 6 (1956), 30–31.

85. V. Komarov, "Chelovek i kosmos," *Sem'ia i shkola* 7 (1959), 40–42.

86. V. Komarov, "Dorogi novye, zvezdnye," *Sem'ia i shkola* 7 (1961), 7–9.

87. Anatolii Belov, "Gzhatskie byli: Rasskaz materi geroia," *Sem'ia i shkola* 8 (1961), 22–26; and Belov, "Gzhatskie byli: V shkole, gde uchilsia geroi," *Sem'ia i shkola* 9 (1961), 17–19.

88. *Komsomol'skaia pravda,* 1 January 1958, 1.

89. *Literaturnaia gazeta,* 1 January 1960, 1.

90. See, for example, "Ves sputnikov," *Komsomol'skaia pravda,* 17 May 1960, 2, published two days after the launching of the first U.S. spaceship.

91. V. Komarov, "Gladia na nebo," *Kul'turno-prosvetitel'naia rabota* 1 (1958), 56–60.

92. Ibid.

93. *Komsomol'skaia pravda,* 17 May 1960, 2.

94. A. Masevich (deputy chairman of the Astronomical Council of the Academy of Sciences), "Nabliudaia za sputnikov," *Pravda,* 4 March 1958, 6.

95. N. Syromiatnikov, "Dva mira—dva poleta," *Komsomol'skaia pravda,* 17 May 1958, 2.

96. S. G. Aleksandrov, "O nauchnykh itogakh issledovaniia kosmicheskogo prostranstva," *Pravda,* 4 October 1962, 2.

97. V. Nikolaev, "Doroga k zvezdam," *Trud,* 13 November 1957, 3. The film is also reviewed in "Put' v kosmos," *NiZ* 1 (1958), 40–41.

98. G. Kudriavtsev and A. Finogenov, "Poema o velikom podvige," *Trud,* 30 September 1961, 4. The film describes the training and personality of the cosmonaut German Titov, who is shown as not only a pilot but also a navigator, radiologist, biologist, astronomer, geographer, and physicist.

99. See, for example, Stuart H. Loory, "Why This Space Race?" *Reporter,* 27 April 1961, 19–22; Lee DuBridge, "A Scientist Calls for Common Sense," ibid., pp. 22–23; and Joseph Krutch, "Why I Am Not Going to the Moon," *Saturday Review,* 20 November 1965, 29–31.

100. V. Liapunov, "Nachalo kosmicheskoi ery," *Ogonek* 41 (1958), 6–7.

101. V. Dobronravov, "Chelovek izuchaet kosmos," *Molodoi kommunist* 11 (1959), 15–22. Also see A. A. Blagonravov, "Zachem my osvaivaem kosmos," *Molodoi bol'shevik* 6 (1960), 99–104.

102. "Ne rano li zaigryvat' s lunoi," *Komsomol'skaia pravda,* 11 June 1960, 1.

103. V. L'vov, "Zavtrashnyi kosmos," *Neva* 7 (1966), 140–142.

104. B. Danilin, "Kto poletit v kosmos—chelovek ili avtomat?" *Molodaia gvardiia* 1 (1961), 204–207. Also see M. Kroshkin, "Pervyi kosmicheskii korabl'," *Komsomol'skaia pravda,* 17 May 1960, 2.

105. Danilin, "Vtorzhenie v kosmos," pp. 6–8.

106. V. Vasilevskii and Iu. Fedotov, "Meditsina i kosmos," *Molodoi kommunist* 12 (1962), 49.

107. O. G. Gazenko and V. B. Malkin, "Biologiia kosmicheskikh poletov," *NiZ* 11 (1958), 17–21.

108. O. G. Gazenko and V. B. Malkin, "Chelovek v kosmose," *NiZ* 12 (1959), 17–23.

109. I. A. Savenko et al., "Kosmicheskie polety i radiatsionnaia opasnost'," *Priroda* 2 (1962), 40–48; and V. V. Antipov et al., "Na trasse zemlia-luna," *Priroda* 4 (1965), 46–49.

110. I. S. Balakhovskii and V. B. Malkin, "Biologicheskie problemy mezhplanetnykh poletov," *Priroda* 8 (1956), 15–21. Danilin also acknowledged the potential danger of meteorites in "Kto poletit v kosmos," p. 207.

111. V. Bazykin, "Zhizn' sputnikov," *Agitator* 2 (1967), 52–53. It will be recalled that several U.S. and Soviet satellites have fallen to the earth without completely burning up, including at least three Soviet satellites with nuclear-powered (radioactive) power packs.

112. A. A. Shternfel'd, "Problemy kosmicheskogo poleta," *Priroda* 12 (1954), 13–14.

113. Ibid., pp. 16–21. For a detailed, straightforward discussion of the early scientific and technical problems facing the Soviet space program, see V. G. Fesenkov, "Problemy astronavtiki," *Priroda* 6 (1955), 11–18, trans. in F. J. Krieger, ed., *Behind the Sputniks: A Survey of Soviet Space Science* (Washington, D.C.: Public Affairs Press, 1958), pp. 83–94; A. G. Karpenko and G. A. Skurindin, "Sovremennye problemy kosmicheskikh poletov," *VAN* 9 (1955), 19–30, trans. in Krieger, *Behind the Sputniks,* pp. 95–111; A. N. Nesmeianov, "Problema sozdaniia iskusstvennogo sputnika zemli," *Pravda,* 1 June 1957, 2; L. I. Sedov, "Problemy kosmicheskikh poletov," *Pravda,* 12 June 1957, 4; E. K. Fedorov, "Issledovaniia verkhnikh sloev atmosfery pri pomoshchi raket i iskusstvennykh sputnikov zemli," *Priroda* 9 (1957), 3–12; and G. A. Skurindin and L. V. Kurnosova, "Nauchnye issledovaniia pri pomoshchi iskusstvennykh sputnikov zemli," *Priroda* 12 (1957), 7–14.

114. Shternfel'd, "Problemy kosmicheskogo poleta," pp. 16–21; B. Liapunov, "Okno v budushchee," *NiZ* 6 (1953), 34; and N. A. Varvarov, "Iskusstvennye zemli," *NiZ* 1 (1957), 17–19.

115. Liapunov, "Okno v budushchee," p. 35. Also see R. G. Perel'man, "Dvigateli galakticheskikh korablei," *NiZ* 7 (1958), 60–64; G. G. Zel'kin, "Fotonnaia raketa," *Priroda* 11 (1960), 69–72; and F. Kedrov, "Fotonnaia raketa," *Oktiabr'* 7 (1957), 169–172. The United States also had an extensive research and development program for atomic-powered rockets, and under the influence of an

administration that strongly supported diverse applications of atomic energy, the United States has embarked on research in this area with renewed faith.

116. Shternfel'd, "Problemy kosmicheskogo poleta," pp. 14–15.
117. B. Liapunov, "Mezhplanetnye puteshestviia," *NiZ* 6 (1953), 34–35.
118. V. L. Ginzburg, "Eksperimental'naia proverka obshchei teorii otnositel'nosti i iskusstvennye sputniki zemli," *Priroda* 9 (1956), 30–39.
119. Vasil'ev, "Dorogi vo vselennuiu," p. 106.
120. See, for example, Chapter 9 of this volume.

7. Biomedical Ethics

1. Strachan Donnelley, "Biomedical Ethics: A Multinational View," *Hastings Center Report* 17, no. 3, special supp. (June 1987), 2.
2. I. N. Smirnov, "Filosofskie izmereniia bioetiki," *Voprosy filosofii*, no. 12 (1987), 83–97.
3. From a computer search using Medlar's on-line, produced by National Library of Medicine's Elhill Retrieval System, Washington, D.C., and Bioethics-on-line, produced by Center for Bioethics, Kennedy Institute of Ethics, Washington, D.C., using keywords *bioethics* and *USSR or Soviet*, completed February 24, 1987.
4. A. Koryagin, "Abuse of Psychiatry: Appeal to Psychiatrists," *Lancet*, 14 November 1981, 1121.
5. Allan Wynn, "The Soviet Union and the World Psychiatric Association," *Lancet*, 19 February 1983, 407.
6. Felicity Barringer, "Soviet Abuse of Psychiatry Said to Linger," *New York Times*, 21 October 1987, 1–2.
7. Ibid., p. 2.
8. *Vedomosti verkhovnovo Soveta SSSR*, no. 2 [2440] (13 January 1988), item 19, pp. 22–27. The complete text in English translation, "Statute on Conditions and Procedures for the Provision of Psychiatric Assistance," appeared in *Current Digest of the Soviet Press* (hereafter cited as *CDSP*), 40, no. 6 (1988), 11–13. For comments, see "O psikhiatricheskoi pomoshchi," *Izvestiia*, 15 January 1988, 6; and S. Staroselskaya, "Look Before You Leap," *Meditsinskaia gazeta*, 13 January 1988, 4; reported in *CDSP* 40, no. 6 (1988), 14–15.
9. See, for example, Peter Gumbel, "Critics Doubt Soviet Psychiatry Reformed: Those Who Ruled Dissidents 'Schizophrenic' Stay in Posts," *Wall Street Journal*, 2 February 1989, A8; and Bill Keller, "U.S. Psychiatrists Fault Soviet Units: Team Finds Inmates Are Still Held for Political Reasons," *New York Times*, 12 March 1989, 1.
10. See Wynn, "The Soviet Union and the WPA," p. 408, which cites information bulletins 1–6 of the International Association on the Political Use of Psychiatry, and the Working Group on the Internment of Dissenters in Mental Hospitals information bulletin (October 1980), charging "the severe harassment of Soviet psychiatrists and others who have complained about these practices."
11. E. Maksimova and I. Martkovich, "Bez zashchity," *Izvestiia*, 11 July 1987, 3.

12. A. Novikov, S. Razin, and M. Mishin, "A Closed Subject," *Komsomol'skaia pravda*, 11 November 1987, 4; reported in *CDSP* 39, no. 46 (16 December 1987), 1.

13. Wynn, "The Soviet Union and the WPA," p. 408.

14. G. Morozov, "We Condemn This Unseemly Activity," *Meditsinskaia gazeta*, 25 March 1983, 3; reported in *CDSP* 35, no. 13 (27 April 1983), 1. For a commentary, see "Why the Soviet Psychiatrists Resigned," *Lancet*, 12 March 1983, 582.

15. See, for example, E. Gorbunova's interview with A. I. Potapov, "In the Shadow of Semiknowledge," *Sovetskaia Rossia*, 20 November 1987, 4; reported in *CDSP* 39, no. 46 (1987), 5.

16. Thomas S. Szasz, "The Mental Health Ethic," in Richard T. De George, ed., *Ethics and Society* (New York: Anchor Books, 1966), pp. 85–110; and Szasz, *The Myth of Mental Illness* (New York: Hoeber-Harper, 1961).

17. Michel Foucault, *Madness and Civilization: A History of Insanity in the Age of Reason*, trans. Richard Howard (New York: Pantheon, 1965).

18. Klara Bronislavovna Segieniece, "Is Induced Abortion Murder by Experts?" *Nauka i tekhnika*, no. 9 (1980), 27–30; reported in *CDSP* 32, no. 5 (1980), 11.

19. Figure cited by V. Kulakov in an interview, "Dva arbuza v odnoi ruke," *Pravda*, 16 December 1988, 3.

20. "With an Eye to Everyday Situations," *Meditsinskaia gazeta*, 12 February 1988, 1; reported in *CDSP* 40, no. 7 (1988), 24.

21. Boris Urlanis, "A Wanted Child," *Nedelia*, 1–7 December 1980, 16; reported in *CDSP* 32, no. 49 (1980), 10–11.

22. Segieniece, "Is Induced Abortion Murder?" p. 11.

23. Vicente Navarro, *Social Security and Medicine in the USSR: A Marxist Critique* (Lexington, Mass.: Lexington Books, 1977), p. 45; and Gordon Hyde, *The Soviet Health Service: A Historical and Comparative Study* (London: Lawrence Wishart, 1974), p. 106.

24. Christopher Tietze, *Induced Abortion: A World Review, 1983* (New York: Population Council, 1983), p. 12.

25. Hyde, *Soviet Health Service*, p. 246.

26. Tietze, *Induced Abortion*, p. 24. Kulakov, "Dva arbuza v odnoi ruke," p. 3, cites 6.8 million abortions for each 5.6 million births registered each year.

27. Larisa Remennik, "The Life That Has Been Killed Inside You," *Nedelia*, no. 38 (21–27 September 1987), 12; reported in *CDSP* 39, no. 44 (1987), 15.

28. Ibid.

29. T. Maiboroda, "Deciding to Become a Mother," *Pravda Ukrainy*, 23 October 1987, 4; reported in *CDSP* 39, no. 44 (1987), 15.

30. "Soviets Reveal Growing Illegitimacy," *International Herald Tribune*, 26 August 1988, 2.

31. "Responses to the Feuilleton: 'About That,'" *Meditsinskaia gazeta*, 10 June 1988, 4; reported in *CDSP* 40, no. 25 (1988), 25.

32. G. Denisova, "About That," *Meditsinskaia gazeta*, 4 March 1988, 4; reported in *CDSP* 40, no. 25 (1988), 24.

33. A. Novikov, "Spid," *Komsomol'skaia pravda*, 1 August 1987, 3–4.

34. "Programme of the Communist Party of the Soviet Union," pt. 2, in *The Road to Communism: Documents of the 22nd Congress of the Communist Party of the Soviet Union* (Moscow: Foreign Languages Publishing House, 1961), pp. 566–567. The 1986 revised program continues to speak of "moral purity"; *Programme of the Communist Party of the Soviet Union: A New Edition* (Moscow: Novosti Press Agency Publishing House, 1986), p. 55. Also, see such books written for the popular market as S. Laptenko, *Moral' i sem'ia* (Minsk: Izdatel'stvo "Nauka i Tekhnika," 1967); V. Chekalin, *Liubov' i sem'ia* (Moscow: Izdatel'stvo Politicheskoi Literatury, 1964); and A. G. Kharchev, *Brak i sem'ia v SSSR* (Moscow: Izdatel'stvo Sotsial'no-ekonomicheskoi literatury "Mysl'," 1964).

35. "Spid bez sensatsii: V. I. Pokrovskii otvechaet na voprosy Kima Smirnova," *Izvestiia*, 16 June 1987, 3.

36. See Richard T. De George, "A Bibliography of Soviet Ethics," *Studies in Soviet Thought* 3 (1963), 83.

37. See Richard T. De George, "Soviet Ethics and Soviet Society," *Studies in Soviet Thought* 4 (1964), 206–217; and Richard T. De George, "Moral Inculcation and Social Control," in *Soviet Ethics and Morality* (Ann Arbor: University of Michigan Press, 1969), chap. 6.

38. See, for example, Ia. A. Mil'ner-Irinin, "Poniatie o prirode cheloveka i ego mesto v sisteme nauki etiki," *Voprosy filosofii*, no. 5 (1987), 71–82; and P. N. Fedoseev, "Sotsialisticheskii gumanizm: Aktual'nye problemy teorii i praktiki," *Voprosy filosofii*, no. 3 (1988), 3–23.

39. For a summary of various articles on these topics, see *CDSP* 39, no. 41 (1987), 15–17, 21.

40. For a history of the development of Soviet medicine and its organization, see Hyde, *Soviet Health Service*.

41. G. Rubanovich, S. Guseva, and A. Cherviakov, "Lying is No Answer," *Nedelia*, 13–19 April 1987, 18; reported in *CDSP* 39, no. 18 (1987), 18.

42. A. Serdiuk, "Belyi khlat i chistaia sovest'," *Izvestiia*, 8 February 1982, 2.

43. Ralph Crawshaw, "Medical Deontology in the Soviet Union," *Archives of Internal Medicine* 134 (September 1974), 592–594.

44. E. Gabrielian, "Dobraia ulybka bracha," *Izvestiia*, 18 May 1975, 3.

45. Michael Ryan, "USSR Letter: Ethics and the Patient with Cancer," *British Medical Journal*, 25 August 1979, 480. Medical deontology is broader than medical ethics. The comparison was the subject of a discussion in 1982 in Moscow; see G. I. Tsaregorodtsev and A. Ia. Ivaniushkin, "Meditsina i etika," *Voprosy filosofii*, no. 9 (1983), 147–152.

46. On this issue see Tsaregorodtsev and Ivaniushkin, "Meditsina i etika," p. 152, and A. D. Naletova, "Problema avtonomii lichnosti v sovremennoi amerikanskoi etike," *Voprosy filosofii*, no. 6 (1982), 121–122.

47. Ryan, "USSR Letter: Ethics and the Patient," pp. 480–481.

48. Michael Ryan, "USSR Letter: Aspects of Ethics (2)," *British Medical Journal*, 15 September 1979, 648–649.

49. Rubanovich, Guseva, and Cherviakov, "Lying," p. 18.

50. Mark D'Anastasio, "Soviet Health System, Despite Early Claims, Is Riddled by Failures," *Wall Street Journal*, 18 August 1987, 8.
51. A. Nikolaev, "Sud vynes reshenie," *Pravda*, 25 January 1982, 7.
52. Serdiuk, "Belyi khlat i chistaia sovest'," p. 2.
53. Y. I. Chazov, "Speech on Soviet Medical Workers' Day," 21 June 1987, cited in D'Anastasio, "Soviet Health System," p. 8.
54. 5 July 1987, 1; reported in *CDSP* 39, no. 27 (1987), 22.
55. Novikov, "Spid," pp. 3–4.
56. "Once More on AIDS," *Komsomol'skaia pravda*, 28 October 1987, 3; reported in *CDSP* 39, no. 45 (1987), 27.
57. V. Belikov, "Chto nado znat' o SPIDe," *Izvestiia*, 15 September 1987, 6.
58. "Spid bez sensatsii," *Izvestiia*, 16 June 1987, 3.
59. N. Boiarkina, "Spid i deti," *Komsomol'skaia pravda*, 28 January 1989, 1; and S. Tutorskaia, "Opasnei virusa prestupnaia khalatnost'," *Izvestiia*, 17 February 1989, 1.
60. For a summary of several articles, see *CDSP* 39, no. 42 (1987), 11–15.
61. Boiarkina, "Spid i deti," p. 1.
62. David Kugultinov, "Emergency Situation: Be People!" *Izvestiia*, 23 February 1989, 3; reported in *CDSP* 41, no. 8 (1989), 23.
63. "Ukaz prezidiuma verkhovnogo Soveta SSSR," *Izvestiia*, 26 August 1987, 2.
64. V. Kalita, "Found Guilty: The Defendant is a Carrier of the AIDS Virus," *Meditsinskaia gazeta*, 8 January 1989, 4; reported in *CDSP* 41, no. 2 (1989), 31.
65. "New Laws in States Target Drugs," *Lawrence Journal-World*, 30 June 1988, 2A.
66. N. S. Malein, "Pravo na meditsinskii eksperiment," *Sovetskoe gosudarstvo i pravo*, no. 11 (1975), 35–41.
67. Tsaregorodtsev and Ivaniushkin, "Meditsina i etika," p. 151.
68. Ibid., pp. 147–152.
69. Leonid Zagal'skii and Valerii Sharov, "Peresadka serdtsa: Problemy i nadezhdy," *Literaturnaia gazeta*, 26 November 1986, 11.
70. D. Sarkisov, "Peresadka serdtsa: Problemy i nadezhdy," *Literaturnaia gazeta*, 24 December 1986, 11.
71. V. A. Negovskii, "Ob odnoi idealisticheskoi kontseptsii klinicheskoi smerti," *Filosofskie nauki*, no. 4 (1981), 55.
72. I. T. Frolov, *Homme, Science, Humanisme: Une nouvelle synthèse* (Moscow: Editions du progrès, 1986), p. 306.
73. V. A. Negovskii, "Nekotorye metodologicheskie problemy sovremennoi reanimatologii," *Voprosy filosofii*, no. 8 (1978), 71.
74. Frolov, *Homme*, p. 307.
75. Viktor Perevedentsev, "Aging," *Moskovskie novosti*, no. 4 (24 January 1988), 12; reported in *CDSP* 40, no. 8 (1988), 24.
76. A. D. Naletova, "Problema avtonomii lichnosti v sovremennoi amerikanskoi etike," *Voprosy filosofii*, no. 6 (1982), 116–123.
77. From an interview with Y. I. Chazov, *Sovetskaia Rossiia*, 5 July 1987, 1; reported in *CDSP* 39, no. 27 (1987), 22.
78. O. Frantsen, "Serdtse Olimpa," *Pravda*, 16 March 1985, 3.

79. L. Zagal'skii, "Po sledam sensatsii: Odno serdtse—dve zhizni," *Pravda*, 11 March 1988, 4.
80. Zagal'skii and Sharov, "Peresadka serdtsa," p. 11.
81. Malein, "Pravo na meditsinskii eksperiment," pp. 35–41.
82. Frolov, *Homme*, p. 165.
83. Yelena Mushkina, "The Happiness of Having a Baby," *Nedelia*, no. 31 (3–9 August 1987), 16; reported in *CDSP* 39, no. 33 (1987), 17.
84. Iu E. Vel'tishchev and M. E. Lobashov, "Eugenics," in *Great Soviet Encyclopedia*, 3rd ed., vol. 8 (New York: Macmillan, 1975), pp. 501–502.
85. "There Will Be No Miracle: Interview of A. A. Sozinov Conducted by Maksim Karpinskii," *Nedelia*, no. 30 (27 July–2 August 1987), 6–7; reported in *CDSP* 39, no. 30 (1987), 23.
86. Ibid.
87. A. A. Sozinov, "Sovremennaia genetika: Problemy i perspektivy," *Kommunist*, no. 4 (1987), 110–121.
88. V. Efroimson, "Rodoslovnaia al'truizma," *Novy mir*, no. 10 (1971), 193–213; and B. Astaurov, "Homo sapiens et humanus—chelovek s bol'shoi bukvy i evoliutsionnaia genetika chelovechnosti," ibid., pp. 214–224.
89. Loren R. Graham, "Reasons for Studying Soviet Science: The Example of Genetic Engineering," in Linda L. Lubrano and Susan Gross Solomon, eds., *The Social Context of Soviet Science* (Boulder: Westview Press, 1980), pp. 205–240.
90. See the following articles, all by N. P. Dubinin: "Filosofskie i sotsiologicheskie aspekty genetiki cheloveka, *Voprosy filosofii*, no. 1 (1971), 36–45; ibid., no. 2, pp. 55–64; "Filosofiia dialekticheskogo materializma i problemy genetiki," *Voprosy filosofii*, no. 4 (1973), 94–107; "O filosofskoi bor'be v biologii," *Filosofskie nauki*, no. 6 (1975), 12–20; "Biologicheskie i sotsial'nye faktory v razvitii cheloveka," *Voprosy filosofii*, no. 2 (1977), 46–57; and "Aktual'nye filosofsko-metodologicheskie problemy sovremennoi biologii," *Voprosy filosofii*, no. 7 (1978), 46–56.
91. See, among others, Frolov, *Homme*, p. 188.
92. A. F. Shishkin, "Etologiia i etika," *Voprosy filosofii*, no. 9 (1974), 111–122.
93. I. T. Frolov, *Global Problems and the Future of Mankind* (Moscow: Progress Publishers, 1982), p. 72.
94. Frolov, *Homme*, p. 141.
95. Sozinov, "Sovremennaia genetika," pp. 110–121.
96. Frolov, *Homme*, p. 158.
97. Frolov, *Global Problems*, p. 74; and *Homme*, p. 217.
98. Frolov, *Global Problems*, p. 74; also see pp. 396–397.
99. Ibid., p. 75.
100. Frolov, *Homme*, pp. 228–229.

8. Fact, Value, and Science

1. For a recent discussion of the effect of personal and cultural values on scientific practice, see Helen Longino, "Beyond 'Bad Science': Skeptical Reflections on

the Value-Freedom of Scientific Inquiry," *Science, Technology, and Human Values* 8, no. 1 (Winter 1983), 7–17.

2. The belief that science and values are separate realms was the explicit ethos of science in western Europe and North America in the early part of this century; see discussion in Loren Graham, *Between Science and Values* (New York: Columbia University Press, 1981), p. 28.

3. "Report by M. S. Gorbachev to the Nineteenth All-Union CPSU Conference, 28 June 1988," *Pravda*, 29 June 1988, 3.

4. *Sed'moi ekstrennyi s"ezd PKP Mart 1918-ogo goda, stenograficheskii otchet* (Moscow: Gosudarstvennoe izd. politicheskoi literatury, 1962), pp. 89, 108.

5. V. I. Lenin, *State and Revolution* (New York: International Publishers, 1932), p. 70.

6. Karl Marx, *The Poverty of Philosophy*, in Richard Dixon et al., trans., *The Collected Works of Marx and Engels* (London: Lawrence and Wishart, 1975), pp. 177–178.

7. The notion of first contemplating reality and then formulating means to achieve some moral end was foreign to Marx's world view. In *The German Ideology* he wrote, "Communism is for us not a state of affairs which is to be established, an ideal to which reality will have to adjust itself. We call communism the real movement which abolishes the present state of things. The conditions of this movement result from the premises now in existence." Karl Marx and Frederick Engels, *Collected Works* (New York: International Publishers, 1976), V, 49. Although Marx detested nineteenth-century capitalism for its victimization of labor, he made use of the Hegelian dialectic to achieve an affirmation of the present: what *is* already contains its own negation, which will produce a supremely valuable negation of the negation.

8. Charles Taylor, "Marxism and Empiricism," in B. Williams and A. Montefiore, eds., *British Analytical Philosophy* (New York: Humanities Press, 1966), pp. 227–246.

9. Robert Tucker, *Philosophy and Myth in Karl Marx* (Cambridge: Cambridge University Press, 1972), esp. pp. 228–231.

10. Susan M. Easton, "Facts, Values, and Marxism," *Studies in Soviet Thought* 17, no. 2 (August 1977), 117–134.

11. W. K. Frankena, "The Naturalistic Fallacy," *Mind* 47 (1939), 464–477.

12. Richard T. De George has characterized Soviet ethical theory as objectivist (it has an absolute objective norm or ideal in terms of which it evaluates the ideals of different societies as more or less progressive, depending on whether or not they tend toward the ideal of communism), naturalistic (it claims that "good" is definable in terms of certain natural properties), and teleological (it is of the self-realization type). See his *Soviet Ethics and Morality* (Ann Arbor: University of Michigan Press, 1969), pp. 27–28. Philip T. Grier provides a useful historical overview of Marxist thought on morality, showing how the naturalist character of Marxism has resulted in a Soviet ethical theory that reduces justification to explanation. See Philip T. Grier, *Marxist Ethical Theory in the Soviet Union* (Dordrecht: D. Reidel, 1978), p. 276.

13. V. I. Lenin, *Materialism and Empiriocriticism* (Peking: Foreign Languages Press, 1972), p. 157.

14. Leszek Kolakowski, *Main Currents of Marxism*, vol. 2 (Oxford: Oxford University Press, 1978), p. 455.

15. Kolakowski notes that Lenin uses the term *partiinost'* in two senses: (1) there can be no middle position between materialism and idealism as Engels defines them; and (2) philosophical theories are not neutral in the class struggle but are instruments of it. Kolakowski, *Main Currents*, pp. 451–452.

16. Steven Lukes, *Marxism and Morality* (Oxford: Clarendon Press, 1985), pp. 1–4; De George, *Soviet Ethics*, p. 3.

17. V. I. Lenin, "Speech at the Third Komsomol Congress, 2 October 1920," in *Collected Works*, vol. 31 (Moscow: Foreign Languages Publishing House, 1960–1963).

18. Karl Popper, *The Poverty of Historicism* (London: Routledge, 1957), p. 54. Popper writes, "The morally good is that which is ahead of its time in conforming to such standards of conduct as will be adopted in the period to come."

19. See John M. McMurtry, *The Structure of Marx's World-View* (Princeton: Princeton University Press, 1978), p. 235.

20. Eugene Kamenka, *Marxism and Ethics* (New York: St. Martin's, 1969), p. 57.

21. Lukes, *Marxism and Morality*, p. 148.

22. For a detailed study, see D. Joravsky, *Soviet Marxism and Natural Science, 1917–1932* (New York: Columbia University Press, 1961).

23. Ibid., pp. 54–70.

24. See Paul R. Gregory and Robert C. Stuart, *Soviet Economic Structure and Performance* (New York: Harper and Row, 1974), p. 97.

25. Quoted by Loren Graham, *Science and Philosophy in the Soviet Union* (New York: Alfred Knopf, 1972), p. 210. Although the point is disputed, Graham argues that Lysenkoism was not necessitated by the premises of historical materialism.

26. Herbert Marcuse, *Soviet Marxism* (New York: Cambridge University Press, 1958), p. 87.

27. See G. Cohen, "Karl Marx and the Withering Away of Social Science," *Philosophy and Public Affairs* 1, no. 2 (Winter 1972).

28. Kamenka, *Marxism and Ethics*, pp. 60–61.

29. Wayne R. LaFave, ed., *Law in Soviet Society* (Urbana: University of Illinois Press, 1965), p. 13.

30. M. A. Dynnik et al., eds., *Istoriia filosofii*, vol. 6, bk. 1, p. 478; cited in Kamenka, *Marxism and Ethics*, p. 61.

31. A. L. Subbotin, "Printsipy gnoseologii Lokka," *Voprosy filosofii*, no. 2 (1955), 199.

32. A. F. Shishkin, "Nekotorye voprosy teorii kommunisticheskoi morali," *Voprosy filosofii*, no. 4 (1956), 3.

33. See De George, *Soviet Ethics*, chap. 5.

34. See James P. Scanlan, *Marxism in the USSR: A Critical Survey of Current Soviet Thought* (Ithaca: Cornell University Press, 1985), pp. 270–271.

35. For an example that received much attention from Western analysts, see V. P. Tugarinov, *O tsennosti zhizni i kultury* (Leningrad, 1960), p. 125.
36. Kamenka, *Marxism and Ethics*, p. 61.
37. For a useful history of Soviet discussions of value theory and the fact-value distinction in the 1960s, see Grier, *Marxist Ethical Theory*, chap. 5.
38. Cited in ibid., p. 114.
39. O. M. Bakuradze, "Istina i tsennost'," *Voprosy filosofii*, no. 7 (1966), 46, 47, 48.
40. A. F. Shishkin and K. A. Shvartsman, *XX vek i moral'nye tsennosti chelovechestva* (Moscow: Mysl', 1968), p. 14.
41. Alexander Vucinich, *Empire of Knowledge: The Academy of Sciences of the USSR (1917–1970)* (Berkeley: University of California Press, 1984), p. 346.
42. Bakuradze, "Istina i tsennost'," p. 48.
43. M. C. Strogovich, *Problemy sudebnoi etiki* (Moscow: Nauka, 1974), p. 22.
44. Shishkin and Shvartsman, *XX vek i moral'nye tsennosti chelovechestva*, p. 19.
45. O. G. Drobnitskii, "Diskussiia po problemam etiki mezhdu sovetskimi i britanskimi filosofami," *Voprosy filosofii*, no. 2 (1969), 145.
46. Scanlan argues that this argument is circular; see Scanlan, *Marxism in the USSR*, p. 291.
47. S. Popov, "Aksiologiia i ee burzhuaznye istolkovateli," *Kommunist*, no. 11 (1969), 97.
48. Alan Montefiore, "Fact, Value, and Ideology," in Williams and Montefiore, *British Analytical Philosophy*, p. 197.
49. George Breslauer, "On the Adaptability of Soviet Welfare-State Authoritarianism," in Karl W. Ryavec, ed., *Soviet Society and the Communist Party* (Amherst: University of Massachusetts Press, 1978), p. 9.
50. See P. Josephson, "Science and Ideology in the Soviet Union: the Transformation of Science into a Direct Productive Force," *Soviet Union/Union Sovietique* 8, pt. 2 (1981), 159–185.
51. Zhores Medvedev, *The Rise and Fall of T. D. Lysenko* (New York: Columbia University Press, 1968), p. 113.
52. See Loren Graham, "The Development of Science Policy in the Soviet Union," in T. Dixon Long and Christopher Wright, eds., *Science Policies of Industrialized Nations* (New York: Praeger, 1975), p. 37.
53. See *Voprosy filosofii*, nos. 1, 5, and 6 (1962).
54. Scanlan, *Marxism in the USSR*, pp. 26–32.
55. I. T. Frolov, *Genetika i dialektika* (Moscow: Nauka, 1968), p. 253. This position had been put forth earlier by other Soviet theorists; see Loren Graham, "Biomedicine and the Politics of Science in the USSR," *Soviet Union/Union Sovietique* 8, pt. 2 (1981), 151.
56. Later Soviet discussions of complementarity (*dopolnitel'nost'*) in physics would emphasize the subjective elements involved even in scientific research.
57. Stephen Fortescue, "Research Institute Party Organizations and the Right of Control," *Soviet Studies*, no. 2 (April 1983), 180.
58. Vucinich, *Empire of Knowledge*, p. 359.

59. Loren Graham, "The Development of Science Policy in the Soviet Union," in Long and Wright, *Science Policies of Industrialized Nations*, esp. p. 37.

60. Jerome M. Gilson, *The Soviet Image of Utopia* (Baltimore: Johns Hopkins University Press, 1975), epilogue.

61. V. N. Sherdakov, "Etika i normativnost'," *Voprosy filosofii*, no. 2 (1982), 86.

62. R. Kosolapov, "Vklad XXIV, XXV i XXVI s"ezdov KPSS v razrabotke teoreticheskikh i politicheskikh problem razvitogo sotsializma i perekhoda k kommunizmu," *Kommunist*, no. 5 (1982), 54–67.

63. V. Kudriavstev, "Osnovnoi zakon strany razvitogo sotsializma," *Kommunist*, no. 4 (1982), 58.

64. Strogovich, *Problemy sudebnoi etiki*, p. 28.

65. "Nauchno-tekhnicheskaia revoliutsiia i ee sotsial'nye aspekty," *Kommunist*, no. 2 (1982).

66. Grier, *Marxist Ethical Theory*, p. 150.

67. R. Petropavlovskii, "Po povodu odnoi knigi," *Kommunist*, no. 8 (1983); discussed by Scanlan in *Marxism in the USSR*, pp. 277–278.

68. For a review of the department's work during its first ten years, see S. F. Anisimov and B. O. Nikolaichev, "Kafedra Marksistsko-Leninskoi etiki: Desiat' let prepodavanii, nauchnoi i obshchestvenno—politicheskoi deiatel'nosti," *Vestnik moskovskogo universiteta*, no. 6 (1979).

69. M. Suslov, "Vstupitel'noe slovo," *Kommunist*, no. 7 (1981), 16.

70. See the series of articles on "Etika i moralevedenie," *Voprosy filosofii*, no. 2 (1982). Elements of the debate had been articulated ten years earlier in *Voprosy filosofii*, no. 4 (1972), 146–149.

71. V. T. Efimov, "Etika i moralevedenie: Polemicheskie zametki," *Voprosy filosofii*, no. 2 (1982), 67.

72. V. Zh. Kelle, "O sotsial'nikh funktsiakh obshchestvennykh nauk pri sotsializme," in *Sotsial'nye problemy nauki* (Moscow, 1974), p. 155.

73. L. M. Arkhangel'skii, "Etika ili moralevedenie," *Voprosy filosofii*, no. 2 (1982), 78.

74. See Graham, *Between Science and Values*, p. 28.

75. L. M. Kosareva and M. K. Petrov, "Formirovanie ideala tsennostno-neutral'nogo nauchnogo znaniia," *Voprosy istorii estestvoznaniia i tekhnika*, no. 1 (1987), 62–72. Also see L. M. Kosareva, "Tsennostnye orientatsii i razvitie nauchnogo znaniia," *Voprosy filosofii*, no. 8 (1987), 44–54.

76. A. A. Guseinov, "Etika—nauka o morali," *Voprosy filosofii*, no. 2 (1982), 82.

77. V. N. Sherdakov, "Etika i normativnost'," *Voprosy filosofii*, no. 2 (1982), 86.

78. A. I. Titarenko, "Predmet etiki: Osnovaniia obsuzhdeniia i perspektivy issledovaniia," *Voprosy filosofii*, no. 2 (1982), 91.

79. See "Razvitie etiki: Panorama idei," *Voprosy filosofii*, no. 6 (1984), 109–132.

80. I. I. Kosarev, "Evoliutsiia meditsinskogo eticheskogo kodeksa v SSSR i za rubezhom (printsipy i ikh realizatsiia)," *Vestnik akademii meditsinskikh nauk SSSR*, no. 4 (1980), 47–52.

81. Fortescue, "Research Institute Party Organizations," p. 181. See his more recent analysis in Stephen Fortescue, *The Communist Party and Soviet Science* (London: Macmillan, 1986), chap. 4.

82. Robert F. Miller, "The Role of the Communist Party in Soviet Research and Development," *Soviet Studies* 37, no. 1 (January 1985), 57.

83. I. T. Frolov and B. G. Iudin, "Etika nauki: Sfera issledovaniia, problemy i diskussii," *Voprosy filosofii*, no. 2 (1985), 74.

84. See Graham, "Biomedicine and the Politics of Science in the USSR," pp. 147–158, esp. p. 157.

85. See *Voprosy filosofii*, nos. 6 and 8 (1973).

86. *Voprosy istorii estestvoznaniia i tekhniki*, no. 4 (1980); ibid., no. 2 (1982).

87. M. S. Gorbachev, *Perestroika* (New York: Harper and Row, 1987), pp. 145–147.

88. Ivan Frolov, *Man-Science-Humanism: A New Synthesis* (Moscow: Progress, 1986), p. 124.

89. *Programma kommunisticheskoi partii sovetskogo soiuza* (Moscow: Politizdat, 1987), p. 10.

90. P. N. Fedoseev, "Sotsialisticheskii gumanizm: Aktual'nye problemy teorii i praktiki," *Voprosy filosofii*, no. 2 (1988), 3–23.

91. See I. T. Frolov, "Vystuplenie akademika I. T. Frolova po tseremonii ofitsial'nogo zakrytiia kongressa," *Voprosy filosofii*, no. 3 (1988), 25. The bureau of the committee includes a range of natural scientists such as the physicist E. Velikhov, the astronomer V. A. Ambartsumian, and social scientists, including D. M. Gvishiani.

92. N. Lobkowicz, "Is the Soviet Notion of Practice Marxian?" *Studies in Soviet Thought* 6, no. 1 (March 1966), 25–35.

93. M. S. Gorbachev, "Report to January 1987 Central Committee Plenum," *Pravda*, 28 January 1987; trans. in *CDSP* 39, no. 4, 1. The term "integral" (*tsel'nyi*) socialism, used in the draft version of the party program, was deleted from the final version.

94. D. P. Gorskii, "O kriteriiakh istiny (k dialektike teoreticheskogo znaniia i obshchestvennoi praktiki)," *Voprosy filosofii*, no. 2 (1988), 28–39.

95. G. A. Arbatov, "Rech' tovarishcha G. A. Arbatova," *Pravda*, 30 June 1988, 6–7.

96. R. Z. Sagdeev, "Science and Perestroika: A Long Way to Go," *Issues in Science and Technology* (Summer 1988), 49.

97. "Dumat' i postupat' po-novomu" (interview with Academician V. I. Goldanskii), *Ogonyok*, 7 November 1987, 16, 25.

98. As one Soviet philosopher wrote, "Statements about social conditions which are not based on a Marxist analysis are of no relevance, even if the statistical data on which they are based is [*sic*] consistent from a formal point of view." See F. Rapp, "Contemporary Soviet Theory of Social Law," in Ervin Laszlo, ed., *Philosophy in the Soviet Union* (Dordrecht: D. Reidel, 1967).

99. M. I. Mikeshin, "Metodologiia nauki: Problemy i resheniia," *Voprosy filosofii*, no. 3 (1988), 121.

100. "Dumat': postupat' po-novomu," pp. 16, 25.

101. Anatolii Rybakov, *Deti arbata*, in *Druzhba narodov*, no. 6 (1987), 49, 86.

102. *Programma*, pp. 53–54.

103. Iakov Mil'ner-Irinin, "Poniatie o prirode cheloveka i ego mesto v sisteme nauki etiki," *Voprosy filosofii,* no. 5 (1987), 71–82.

104. See M. S. Kagan, "O dukhovnom," *Voprosy filosofii,* no. 9 (1985). Kagan divides the spiritual into cognitive, valuative, and projectional spheres. See also V. Brozhik, *Marksistskaia teoriia otsenki* (Moscow, 1982).

105. O. M. Bakuradze, *Priroda moral'nogo suzhdeniia* (Tbilisi: Izd. Tbilisi University, 1982).

106. Iu. D. Granin, "O gnoseologicheskom soderzhanii poniatiia 'otsenka'," *Voprosy filosofii,* no. 6 (1987), 59.

107. *Voprosy filosofii,* nos. 9 and 10 (1987); ibid., no. 3 (1988).

108. Ibid., no. 8 (1987), 67.

109. Gorbachev, Party Conference, p. 4.

110. Aleksandr N. Iakovlev, "Briefing on the 19th Party Conference, 28 June 1988," trans. in *FBIS,* 29 June 1988, 38.

111. Gorbachev, *Perestroika,* p. 222.

112. T. Zaslavskaia, "Chelovecheskii faktor i sotsial'naia spravedlivost'," *Sovetskaia kul'tura,* 23 January 1986, 3.

113. Personal interview, September 1987.

114. Frolov, *Man-Science-Humanism,* pp. 329–330.

115. See Vucinich, *Empire of Knowledge,* pp. 348–353.

116. O. S. Soina, "L. Tolstoi o smysle zhizni: Eticheskie iskaniia i sovremennost'," *Voprosy filosofii,* no. 11 (1985), 124–132.

117. Frolov, *Man-Science-Humanism,* p. 333.

118. Gorbachev, Party Conference, p. 4.

119. "Filosofiia i zhizn'," *Voprosy filosofii,* no. 2 (1988), 111.

120. Iakovlev, "Briefing," p. 39.

121. G. I. Marchuk, "Speech to 19th All-Union CPSU Conference," *Pravda,* 1 July 1988, 2.

122. See *Pravda,* 11 February 1988, 2.

123. *Vedomosti verkhovnogo soveta SSSR,* no. 2 (13 January 1988), 22–27.

124. S. N. Fyodorov, "Speech to 19th CPSU Conference," trans. in *FBIS,* 5 July 1988, 51.

125. *Programma,* p. 58.

126. Marchuk, "Speech," p. 3.

9. The Changing Image of Science and Technology in Soviet Literature

1. H. G. Wells, *Russia in the Shadows* (1921; reprinted, Westport, Conn.: Hyperion Press, 1973), p. 17; and L. Trotsky, *Literature and Revolution* (Ann Arbor: University of Michigan Press, 1968), pp. 252–253.

2. V. I. Lenin, "Odna iz velikikh pobed tekhniki," *Pravda,* 21 April 1913.

3. V. I. Lenin, "Nasha programma" (1899); reprinted in *Polnoe sobranie sochinenii* (Moscow, 1958), IV, 182.

4. Trotsky, *Literature and Revolution,* p. 252.

5. Ibid., pp. 249, 251.

6. Ibid., pp. 254, 256.

7. Russian translations of the More and Campanella utopias had been published in 1905 and 1907 respectively.

8. See N. A. Trifonov, "Vstupitel'naia stat'ia," in A. V. Lunacharskii, "Nezakonchennaia p'esa 'Solntse' (3-ia chast' dramaticheskoi trilogii 'Foma Kampanella')," *Iz istorii sovetskoi literatury 20–30 godov, Literaturnoe nasledstvo,* vol. 93 (Moscow: Nauka, 1983), pp. 125–126.

9. K. Marks and F. Engels, *Sochineniia,* 2nd ed. (Moscow, 1928), II, 142.

10. Thomas Campanella, *City of the Sun: A Poetical Dialogue Between a Grandmaster of the Knights Hospitallers and a Genoese Sea-Captain, His Guest,* in *The World's Greatest Literature,* vol. 32, rev. ed., *Ideal Commonwealths* (New York: P. F. Collier, 1901), p. 151.

11. V. I. Lenin, "Iz doklada Vserossiiskogo tsentral'nogo ispolnitel'nogo komiteta i Soveta narodnykh komissarov o vneshnei i vnutrennei politike 22 dekabria," *Polnoe sobranie sochinenii,* XLII, 145–157.

12. Trotsky also talks of this in *Literature and Revolution;* see p. 249.

13. Campanella, *City of the Sun,* p. 158.

14. See Katerina Clark, "Utopian Anthropology as a Context for Stalinist Literature," in Robert C. Tucker, ed., *Stalinism: Essays in Historical Interpretation* (New York: W. W. Norton, 1977), pp. 185–189.

15. Since it generally takes some time for literary works to be written and to go through the process of publication, literature that can be identified with a given period in Soviet history (such as War Communism or the First Five-Year Plan) often appeared after the period had ended.

16. Vladimir Kirillov, "My," *Literaturnyi al'manakh: Zhurnal Proletkul'ta* (1918), 10.

17. See, for example, N. Kliuev, "My rzhannye, tolokonnye . . . Vy—chugunnye, betonnye" (a poem of 1918 addressed to Kirillov), and "Tvoe prozvishche—russkii gorod" (a poem of about the same time addressed to Kirillov and Gastev), in his *Sochineniia,* ed. G. P. Struve and B. A. Filippov, vol. 1 (Munich: Neimanis, 1969), pp. 402–404.

18. Robert A. Maguire, *Red Virgin Soil: Soviet Literature in the 1920s* (Princeton: Princeton University Press, 1968), p. 124.

19. Chaianov published part 1, *Puteshestvie moego brata Alekseia v stranu krest'ianskoi utopii,* in 1920 under the pseudonym, Iv. Kremnev (Moscow: Gos. izd., 1920). Part 2 was never passed by the censor.

20. Compare Iv. Kremnev, *Puteshestvie,* p. 18; and A. Solzhenitsyn, *Pis'mo vozhdiam sovetskogo soiuza* (Paris: YMCA Press, 1974), pp. 29–33.

21. For example, Nikolai Kliuev, "Krasnyi kon'," *Griadushchee,* nos. 5/6 (1919), 14–15; and "Ognennaia gramota," ibid., nos. 7/8 (1919), 17.

22. When V. Kaverin wrote "The Eleventh Axiom" ("Odinnadtsataia aksioma," 1921), for instance, he tried to apply Lobachevskii's theories to plot structure.

23. This was a favorite distinction of the writers E. Zamiatin and V. Khlebnikov, but it was also used by champions of "left art"; see, for example, N. P. [Punin], "Nashi zadachi i profsoiuzy khudozhnikov," *Iskusstvo kommuny,* no. 3 (22 December 1918), 2.

24. "On Literature, Revolution, Entropy, and Other Matters," in Yevgeny Zamia-

tin, *A Soviet Heretic: Essays,* trans. Mirra Ginzburg (Chicago: University of Chicago Press, 1975), pp. 107–108.

25. For example, A. Piotrovskii, "Diktatura," *Zhizn' iskusstva,* nos. 584/585 (1920), 2.

26. My observations here do not apply to science fiction, as is clear in Richard Stites's contribution to this book (see Chapter 10).

27. For example, F. Gladkov, *Tsement, Krasnaia nov',* no. 3 (1925), 56. Gladkov was a member of the proletarian literary organization the Smithy, which had broken away from Proletcult.

28. Gladkov, ibid., no. 2 (1925), 107–109.

29. Gladkov, ibid., no. 1 (1925), 105.

30. Gladkov, ibid., no. 5 (1925), 73.

31. V. Kataev, *Vremia, vpered!, Krasnaia nov',* no. 1 (1932), 16.

32. For example, V. Il'enkov, *Vedushchaia os', Oktiabr',* nos. 11–12 (1931), 59.

33. For example, F. Gladkov, *Energiia, Novyi mir,* no. 1 (1932), 8; ibid., no. 2, p. 112; and ibid., nos. 7–8, p. 105.

34. Consider the fates of Kliuev, Pilniak, and the many other antiurbanist writers who suffered various forms of repression in these years.

35. V. Kataev's *Time, Forward!* (New York: Farrar and Rinehart, 1933) represents a happy exception to this rule, although it also slows down in the sections presenting Soviet-Western philosophical dialogues.

36. Science fiction, which had been prominent in the early to mid-twenties, faded out almost completely in the thirties. As early as 1934, the writer K. Fedin claimed in a speech to the First Writers' Congress that Soviet science fiction had died and been laid to rest; *Pervyj s"ezd pisatelei: Stenograficheskii otchet* (Moscow, 1934). Actually, this was something of an exaggeration, although science fiction of the thirties and forties tended to be concerned with present or future construction plans for the frozen Arctic (a place where the struggle with nature is played out most dramatically), rather than with life in some fantastic *other* time or place. See Rafail Nudel'man, "Sovetskaia nauchnaia fantastika i ideal sovetskogo obshchestva," *Canadian-American Slavic Studies* 28, nos. 1–2 (Spring–Winter 1984), 12–14.

37. For more on this see Katerina Clark, *The Soviet Novel: History as Ritual* (Chicago: University of Chicago Press, 1985), esp. chaps. 4 and 5.

38. An important exception is A. Malyshkin's *Liudi iz zakholust'ia* (People of the backwoods) (Moscow: Sovetskii pisatel', 1938), about the building of Magnitogorsk; this work, however, largely belongs to the period of the First Five-Year Plan, when it was begun (it took Malyshkin a long time to write it because he was preoccupied with editorial duties on *Novyi mir*).

39. So named for A. A. Zhdanov, whose speech attacking the journals *Zvezda* and *Leningrad* in 1946 signaled a return to socialist realist conformism and an end to hopes that victory in the Second World War might bring a more liberal climate.

40. For example, S. Babaevskii, *Kavaler zolotoi zvezdy* (Moscow, 1952); and his *Svet nad zemlei* (Moscow, 1951).

41. V. Azhaev, *Daleko ot Moskvy, Novyi mir,* no. 7 (1948), 11, 20.

42. V. Azhaev, *Daleko ot Moskvy, Novyi mir,* no. 8 (1948), 29; S. Babaevskii, *Kavaler zolotoi zvezdy,* pt. 2, *Oktiabr',* no. 4 (1948), 21–22; and ibid., no. 5 (1948), 46.

43. See, for example, Pavel Nilin, *Znakomstvo s Tishkovym, Znamia,* no. 9 (1954), 67; S. Antonov, *Delo bylo v Pen'kove, Oktiabr',* no. 6 (1956); and G. Nikolaeva, *Bitva v puti, Oktiabr',* nos. 3–7 (1957).

44. J. V. Stalin, *The Economic Problems of Socialism* (Moscow: Foreign Languages Publishing House, 1952), pp. 76–77.

45. J. V. Stalin, *Concerning Marxism and Linguistics* (London: Soviet News, 1950), p. 22.

46. J. V. Stalin, *Marksizm i voprosy iazykoznaniia* (Moscow: Gos. izd. polit. lit., 1952), p. 32.

47. V. Kaverin, *Dr. Vlasenkova,* in *Otkrytaia kniga* (Moscow: Molodaia gvardiia, 1959); and L. Leonov, *Russkii les, Znamia,* no. 11 (1953).

48. This label is, of course, imprecise in that Khrushchev did not emerge as the sole leader until 1955.

49. N. De Witt, "Table V1-A-1, Aggregate Number of Professionals and Semi-Professionals Employed in the Civilian Economy of the USSR, Selected Years, 1913–1959," *Education and Professional Employment in the USSR* (Washington, D.C.: National Science Foundation, 1961), p. 779.

50. F. Abramov joined the Party in 1945 and worked as a journalist until 1956. Ch. Aitmatov was a veterinarian. V. Dudintsev had been an economic journalist for *Pravda* and *Komsomol'skaia pravda* in the late forties. D. Granin joined the Party in 1942 and was trained as an electrical engineer. V. Nekrasov was trained as an architect and joined the Party in 1944. P. Nilin had worked as an economic journalist since the thirties, when he wrote for *Nashi dostizheniia;* he joined the Party in 1944. V. Ovechkin joined the Party in 1929 and worked for many years as an economic journalist, primarily on agricultural topics and in the provincial Party press. V. Tendriakov joined the Party in 1948 and worked in the provincial Komsomol hierarchy. G. Troepolskii was trained, and worked, as an agronomist. S. Zalygin was trained as a hydrologist.

51. Ovechkin's first work was "Savelev," published in 1927. See his collected works, *Izbrannoe* (Moscow, 1955). In 1962 Ovechkin submitted a memorandum to the Central Committee about reforming collective farms on the Yugoslav model. As a result, he was confined to a mental hospital where he tried to commit suicide. He died in 1968. D. Pospielovsky, "The 'Link' System in Soviet Agriculture," *Soviet Studies* 21 (1970), 415.

52. The others were "Na perednem krae" (1953), "V tom zhe raione" (1954), "Svoimi rukami" (1954), "Trudnaia vesna" (1956), and also a separate sketch, "Ob initsiative i talantakh" (1956).

53. "Na perednem krae," *Pravda,* 20 and 23 July 1953; "V tom zhe raione," *Pravda,* 28 February and 1 March 1954; supporting article, *Pravda,* 8 March 1954; "Svoimi rukami," *Pravda,* 27 and 30 August 1954; supporting article, *Pravda,* 12 September 1954; and "Trudnaia vesna," *Pravda,* 10 and 15 August 1956.

54. These include his recommendation of new plowing techniques ("Na perednem

krae"), the decentralization of the bureaucracy ("Svoimi rukami"), the granting of more power to MTS (Machine Tractor Stations) officials ("Na perednem krae")—and even Khrushchev's panacea of growing more maize ("Trudnaia vesna").

55. V. Tendriakov, "Sredi lesov," *Nash sovremennik,* no. 2 (1953); and G. Troepol'skii, "Iz zapisok agronoma," *Novyi mir,* no. 3 (1953).

56. A. Nove, "Some Notes on the 1953 Budget and the Peasants," *Soviet Studies* 5, no. 3 (1953–54), 228–229.

57. See, for example, Vladimir Kavtorin, *Gorod bez nazvaniia, Zvezda,* nos. 10–12 (1983). Fiction of this period also questioned the notion that man should become machinelike, or that humans should be to society as parts to the machine. See, for example, A. Iashin, "Rychagi," *Literaturnaia Moskva* 2 (1956).

58. V. Zhabinskii, *Prosvety: Zametki o sovetskoi literature 1956–57* (Munich: Institute for the Study of the USSR, 1958), p. 84.

59. See, for example, a signal statement of this policy from the time of the Stalinist "Great Breakthrough" (1929) in the "Resolution of the Second All-Union Conference of Marxist-Leninist Research Institutions," April 1929, in R. V. Daniels, *A Documentary History of Communism* (New York: Vintage Books, 1962), II, 7–8. Compare *Istoriia Vsesoiuznoi kommunisticheskoi partii (Bol'shevikov): Kratkii kurs* (Moscow: Ogiz, 1946), p. 174.

60. For example, V. Nekrasov, *V rodnom gorode, Novyi mir,* no. 11 (1954), 16.

61. For example, D. Granin, *Iskateli* (Leningrad, 1955).

62. The same topos was used, however, by the conservative writer V. Kochetov in arguing for the old view that the interests of science should be completely subordinated to those of industry. See his *Molodost' s nami, Zvezda,* no. 9 (1954), 33.

63. For example, D. Granin, "Variant vtoroi," *Zvezda* (1949).

64. V. Dudintsev, *Ne khlebom edinym, Novyi mir,* no. 8 (1956), 87; and ibid., no. 10 (1956), 92.

65. See, for example, V. Kaverin, *Otkrytaia kniga, Novyi mir,* no. 10 (1949), 117, 118, 165–166.

66. Compare, in the earlier sections of Kaverin, *Doktor Vlasenkova,* pp. 360, 368, 370–372, and 376 with, in the later sections, pp. 425, 427, 460, 485, and 495.

67. Ibid., p. 482; emphasis added.

68. Ibid., p. 485.

69. V. Kaverin, *Poiski i nadezhdy, Literaturnaia Moskva,* no. 2 (1956), 85.

70. For example, I. Grekova, *Porogi, Oktiabr',* nos. 10–11 (1984); and Andrei Molchanov, "Novyi god v oktiabre," *Iunost',* no. 2 (1983).

71. Dudintsev, *Ne khlebom edinym, Novyi mir,* no. 8 (1956), 93, 98; no. 9 (1956), 96; and no. 10 (1956), 69.

72. See note 36.

73. I. Efremov, "Ot avtora," *Tumannost' Andromedy* (Moscow: Molodaia gvardiia, 1958), pp. 3–4.

74. An exception would be M. Bremener, *Pust' ne soshlos' s otvetom, Iunost',* no. 10 (1956).

75. Trotsky, *Literature and Revolution,* p. 252.

76. Leonov, *Russkii les,* p. 52. Although technically this book was published after Stalin's death (in October–December 1953), it is clear that much of it was written before Stalin died (Leonov dates commencement of the book to January 1950). Indeed, the book is at least in part dedicated to praising Stalin's afforestation scheme. Therefore, I see it as an indicator of a change of attitudes to be felt even under Stalin.

77. For example, Fedor Abramov, who was originally identified with the Ovechkin school, was actually anomalous there because he was a professional literary academic and served as head of the department of Soviet literature at Leningrad University from 1956 to 1960, after which he abandoned his academic career to write fulltime. Vasilii Belov trained at the Gorky Literary Institute from 1961 to 1963. Valentin Rasputin graduated from the Historico-Philological Faculty of Irkutsk University. Vasilii Shukshin trained at the Gorky Cinematic Institute in Moscow, starting in 1954. Vladimir Soloukhin attended the Gorky Literary Institute from 1945 to 1951.

78. V. Rasputin, *Proshchanie s Materoi, Nash sovremennik,* no. 10 (1976), 63.

79. Ibid., p. 43.

80. V. Rasputin, "Eto stanovitsia traditsiei (Valentin Rasputin v MGU)," *Vestnik Moskovskogo universiteta,* Seriia filologii, no. 3 (1977), 81.

81. See Thane Gustafson, "Environmental Disputes in the USSR," in D. Nelkin, ed., *Controversy: The Politics of Technological Decisions* (Beverly Hills: Sage Publications, 1979).

82. For example, in 1979 and 1980 *Literaturnaia gazeta* ran a long series of articles under the pointed general heading "Village Prose: High Roads and Back Roads," implying that village prose was taking the country along back roads not only literally but also metaphorically.

83. Postanovlenie TsK KPSS, "O tvorcheskikh sviazakh literaturno-khudozhestvennykh zhurnalov s praktikoi kommunisticheskogo stroitel'stva," *Pravda,* 30 July 1982, 1.

84. See, for example, the bitter remarks made by science fiction writers at a *Literaturnaia gazeta* roundtable discussion of February 1980, "Nauchnaia fantastika—zhanr v krizise?" *Literaturnaia gazeta,* 27 February 1980, 4.

85. For example, O. S. Miauer, comp., *Derevenskii dnevnik: Sel'skie ocherki 50-60-kh gg.* (Moscow: Sovetskii pisatel', 1984); and *Vospominaniia o V. Ovechkine* (Moscow: Sovetskii pisatel', 1982).

86. Igor' Vinogradov, "Dvadtsat' let spustia," *Literaturnoe obozrenie,* no. 2 (1985), 61.

87. For example, Ivan Vasil'ev, *Dopusk na initsiativu: Derevenskii ocherk* (Moscow: Sovremennik, 1983); Vladimir Leonov, *Grachi moi grachiki* (Moscow: Moskovskii rabochii, 1982); Leonid Ivanov, *Berezovskie orientiry* (Moscow: Sovetskii pisatel', 1983); and Iosif Gerasimov, *Probel v kaledare, Novyi mir,* no. 3 (1983).

88. G. Markov, "Sovetskaia literatura v bor'be za kommunizm," p. 2.

89. Chingiz Aitmatov, *I dol'she veka dlitsia den', Novyi mir,* no. 11 (1980), 53–57.

90. For example, Anatolii Kim, *Lotos, Druzhba narodov,* no. 10 (1980); Vladimir Krupin, *Zhivaia voda* (Moscow: Sovetskii pisatel', 1980); Ivan Makanin, *Pred-*

techa (Moscow: Sovetskii pisatel', 1983); and V. Orlov, *Al'tist Danilov* (Moscow: Sovetskii pisatel', 1981).

91. Anatolii Kim, *Nefritovyi poias: Povesti* (Moscow: Molodaia gvardiia, 1981), p. 240.

92. For example, in 1983 S. Semenova's book on Fyodorov was withdrawn while actually in press.

93. For example, *Airport* appeared in *Inostrannaia literatura* in Russian translation in 1971.

94. A. Bocharov, "Ekzamenuet zhizn'," *Novyi mir,* no. 8 (1982), 228–229.

95. See, for example, Vinogradov, "Dvadtsat' let spustia," p. 60. Other literature making a similar point includes Gennadii Nikolaev, *Gorod bez nazvaniia, Zvezda,* nos. 10–11 (1983).

96. A sign of Bondarev's rapid rise to prominence is the fact that three new books about him were published between 1983 and 1984, as compared with two for all his preceding twenty-two-year literary career.

97. Iu. Bondarev, *Igra, Novyi mir,* no. 1 (1985), 70–71.

98. Ibid., pp. 72, 64.

99. For example, ibid., p. 64.

100. Ibid., p. 72.

101. See, for example, V. Gubarev's play of 1986, *The Sarcophagus,* which has been produced both in the Soviet Union and in the West, and his sketches written as science correspondent for *Pravda,* collected in *Zarevo nad Pripiatiu* (Moscow: Molodaia gvardiia, 1987).

102. See, for example, "Pisatel'," "Vremia, vpered!" *Literaturnaia gazeta* (July 1985), which presents a revised list of the classics of socialist realism as models for literature, ignoring those that deal with revolutionary struggle, and fore-grounding those that deal with modernizing industry and agriculture, many of them from the First Five-Year Plan: Gorky's collection of sketches from the plan years, *Around the Land of the Soviets;* Kataev's *Time, Forward!* (1932); I. Ehrenburg's *On the Second Day* (1934); M. Sholokhov's *Virgin Soil Upturned*; A. Malyshkin's *People from the Backwoods* (1938); Y. Krymov's *Tanker Derbent* (1938); V. Azhaev's *Far Away from Moscow* (1948); Ovechkin's *District Routine* (1952–1956); G. Nikolaeva's *Battle en Route* (1957); and even M. Shaginian's *Hydrocentral* (1934), about building a hydroelectric station during the First Five-Year Plan, which had long since faded from this sort of prominence.

103. See, for example, in Ch. Aitmatov's *Executioner's Block (Plakha)* (Moscow: Molodaia gvardiia, 1987), the slaughter of antelope in a paramilitary raid with helicopters and machine guns, initiated merely in order to make up for shortfall in the monthly plan for meat production.

104. See, for example, S. Zalygin's novel *After the Storm (Posle buri), Druzhba narodov,* nos. 4–5 (1980), nos. 7, 8, 9 (1985). The appearance of this novel well before Gorbachev's accession and its continuing publication afterward provides further evidence that these sorts of themes are characteristic of literature in the eighties generally, and not just in the period of glasnost. In 1986 Zalygin was made editor of the prestigious and outspoken literary journal *Novyi mir;*

hence his position in *After the Storm,* while representing one pole of intellectual debate, is far from dissident.

105. For example, Aitmatov, *The Executioner's Block;* V. Dudintsev, *The White Raiments (Belye odezhdy), Neva,* no. 1–4 (1987).

106. For example, V. Rasputin, *Fire (Pozhar), Nash sovremennik,* no. 7 (1985).

107. See, for example, D. Granin, *The Aurochs (Zubr), Novyi mir,* no. 1 (1987).

108. The case was also complicated because Timofeev-Resovskii decided not to return to the Soviet Union from Germany in the 1930s, and did so only in 1945, after the war.

109. It is stated inside the front cover of the book edition that the subject is Timofeev-Resovskii. *Zubr* (Leningrad: Sovetskii pisatel', 1987).

110. D. Granin, *Zubr, Novyi mir,* no. 1 (1987), 24–25.

111. See, for example, Ch. Aitmatov, *Plakha* (Moscow: Sovremennik, 1986), p. 598.

112. For example, V. Zinchenko, "Chelovecheskii intellekt i tekhnokraticheskoe myshlenie," *Kommunist,* no. 3 (1988), 96–104.

10. World Outlook and Inner Fears in Soviet Science Fiction

1. For background, see Richard Stites, "Hopes and Fears of Things to Come: The Anticipation of Totalitarianism in Russian Fantasy and Utopia," *Nordic Journal of Soviet and East European Studies* 3, no. 1 (1986), 1–20; Richard Stites, "Utopias in the Air and on the Ground: Futuristic Dreaming in the Russian Revolution," *Russian History* 11, nos. 2–3 (Summer-Fall 1984), 236–257; and Alexander Bogdanov, *Red Star: The First Bolshevik Utopia,* ed. Loren Graham and Richard Stites (Bloomington: Indiana University Press, 1984).

   Concerning the material in this chapter, I wish to thank Tanya Stites for research assistance, Jarri Koponen of Helsinki for use of his science fiction collection, and Mark Adams and Alexander Batchan for commentary.

2. This experimental atmosphere is the subject of my book *Revolutionary Dreams: Utopian Vision and Experimental Life in the Russian Revolution* (New York: Oxford University Press, 1989).

3. For the relationship among these visions, see Richard Stites, "Utopias of Time, Space, and Life in the Russian Revolution," *Revue des études slaves* 56, no. 1 (1984), 141–154.

4. For an introduction to the genre, see A. F. Britikov, *Russkii-sovetskii nauchno-fantasticheskii roman* (Moscow: Nauka, 1979); Darko Suvin, "The Utopian Tradition of Russian Science Fiction," *Modern Language Review* 66 (1979), 139–159; and Leonid Geller, *Vselennaia za predelom dogmy* (London: Overseas Publications, 1985). My thanks to Mark Adams for this book.

5. Ya. Okunev, *Griadushchii mir: Utopicheskii roman* (Petrograd: Priboi, 1923), p. 44.

6. See, for example, V. Itin, *Strana gonguri* (Kansk: Gos. izd., 1922); and V. D. Nikolsky, *Cherez tysiachu let: Nauchno-fantasticheskii roman* (Leningrad: Soikin, 1927).

7. See A. R. Palei, *Golfstrem* (Moscow: Ogonek, 1928); and A. R. Belyaev, *Bor'ba v efire* (Moscow: Molodaia gvardiia, 1928).

8. Evgenii Zamiatin, *My: Roman, povesti, rasskazy* (Moscow: Sovremennik, 1989).

9. Zamiatin, *My* (1920; reprinted, New York: Interlanguage Literary Associates, 1967); A. Karelin, *Rossiia v 1930 godu* (Moscow: Vserossisakaia Federatsiia Anarkhistov, 1920); and P. N. Krasnov, *Za chertopolokhom: Fantasticheskii roman* (Berlin: Diakov, 1922).

10. Ian Larri, *Strana schastlivykh* (Leningrad: Leningradskoe oblastnoe izd., 1931).

11. See Ilmari Susiluoto, *The Origins and Development of Systems Thinking in the Soviet Union* (Helsinki: Suomalainen Tiedeakatemia, 1982).

12. Darko Suvin, *Metamorphoses of Science Fiction* (New Haven: Yale University Press, 1979), p. 269; Ivan Efremov, *Tumannost' Andromedy*, trans. G. Hanna as *Andromeda: A Space-Age Tale* (Moscow: Progress, 1980). Also see G. V. Grebens, *Ivan Efremov's Theory of Soviet Science Fiction* (New York: Vantage, 1978); and Ben Hellman, "Paradise Lost: The Literary Development of Arkadii and Boris Strugatskii," *Russian History* 11, nos. 2–3 (Summer-Fall 1984), 313.

13. Suvin, "Utopian Tradition," p. 153; Vladimir Gakov, ed., *World Spring*, trans. R. de Garis (New York: Collier, 1981), p. ix; and Patrick McGuire, *Red Stars: Political Aspects of Science Fiction* (Ann Arbor: UMI Research Press, 1977), pp. xiv, 18–23.

14. McGuire, *Red Stars*, p. 23.

15. Ibid., pp. 93–94.

16. Ibid., pp. 85–92, 95–99; Peter Nicholls et al., eds., *Encyclopedia of Science Fiction* (London: Granada, 1979), p. 212; Ariadne Gromova, "At the Frontiers of the Present Age," in C. G. Bearne, ed., *Vortex: New Soviet Science Fiction* (London: Pan, 1970), pp. 9–26. *Fantastika* is an annual of about 350 pages, printing twenty to twenty-five stories and articles.

17. S. Gorschev [Gorshchev] and M. Vassilev [Vasilev], eds., *Russian Science in the Twenty-first Century* (New York: McGraw, 1959); and L. E. Etingen, *Chelovek budushchego: Oblik, struktura, forma* (Moscow: Sovetskaia Rossiia, 1976).

18. Gromova, "At the Frontiers."

19. Vladimir Savchenko, *Otkrytie sebia: Roman* (Moscow: Molodaia gvardiia, 1967), trans. A. Bouis as *Self-Discovery* (New York: Collier, 1979). It is worth noting that the Soviet film *Letter from a Dead Man* has as its premise a Soviet-caused accidental nuclear holocaust.

20. "The Ban," in Dmitrii Bilenkin, *The Uncertainty Principle*, trans. A. Bouis (New York: Collier, 1978), pp. 46–54; "Final Exam," ibid., pp. 23–28; "Strangers' Eyes," ibid., pp. 63–73; and "The Painter," ibid., pp. 132–137.

21. Pavel Amnuel, "Segodnia, zavtra, i vsegda," in his *Segodnia, zavtra, i vsegda: Sbornik nauchno-fantasticheskikh rasskazov* (Moscow: Znanie, 1984), pp. 4–63 (esp. p. 23); and Savchenko, *Self-Discovery*, pp. 36, 119–120.

22. Vladimir Savchenko, "Success Algorithm," in H. S. Jacobson, trans., *New Soviet Science Fiction* (hereafter cited as *NSSF*) (New York: Collier, 1979), pp. 140–191. Also see Anatolii Dneprov, "Formula for Immortality," ibid., pp.

113–140, a blast against "apolitical" scientists whose research is used for destructive purposes.

23. For one example of internationalism in science, see M. Emstev and E. Parnov, "The Pale Neptune Equation," in *NSSF*, pp. 192–231.

24. Mikhail Greshnev, "Sagan-Dalin," in Spartak Akhmetov, ed., *Fantastika 84: Sbornik nauchno-fantasticheskikh povestei, rasskazov, i ocherkov* (Moscow: Molodaia gvardiia, 1984), pp. 6–18. For Gastev's utopia, see Charles Rougle, "'Express': The Future According to Gastev," *Russian History* 11, nos. 2–3 (Summer-Fall 1984), 258–268.

25. For a sampling of science fiction gadgetry, see M. Emstev and E. Parnov, *World Soul*, trans. A. Bouis (New York: Collier, 1978); and the discussion of robots by E. Voiskunskii, in "'Stremitsia chelovechosti,'" his introduction to *NF: Sbornik nauchnoi fantastiki*, 23 (Moscow: Znanie, 1980), pp. 3–6. For those examples cited, see Savchenko, *Self-Discovery*, p. 211; and "Success Algorithm." Those who know Soviet urban life from the inside (or who have read the works of Iuri Trifonov) will appreciate the pathos of such inventions.

26. "What Never Was," in Bilenkin, *Uncertainty Principle*, pp. 55–62; Bilenkin, "Personality Probe," in *NSSF*, pp. 54–74; and Vladlen Bakhnov, "Cheap Sale," ibid., pp. 97–104.

27. Savchenko, *Self-Discovery*, pp. 168–169, 178–183, passim.

28. "Time Bank," in Bilenkin, *Uncertainty Principle*, pp. 159–163. For the historical antecedents in Soviet Taylorism, see "Man the Machine," in Stites, *Revolutionary Dreams;* Il'ia Varshavsky, "The Duel," in *NSSF*, pp. 9–14; and Vladlen Bakhnov, "Beware of the Ahs!" ibid., pp. 105–114.

29. Viktor Pronin, "Sila slova," in Ivan Chernykh, ed., *Fantastika 85* (Moscow: Molodaia gvardiia, 1985), pp. 57–67.

30. Emstev and Parnov, *World Soul*, passim.

31. For a sober analysis of women's status in the Soviet Union, see Gail Lapidus, *Women in Soviet Society* (Berkeley: University of California Press, 1978); and Barbara Holland, ed., *Soviet Sisterhood* (Bloomington: Indiana University Press, 1985). For women in science fiction, see McGuire, *Red Stars*, pp. 46–47; and the manuscript by Diana Greene, "Male and Female in *The Snail on the Slope* by the Strugatskys," cited with the author's permission. Some examples are Amnuel, "Segodnia" (Irina), and "Zveno i tsepy" (Alena), in *Segodnya*, pp. 125–155; and Svetlana Yagunova, "Bereginia" (Lyudmila), in Akhmetov, *Fantastika 84*, 124–149.

32. Patrocinio Schweichart, "What If . . . Science and Technology in Feminist Utopias," in Joan Rothschild, ed., *Machina ex dea: Feminist Perspectives on Technology* (New York: Pergamon, 1983), pp. 198–224.

33. See McGuire's discussion of the state in *Red Stars*, pp. 36–49. The issue of free speech is taken up in Iuri Moiseev, "Pravo na giperbolu," in Akhmetov, *Fantastika 84*, pp. 63–68; that of interference in the domestic affairs of another society by Stanislav Gagarin, "Agasfer iz Sozvezdiia Lebedia," ibid., pp. 91–105, as well as in many of the Strugatskiis' novels.

34. Emstev and Parnov, *World Soul*, p. 96; Spartak Akhmetov, "Shok," in Akhmetov, *Fantastika 84*, pp. 105–118; Liudmila Sveshnikova, "Kak perekhitrit'

bol'," ibid., pp. 182–190; and Aleksandr Potupa, "Effekt Lakimena," ibid., pp. 190–197. The Mars tale is L. B. Afanas'ev, "Puteshestvie na Mars," *Niva* (January and March 1901).

35. Emstev and Parnov, *World Soul*, p. 76; Pavel Amnuel, "Nevinoven," in *Segodnya*, pp. 155–162. Iurii Moiseev, "Angel-eko," in Chernykh, *Fantastika 85*, pp. 81–90.

36. Bulychev's *glupye* are translated "gloopies" in "Half a Life"; *Half a Life and Other Stories*, trans. H. S. Jacobson (New York: Macmillan, 1977), pp. 1–49. Il'ia Varshavsky, "Escape," in *NSSF*, pp. 28–41.

37. Savchenko, "Success Algorithm."

38. Savchenko, *Self-Discovery*, p. 121.

39. Bilenkin, "Personality Probe."

40. For one example, see Emstev and Parnov, *World Soul*, passim.

41. Pavel Amnuel, "Vyshe tuch, vyshe gor, vyshe neba," in *Segodnya*, pp. 162–191; see esp. pp. 169, 190.

42. For perspectives on village prose, see G. Hosking, *Beyond Socialist Realism* (London: Granada, 1980); and Katerina Clark's contribution to this volume (Chapter 9). Iurii Medvedev, "Liubov' k Paganini," in Akhmetov, *Fantastika 84*, pp. 18–32.

43. Aleksandr Petrin, "Vasil Fomich i EVM," in Akhmetov, *Fantastika 84*, pp. 174–176; Liudmila Ovsiannikova, "Mashina schastya," ibid., pp. 62–63; and a brief discussion of the issue in Dmitrii Bilenkin, "Proverka fantastikoi," in *NF sbornik nauchnoi fantastiki*, 22 (Moscow: Znanie, 1980), pp. 3–6.

44. Valerii Briusov, *Zemnaia os'*, 2nd ed. (Moscow: Skorpion, 1910), contains these and other stories. They have been widely translated. For a good translation of one story set in the context of other Russian science fiction works, see Leland Fetzer, ed., *Pre-Revolutionary Russian Science Fiction: An Anthology* (Ann Arbor: Ardis, 1982), pp. 229–243.

45. Il'ia Varshavsky, "The Violet," in *NSSF*, pp. 1–8; and Emstev and Parnov, *World Soul*, p. 124 and passim.

46. Hellman, "Paradise Lost," pp. 314–315; and Darko Suvin, "Criticism of the Strugatskii Brothers' Work," *Canadian-American Slavic Studies* 1, no. 2 (Summer 1972), 286–307.

47. A. and B. Strugatskii, *Gadkie lebedi: Povest'* (Frankfurt: Posev, 1972). For varying perspectives on the Strugatskiis, see the works of Britikov and Geller, already cited, and T. Chernysheva, "Utopiia i ee evoliutsiia v XX v.: Na primere tvorchesta A. i B. Strugatskikh," *Canadian-American Slavic Studies* 18, nos. 1–2 (Spring-Summer 1984), 76–84.

48. Anthony Alcott, "Glasnost and Soviet Culture," in M. Friedberg et al., eds., *Soviet Society under Gorbachev* (Armonk, N.Y.: M. E. Sharpe, 1987), pp. 101–130.

49. *NF sbornik nauchnoi fantastiki*, 32 (Moscow: Znanie, 1988), p. 204.

50. V. Lakshin, "Antiutopiia Evgeniia Zamiatina," *Znamia* 4 (April 1988), 126–130.

51. Eremei Parnov, "O divnyi novyi mif," *Sovetskaia kul'tura*, 26 July 1988, 6.

52. D. Bilenkin, "Realizm fantastiki," in *NF*, 32 (1988), 184–203.

53. *Washington Post,* 8 September 1988, C8.
54. Unless the impact of *We* is too irresistible.

11. Bridging the Two Cultures

1. Daniil Danin, "Zhazhda iasnosti," *Novyi mir,* no. 3 (March 1960), 207–226. Also see Daniil Danin, *Perekrestok* (Moscow: Sovetskii pisatel', 1974), pp. 7–84.
2. Maxim Gorky, *Sobranie sochinenii,* vol. 27 (Moscow: Gosizdat, 1956), p. 107. Also see Danin, *Perekrestok,* p. 15.
3. Matvei Bronstein, *Solnechnoe veshchestvo* (Leningrad: Detskaia literatura, 1936).
4. Mikhail Il'in, *Rasskaz o velikom plane* (Moscow: Molodaia gvardiia, 1931); first American edition, Mikhail Il'in, *Russia's Primer: The Story of the Five Year Plan* (Boston: Houghton Mifflin, 1931).
5. Lidiia Chukovskaia, *Zapiski ob Anne Akhmatovoi* (Paris: YMCA, 1984), pp. 266–267.
6. Danin, *Perekrestok,* p. 12.
7. Ibid., p. 7.
8. Ibid., pp. 42–43.
9. Daniil Danin, *Neizbezhnost' strannogo mira* (Moscow: Molodaia gvardiia, 1961).
10. T. Nemchuk [Mark Kuchment], *Novyi mir,* no. 1 (1962), 262–264.
11. Daniil Danin, "Neizbezhnost' strannogo mira," *Puti v neznaemoe* 1 (1960), 240.
12. Daniil Danin, *Bol'shaia Sovetskaia entsiklopediia,* vol. 7 (Moscow: Entsiklopediia Tret'e Izdanie, 1972), p. 142.
13. Il'ia Ehrenburg (1891–1967) was a major Soviet writer and journalist and an influential public figure.
14. Personal conversation with Danin, spring 1962.
15. Cosmopolitanism involved a political campaign against nonpatriots—for example, non-Russians—in the USSR in the 1940s and 1950s. See Alexander Nekrich and Mikhail Geller, *Utopia in Power* (New York: Summit Books, 1986), p. 430.
16. Alexander Nekrich, *Otreshis' ot strakha* (London: Overseas Publications Interchange, 1979), p. 39. Also see Joshua Rubenstein and Mark Kuchment, "Forbidden History," *New York Review of Books,* 12 May 1983, 46–49.
17. Daniil Danin, "Material i stil'," *Novyi mir,* no. 1 (1961), 167–185.
18. Ibid., p. 167.
19. Vladimir I. Lenin, *Polnoe sobranie sochinenii,* vol. 5 (Moscow: Piatoe Izdanie, 1960), p. 11.
20. *Puti v neznaemoe: Pisateli russkazyvaiut o nauke,* 20 vols. (Moscow: Sovetskii pisatel', 1960–).
21. Oleg Pisarzhevskii, *New Paths of Soviet Science* (London: Soviet News, 1954), pp. 53–55. These pages contain some astonishing statements: "Scientists . . . proved that the transformation of non-living into living substance continues uninterruptedly in our time in accordance with the laws of nature. As the latest

works of Lysenko and his colleagues have shown, a radical change in heredity leads to the transformation of one species into another."

22. Oleg Pisarzhevskii, "Pust' uchenye sporiat," *Literaturnaia gazeta,* 17 November 1964, 2–22. On the same day, "Kriterii-praktika," in *Komsomol'skaia pravda,* 17 November 1964, 3, a strong attack on Lysenko, was published by a prominent biologist, Professor V. Efroimson, and Roy Medvedev, who later became a famous dissident.

23. Pisarzhevskii's obituary appeared in *Izvestiia,* 21 November 1964, 4. The date of his death was given as 19 November 1964.

24. *Bol'shaia Sovetskaia entsiklopediia,* vol. 1 (Moscow: Tret'e Izdanie, 1970), p. 499.

25. See letter to Boris Agapov, 1 May 1936, in Gorky, *Sobranie sochinenii,* vol. 30, pp. 440–441.

26. I was Agapov's guest in this same apartment in 1963–64.

27. Still, as a member of the editorial board of *Novyi mir,* he was among those who signed the letter to Boris Pasternak rejecting his novel *Doctor Zhivago.* See *Literaturnaia gazeta,* 25 October 1958, 3–5.

28. Personal conversation, 1964.

29. Ibid.

30. Daniil Danin, "Gody svershivshikhsia mechtanii," *Puti v neznaemoe,* no. 13 (1977), 142–157.

31. Yuli Medvedev, "Otkrytie," *Puti v neznaemoe,* no. 17 (1983), 39–83.

32. Boris Volodin, "Ochen' bol'shoe uvelichenie," *Puti v neznaemoe* no. 5 (1965), 5–89.

33. Margarita Aliger, "Vechnyi prazdnik," *Puti v neznaemoe,* no. 6 (1966), 305–326.

34. Iurii Davydov, "Neistovyi," *Puti v neznaemoe,* no. 16 (1982), 322–359.

35. Natan Eidelman, "Posle 14 dekabria," *Puti v neznaemoe,* no. 14 (1978), 269–320.

36. Viktor Frenkel and Aleksandr Iarov, "Patentnyi ekspert Einshtein," *Puti v neznaemoe,* no. 17 (1983), 415–439.

37. Bruno Pontecorvo, "Zagadochnoe neutrino," *Puti v neznaemoe,* no. 3 (1963), 580–586.

38. Elevter Andronikashvili, "Vospominania o gelii II," *Puti v neznaemoe,* no. 17 (1983), 440–479.

39. Lev Artsimovich, *Novyi mir,* no. 1 (1967), 130–205.

40. Ibid., pp. 202–203.

41. In Osip Mandelstam, *O poezii* (Moscow: Academia, 1928), p. 124.

42. Mark Kuchment, "Landau," *ISIS* (March 1983), 141–143.

43. Loren Graham, *Between Science and Values* (New York: Columbia University Press, 1981), pp. 364, 414. Also see Aleksandr Khinchin, "O vospitatel'nom effekte urokov matematiki," *Matematicheskoe prosveshchenie,* no. 6 (1961).

44. Boris Meilakh, *Na rubezhe nauki i iskusstva* (Leningrad: Nauka, 1971), pp. 60–73.

45. Il'ia Ehrenburg, *Sochineniia,* vol. 3 (Moscow: Khudozhestvennaia literatura, 1967), pp. 760–766.

46. G. Ozerov, *Tupolevskaia sharaga* (Frankfurt: Possev Verlag, 1971).
47. Vladimir Shubkin, "Predeli," *Novyi mir,* no. 2 (1978), 187–217.
48. Ibid., p. 210.
49. Ibid., pp. 210–212.
50. Anatolii Chernoisov, "Biznes s navarom," *Nash sovremennik,* no. 1 (1985), 130–137. According to a short biography published by the same magazine, Anatollii Trofimovich Chernoisov was born in 1937 in Siberia, graduated from Omsk Polytechnical Institute, worked at a plant, taught in a technical college, and, as of 1985, was a member of the Union of Soviet Writers, living in Novosibirsk.
51. Daniil Danin, "Uletavl'," *Puti v neznaemoe,* no. 17 (1983), 340–365.

12. The Response to Science and Technology in the Visual Arts

1. Notably Rosalind J. Marsh, *Soviet Fiction since Stalin: Science, Politics, and Literature* (Totowa, N.J.: Barnes and Noble, 1986).
2. This drawing, captioned "A Temple of Machine Worshipers," is reproduced in René Fülöp-Miller, *Geist und Gesicht des Bolschewismus: Darstellung und Kritik des kulturellen Lebens in Sowjet-Russland* (Zurich, 1926), p. 31. Other drawings with related themes are found on pp. 17, 25, 273, 291, and 293.
3. Several impressive textile designs from the late 1920s with industrial motifs such as factories, airplanes, gears, and so on are reproduced in *Art into Production: Soviet Textiles, Fashion, and Ceramics, 1917–1935,* exhibition catalogue (Oxford: Museum of Modern Art, 1985).
4. For examples of Glazunov's work, see Vasilii Dmitrievich Zakharenko, *Il'ia Glazunov: Fotoal'bom* (Moscow, 1978).
5. For a good introductory survey of these developments, see Camilla Gray, *The Russian Experiment in Art, 1863–1922,* revised and enlarged edition by Marian Burleigh-Motley (New York: Thames and Hudson, 1986).
6. This and other aspects of my topic have been well treated by Hubertus Gassner's essay "Von der Utopie zur Wissenschaft und Zurück," in *Kunst aus der Revolution: "Kunst in die Produktion!" Sowjetische Kunst während der Phase der Kollektivierung und Industrialisierung, 1917–1933,* exhibition catalogue (Berlin: Neue Gesellschaft für Bildende Kunst, 1977), II, 51–101.
7. Quoted by Christina Lodder, in *Russian Constructivism* (New Haven: Yale University Press, 1983), p. 79. In addition to the crucial debates discussed later in this chapter, other problems raised at INKhUK "ranged over such widely disparate fields as an examination of the basic painterly elements by Kandinskii, the elements of sculpture by Korolev, the elements of poetry by Rozanov, the elements of music, and the similarity of plastic and musical experiences" (ibid.). A similar spirit dominated the UNOVIS group, which Malevich founded at the Vitebsk Art School in 1919–20. There El Lissitzky and other followers of Malevich more or less systematically applied the master's suprematism to various fields of creative activity.
8. German Karginov, *Rodchenko* (London, 1979), pp. 90–91. The mathematical model invoked in Rodchenko's statement is also reflected in the title of a semi-

nal exhibition of 1921, which marked a crucial transition for many artists on the way to embracing the applied arts. Five artists (including Rodchenko) each exhibited five works in a show entitled *5 × 5 = 25*. Art criticism, too, was not spared the temptation of achieving the precision and accuracy of mathematics. Nikolai Punin, one of the most radical and innovative writers on art in the years after the Revolution, claimed to have discovered the formula for the creative process:

$$S(Pi + Pii + Piii + \ldots P)Y = T$$

"where *S* equals the sum of the principles (*P*), *Y* equals intuition, and *T* equals artistic creation." Quoted in John E. Bowlt, ed., *Russian Art of the Avant-Garde: Theory and Criticism, 1902–1934* (New York, 1976), p. 171.

9. *The Avant-Garde in Russia, 1910–1930: New Perspectives,* exhibition catalogue (Los Angeles: Los Angeles County Museum of Art, 1980), p. 222.

10. Something of this association between avant-garde abstract art and science is reflected in the fact that several pioneering exhibitions of this art were held in scientific institutes in the USSR from the 1960s onward. Works by Popova, for example, were exhibited (with works by Kliun) at the Kurchatov Atomic Energy Institute in June 1972.

11. This was the title of Nikolai Tarabukin's theoretical treatise of 1923, which spelled out much of the ideology underlying the transformation of the artist into an engineer.

12. Bowlt, *Russian Art,* p. 206.

13. For details, see John Milner, *Vladimir Tatlin and the Russian Avant-Garde* (New Haven: Yale University Press, 1983), pp. 217ff., and Lodder, *Russian Constructivism,* chap. 7.

14. As quoted in Milner, *Vladimir Tatlin,* p. 224.

15. Tatlin, quoted ibid., p. 220.

16. A fascinating analysis of a very broad range of literary and artistic material on the theme of flight in the early decades of this century is provided by Felix Philipp Ingold, *Literatur und Aviatik: Europäische Flugdichtung, 1909–1927: Mit einem Exkurs über die Flugidee in der modernen Malerei und Architektur* (Frankfurt am Main: Birkhäuser, 1980).

17. W. Sherwin Simmons, *Kasimir Malevich's 'Black Square' and the Genesis of Suprematism, 1905–1915* (New York, 1981), pp. 109–110.

18. Ibid., pp. 173–174. Malevich reexamined some of these themes in a painting of 1914–15, *The Aviator,* also discussed ibid., pp. 172–194.

19. Later, around 1927, Malevich wrote, "The aeroplane has appeared, not on account of the socio-economic conditions being an expedient cause, but only because the sensation of speed and movement looked for an outlet and in the end took the form of an aeroplane." K. S. Malevich, *The Artist, Infinity, Suprematism: Unpublished Writings, 1913–1933,* vol. 4 (Copenhagen: Borgen, 1978), p. 145.

20. For Malevich and his pupils, see Andrei B. Nakov, "Suprematism after 1919," in *The Suprematist Straight Line: Malevich, Suetin, Chashnik, Lissitzky,* exhi-

bition catalogue (London: Annely Juda Fine Art, 1977). Rodchenko apparently made five such hanging constructions in total, all dating from 1920–21. Only one, *Oval*, has survived and is now in the collection of the Museum of Modern Art, New York. Lissitzky's inscription is quoted from my translation in Peter Nisbet, *El Lissitzky (1890–1941)*, exhibition catalogue (Cambridge, Mass.: Busch-Reisinger Museum, 1987), p. 77.

21. Robin Higham and Jacob W. Kipp, eds., *Soviet Aviation and Air Power: A Historical View* (Boulder: Westview Press, 1977), is a useful collection of essays surveying the history of aviation in the USSR.

22. Robert A. Lewis, *Science and Industrialisation in the USSR* (New York: Holmes and Meier, 1979), p. 139.

23. Ibid., p. 140.

24. Neil M. Heyman, "NEP and the Industrialization to 1928," in Higham and Kipp, *Soviet Aviation*, p. 37.

25. Karginov, *Rodchenko*, pp. 101, 110; *Alexander Rodtschenko und Warwara Stepanowa*, exhibition catalogue (Duisburg: Wilhelm-Lehmbruck-Museum, 1982), p. 54.

26. David Elliott, *Rodchenko and the Arts of Revolutionary Russia* (New York: Pantheon Books, 1979), p. 96.

27. Karginov, *Rodchenko*, p. 155.

28. *Von der Malerei zum Design: Russische konstruktivistische Kunst der Zwanziger Jahre*, exhibition catalogue (Cologne: Galerie Gmurzynska, 1981), p. 231, illustrates the collage for this famous cover. Echoes of Rodchenko's affection for the airplane can be found later both in covers for such run-of-the-mill volumes as L. Marx's 1925 book on airplane engines, *Aviatsionnye dvigateli* (see ibid., p. 220), and in his advocacy of the aerial viewpoint in his later photography, although this was part of a more general thesis about the value of the unusual standpoint, whether from above or below.

29. For full details on this group, see V. Kostin, *OST: (Obshchestvo stankovistov)* (Leningrad, 1976). In fact, an interest in scientific and technological subjects was prevalent among members of the OST group as a whole. "Almost all the future members of OST, while still students, at the time of the formation of their world view and artistic interests and while still under the influence of the most left-wing movements in art, displayed an exclusive interest in the discoveries of science and various branches of physics, in technical achievements, especially in the area of electricity, radio, telephone, communications in general, information, aerodynamics, etc." (p. 25). To take but the most apposite of many examples, A. A. Labas had a long-term interest in the theme of flight, as expressed in such later works as *In the Aeroplane Cabin* of 1928 (oil on canvas, 77.4 × 91.6 cm, State Tretiakov Gallery, Moscow; see ibid., p. 129, and E. I. Butorina, *Aleksandr Arkad'evich Labas* [Moscow, 1979], p. 107); one of the aviation series which Labas exhibited at the fourth OST exhibition of 1928, in which he attempted to convey the feeling of flight (see Kostin, *OST*, pp. 110–113); and *During the Flight* of 1935 (oil on canvas, 52 × 63 cm, location uncertain; see Butorina, *Labas*, p. 114, and Hubertus Gassner and Eckhart Gillen, eds., *Zwischen Revolutionskunst und Sozialistischem Realismus: Dokumente*

und Kommentare: Kunstdebatten in der Sowjetunion von 1917 bis 1934 [Cologne, 1979], color plate 7, p. 24).

30. Oil on canvas, 89 × 71 cm, private collection, Moscow; see Bowlt, *Russian Art*, p. 280; Kostin, *OST*, p. 91.

31. Oil on canvas, 73.4 × 55.6 cm, Museum Ludwig, Cologne, no. 1357.

32. Gassner and Gillen, *Revolutionskunst*, p. 359.

33. Oil on canvas, 121 × 95 cm, State Russian Museum, Leningrad, no. 4487; see Gassner and Gillen, *Revolutionskunst*, p. 508; *Russische und Sowjetische Kunst—Tradition und Gegenwart: Werke aus Sechs Jahrhunderten*, exhibition catalogue (Düsseldorf: Kunstverein für die Rheinlande und Westfalen and Städtische Kunsthalle, 1984), p. 164.

34. Lewis, *Science and Industrialisation*, pp. 140–141.

35. Ibid.

36. Oil on canvas, 134 × 161 cm, State Tretiakov Gallery, Moscow, inv. no. 27654; see Gassner and Gillen, *Revolutionskunst*, p. 508; Vladimir Sysoev, *Alexander Deineka: Paintings, Graphic Works, Sculptures, Mosaics, Excerpts from the Artist's Writings* (Leningrad, 1982), plate 91, no. 206, p. 292; *Russische und Sowjetische Kunst*, exhibition catalogue (1938), no. 90, p. 79. In subject, mood, and composition the work is obviously a reprise of a very famous Victorian painting, Sir John Everett Millais's *Boyhood of Raleigh* (1870), Tate Gallery, London, in which the future explorer hears inspiring tales of the vast lands overseas and the many discoveries to be made. That Deineka should look back to such a model for his painting should not surprise anyone familiar with the principles and practices of so-called socialist realism of the 1930s. Overtly moralizing, didactic, and narrative painting was the ideal.

37. Deineka's treatment of an aviation subject was no exception within his own oeuvre. In 1932 he completed an important mural for a factory canteen on the subject of civil aviation; see B. M. Nikiforov, *Aleksandr Deineka* (Moscow, 1937), p. 73. In the mid-1930s he completed several other works at Sebastopol with aviation themes, such as the 1934 watercolors *Torpedo Boat Sebastapol* (29.6 × 49 cm, State Tretiakov Gallery, Moscow, inv. no. 13598; see Sysoev, *Alexander Deineka*, no. 134, p. 2, and *Alexander Deineka: Malerei, Graphik, Plakat*, exhibition catalogue [Düsseldorf: Städtische Kunsthalle, 1982], no. 89, p. 78) and *Hydroplane;* see Nikiforov, *Deineka*, p. 95. In 1966–1968 Deineka painted a mural for the airport in Moscow.

38. K. E. Bailes, "Technology and Legitimacy: Soviet Aviation and Stalinism in the 1930s," *Technology and Culture* 17, no. 1 (January 1976), 58.

39. Ibid., p. 60.

40. See the relevant entries in my catalogue raisonné of Lissitzky's typographical work, in *El Lissitzky*, pp. 196–198.

41. Oil on canvas, 99.5 × 120 cm, State Russian Museum, Leningrad, no. 1057; see *Russische und Sowjetische Kunst*, p. 173.

42. The image of the dirigible could form a subset of its own within my general theme. It emerges rather suddenly as a preoccupation of artists around 1930–31, although Lissitzky was already working on the interior design for Konstantin Tsiolkovskii's airship in 1926; see *El Lissitzky*, p. 51, n. 86. Deineka's poster

of 1930, *Let Us Construct the Powerful Soviet Dirigible 'Klim Voroshilov'* (77 ×
108 cm; see *Alexander Deineka*, exhibition catalogue, no. 52, p. 52), showing a
massive airship floating over a row of tractors in a field, with a freight train and
factory on the horizon, firmly established the intimate link between the new
flight technology and the industrial and agricultural priorities of the collectivization era. Rodchenko also used two photographs of dirigibles for his cover design
for the magazine *Za Rubezhom*, no. 2 (August 1930); see Karginov, *Rodchenko*, p. 157. Moreover, dirigibles appeared on the cover of the journal *Bri-
gada Khudozhnikov*, one of the leading forums of the constructivist avant-garde
in these years, in 1931. Labas painted several works with dirigibles in 1930–
31, such as his *First Soviet Dirigible: Sketch* (oil on canvas, 76 × 102 cm, State
Tretiakov Gallery, Moscow; see Butorina, *Labas*, pp. 104–106). The design of
a dirigible also became a lifelong obsession for the artist Piotr Miturich. Lodder,
Russian Constructivism, chap. 7, gives a very detailed account.
43. Bailes, "Technology and Legitimacy," p. 71.
44. Otto Preston Chaney, Jr., and John T. Greenwood, "Patterns in the Soviet
Aircraft Industry," in Higham and Kipp, *Soviet Aviation*, p. 272. It should be
noted that the airplane is dropping propaganda banners by parachute (perhaps
as part of an imagined air show). The practice of using parachute jumps for
agitational purposes to arouse the enthusiasm of the general public for aviation
and the air force was common. It may indeed be reflected in the subject matter
of paintings such as Aleksandr Drevin's *Parachute Landing* of 1932 (oil on canvas, 89 × 107 cm, State Russian Museum, Leningrad, no. 8571; see *Russische
und Sowjetische Kunst*, p. 171, color plate, and Deineka's *Parachutist* of 1934
(oil on canvas, Picture Gallery of the Khirgiz SSR, Frunze, inv. no. 1688/5875;
see Nikiforov, *Deineka*, p. 91; Sysoev, *Alexander Deineka*, plate 97, no. 128,
p. 289). As one historian has noted, "The parachute troops, an idea which Tu-
khachevsky pushed vigorously, were first organised in 1931 in two small
units." Kenneth R. Whiting, "Soviet Aviation and Air Power under Stalin,
1928–1941," in Higham and Kipp, *Soviet Aviation*, p. 52. Rodchenko designed
one of his most adventurous layouts for the issue of *USSR in Construction*
devoted to the sport of parachute jumping (December 1936).
45. Gassner and Gillen, *Zwischen Revolutionskunst*, pp. 415–416. Arvatov's scath-
ing remarks are directed at the Association of Revolutionary Artists.
46. Oil on canvas, 209 × 200 cm, State Tretiakov Gallery, Moscow, inv. no.
11977; see Sysoev, *Alexander Deineka*, plate 1, no. 35, p. 284.
47. Oil on canvas, 171 × 195 cm, State Russian Museum, Leningrad, inv. no. B
988; see Kostin, *OST*, p. 56; Sysoev, *Alexander Deineka*, plate 6, no. 39, p.
285.
48. Deineka returned to this theme in his *Donets Basin* of 1947 (tempera on canvas, 179 × 197 cm, State Tretiakov Gallery, Moscow; see *Alexander Deineka*,
exhibition catalogue, no. 100, p. 88), with two women (again one active, the
other proudly standing tall) at work in the railway yards, loading coal.
49. An example from each of these three categories: A. N. Samokhvalov's *Putting
the Shot* of 1933 (oil on canvas, 124.5 × 65.8 cm, State Tretiakov Gallery,
Moscow, inv. no. 17351); Deineka's *Mother* of 1932 (oil on canvas, 120 × 159

cm, State Tretiakov Gallery, Moscow; see Sysoev, *Alexander Deineka*, plate 108); and S. V. Riangina's *Higher, Ever Higher* of 1934 (oil on canvas, 149 × 100 cm, Kiev Museum of Russian Art, inv. no. Zh. 333). This last painting, showing an ecstatic girl on an electrical pylon, climbing higher to a man above her, is intimately linked to our theme by the view of an electric train and more pylons in the mountain valley below. For a rich and differentiated discussion of the image of women in Soviet art of this period, see Christiane Bauermeister-Paetzel and Sylvia Wetzel, "Ein Huhn ist kein Vogel, ein Weib ist kein Mensch: Die neue Frau in der sowjetischen Kunst zur Zeit der Industrialisierung und Kollektivierung," in *Kunst aus der Revolution*, II, 24–48.

50. Oil on canvas, 170 × 200 cm, Ministry of Culture, Moscow, no. 77067; see *Schrecken und Hoffnung: Künstler sehen Frieden und Krieg*, exhibition catalogue (Hamburg: Hamburger Kunsthalle, 1987), no. 279 (as in State Picture Gallery of Armenian SSR).

51. Oleg Sopotinsky, comp., *Art in the Soviet Union: Painting, Sculpture, Graphic Arts* (Leningrad, 1977), no. 314, p. 447. This work and the picture *Before the Dawn* won the artist a State Prize.

52. Oil on canvas, 54 × 69 cm, State Tretiakov Gallery, Moscow; see M. F. Kiselov, *Georgii Nisskii* (Moscow, 1972), p. 21.

53. Oil on canvas, 100 × 187 cm, State Russian Museum, Leningrad, inv. no. Zh-8748; see ibid., p. 125. Such a painting also fits into the tradition of representing flight, as I have described it. The emphatic harmony of machine and landscape is similar to Deineka's 1937 flying boats, and to his 1934 painting *Over the Endless Plains* (oil on canvas, 119 × 221 cm, Ciurlionis Art Museum, Kaunas; see V. Sysoev, *Aleksandr Deineka* [Leningrad, 1971], plate 13).

54. Oil on canvas, 120 × 194 cm, State Tretiakov Gallery, Moscow.

55. Sopotinsky, *Art in the Soviet Union*, no. 170, p. 452.

56. The search for paintings in which technology is overtly threatening or dangerous is likely to be long and unrewarding. A work such as Piotr Villiams' *Auto Race* of 1930 (oil on canvas, 151 × 213 cm, State Tretiakov Gallery, Moscow; see *Russische und Sowjetische Kunst*, p. 129), showing sleek cars racing around a track with factory buildings at the center and a boy trying to calm a frightened horse in the foreground, is not so much criticizing the demonic power of modern technology to disrupt nature as it is mildly commenting on the difference between old and new "horsepower." A recent equivalent would be K. A. Mullashev's *Morning* of 1976 (oil on canvas, 122 × 150 cm, location unknown; see M. T. Kuz'mina, *Molodye sovetskie khudozhniki* [Moscow, 1979], no. 65, p. 155), showing a space capsule gently parachuting to earth over a wide open plain with a galloping herd of (frightened?) white horses.

Some more recent paintings are less sanguine. Sergei Basilev's *Catastrophe* of 1986–87 (oil on canvas, 200 × 300 cm, Ministry of Culture of USSR; see *Schrecken und Hoffnung*, exhibition catalogue, no. 254, p. 305), and Maksim Kantor's *Chernobyl* of 1987 (eleven-part mixed-media polyptych, 400 × 400 cm, Ministry of Culture of USSR; see ibid., no. 261, p. 309) both deal relatively sternly with the infamous nuclear reactor disaster of 1986.

57. Gassner and Gillen, *Revolutionskunst*, p. 361. In this sense the women domi-

nating scenes of industrial production are also a special instance of another fundamental pattern, the compositional disjunction between a heroic worker figure in the foreground, towering over a background where modern technology is being applied. Sometimes this effect is achieved by elevating the massive worker above the site, thereby also drawing on the associations of flight and transcended gravity, which, as we have seen, are integral to art's dealings with technology in the Soviet Union. An example from the 1920s would be Konstantin Vialov's *Telegraph Pole Workers* (see ibid., p. 363), depicting men at the top of masts, in a formal arrangement of wires and poles, high above the earth, where trains and railroad tracks disappear into the distance. Much the same effect is achieved by Oleg Vukulov's *Moscow* triptych of 1982–83 (oil on canvas, 180 × 105 cm, 180 × 210 cm, 180 × 105 cm, State Tretiakov Gallery, Moscow; see *Russische und Sowjetische Kunst*, p. 203). In the central panel of this painting a crane platform and cage holding four workers hovers high above the Moscow cityscape, implying a link between the heroic construction work of the team and the metropolitan scale of their achievements. For an example in which the elevated worker is a woman, see R. N. Baranov's *From the Heights of the Dockyard Crane* of 1973 (medium, dimensions, and location unknown; see Kuz'-mina, *Khudozhniki*, no. 12).

58. Marsh, *Soviet Fiction*, p. 213. As Marsh notes, this was "a subject which had preoccupied Soviet writers in the LEF disputes of the 1920s."

59. 18 June 1960, quoted by Marsh, ibid., p. 214.

60. "Much of the work undertaken by the unofficial painters and sculptors is conventional, at least in form, and repeats the methods of surrealism, expressionism, and pop art. Even though the artists' search for a more mystical, subjective interpretation of reality does, in itself, constitute an ideological and social protest, the result is often a quaint confusion of established styles"; John E. Bowlt, "Moscow: The Contemporary Art Scene," in Norton Dodge and Alison Hilton, eds., *New Art from the Soviet Union: The Known and the Unknown* (Washington, D.C.: Acropolis Books, 1977), p. 25.

61. Cf. the comment that "to some extent . . . the very choice of village life as a literary subject by so many writers in the 1970s implies some rejection of the 'NTR' and modern urban Soviet values"; Marsh, *Soviet Fiction*, p. 191. Glazunov, as I have remarked, might be a parallel to such "ruralist" writers.

62. Lev Nusberg, "What is Kineticism?" *Form* 4 (15 April 1967), p. 20.

63. *Lew Nussberg und die Gruppe Bewegung, Moskau, 1962–1977*, exhibition catalogue (Bochum: Museum Bochum, 1978), p. 7.

64. Nusberg writes that just before he founded the group in 1962, he made "mobile 3-D constructions, controlled by a program using various optical and coloured effects, noises and electronic music" and specifically credits Naum Gabo as an inspiration; ibid., p. 17. For illustrations of objects included in the group's 1965 Leningrad exhibition, such as Francisco Infante's *Kinetic Object: Crystal* (metal, nylon thread with motor and programmed control system), see Nusberg, "Kineticism," p. 21.

65. Media, dimensions, and location unknown; see Vladimir Petrovich Sysoev, *Iskusstvo molodykh khudozhnikov* (Moscow, 1986), p. 9.

Contributors

Mark B. Adams
Department of the History and Sociology of Science
University of Pennsylvania
Philadelphia, Pennsylvania

Bruce J. Allyn
Center for Science and International Affairs
Kennedy School of Government
Harvard University
Cambridge, Massachusetts

Harley Balzer
Department of Government and Russian Area Studies Program
Georgetown University
Washington, D.C.

Katerina Clark
Departments of Comparative Literature and Slavic
Languages and Literature
Yale University
New Haven, Connecticut

Richard T. De George
Department of Philosophy
University of Kansas
Lawrence, Kansas

Seymour Goodman
Department of Management and Information Science
University of Arizona
Tucson, Arizona

Loren R. Graham
Program in Science, Technology, and Society
Massachusetts Institute of Technology
Russian Research Center
Harvard University
Cambridge, Massachusetts

Paul R. Josephson
Department of Social Sciences
Sarah Lawrence College
Bronxville, New York

Mark Kuchment
Russian Research Center
Harvard University
Cambridge, Massachusetts

Peter Nisbet
Busch-Reisinger Museum
Harvard University
Cambridge, Massachusetts

S. Frederick Starr
President, Oberlin College
Oberlin, Ohio

Richard Stites
Department of History
Georgetown University
Washington, D.C.

Douglas R. Weiner
Department of History
University of Arizona
Tucson, Arizona

Index

Space culture *(cont.)*
  Sputnik, 176–180, 282–283; in journals, literature, and films, 177–178, 180–187, 283; popular perceptions, 180–190, 191; *Vostok III* and *IV,* 181, 187; opposition to, 185–187
Sputnik, 176–179, 180, 182, 184, 185, 189–190, 282–283, 304, 339
Stakhanovite movement, 156, 263
Stalin, J.: support for Lysenko, 2; communications policy, 28, 30, 49, 205; Cultural Revolution, 28; science policy, 73, 280–281; mobilization phase, 231–232, 240; "Marxism and Linguistics," 234, 238, 276, 280–281; *Economic Problems of Socialism,* 276; interest in aviation, 350, 352; afforestation scheme, 413n76
Stalinism: irrational political order, 15, 325, 334, 339; control through terror, 51–52; Bolshevization of Academy of Sciences, 101; engineering and, 154–159; gigantomania, 169–171, 190
Stanchinskii, V. V., 73–74
State Film Agency (Goskino, later Sovkino), 28–29
State Plan for Electrification (GOELRO), 169
Strogovich, M. C., 236, 242
Strugatskii brothers, 305, 308, 322; *The Snail on the Slope,* 313; *Homecoming,* 320; *The Ugly Swans,* 320
Strunnikov, V. A., 126
Studitskii, A. N., "Fly-Lovers and Man-Haters," 109–110
Sukachev, V. N., 98, 109, 110; attacks on Lysenko, 76; role in *zapovedniki,* 76–77
Suslov, M., 243–244
Suvorin, A., *Novoe vremia,* 24
Sveshnikova, L., "How to Outwit Pain," 315
Svobodnaia mysl', 36
Sytin, I., *Russkoe slovo,* 24
Szasz, T., 198

Taliev, V. I., 89
Tamizdat, 281
Tamm, I., 110
Tammik, R. E., *In the Studio,* 357
Tatlin, V.: "machine art," 3; Monument to the Third International, 338, 346; *Letatlin,* 338–339, 347
Taylor, C., 228

Taylor, F. W., 266, 300, 311–312, 322
Taylorism. *See* Taylor, F. W.
Technologism, 297–298
Technotronic glasnost, 41–50; rise of horizontal links, 42, 44; privatized information and public opinion, 42; internationalization of information, 42–43; individuation and small technologies, 43–44; growth of networks and groups, 44–45; difficulties of surveillance, 45–46; undermining Party's role as culture maker, 46–48. *See also* Communications technologies
Telegraph, 25–26, 27
Telephone, 25–26, 32
Television, 34–35, 373n13
Tendriakov, V., 277–278
Terent'ev, P. V., 79
Tereshkova, V., 181
Timofeev-Resovskii (Timoféeff-Ressovsky), N. V., 91, 103, 117, 118, 121–122; radiation genetics, 79, 121, 129, 296; prison and exile, 109, 121, 131; Berlin-Buch, 121; campaign to rehabilitate, 122, 131, 296; Granin's biography of, 131; film on, 131; Granin's documentary-novella about, 295–297
Titarenko, A. I., 245
Titov, G., 179, 181
Tokarev, V., 42
Toll, Baron, 22
Tolstoi, L., 252
Topchiev, A. V., 177
"Transrational" art, 344
Troepol'skii, G., 277–278
Trotsky, L., 260, 271, 285
Tsiolkovskii, K. E., 187; *Beyond the Earth,* 260; work in rocketry, 261
Tsvetaeva, M., 330
Tucker, R., 228
Tugendkhold, Ia., 350, 355
Tupelov, A. N., 335, 352
Tyshler, A.: *Woman and Aeroplane,* 349–350; *Lyrical Series,* 350

Ukrainsky visnik, 36
Union of Engineers, 144
Union of Soviet Writers, 297, 322, 329, 339
UNOVIS group, 421n7
Urbanist fiction, 272, 285
Urbanization, 28, 31–32, 85